I0034066

# Permanent Magnet Synchronous Machines and Drives

Permanent magnet synchronous motors (PMSMs) are popular in the electric vehicle industry due to their high-power density, large torque-to-inertia ratio, and high reliability. This book presents an improved field-oriented control (FOC) strategy for PMSMs that utilizes optimal proportional-integral (PI) parameters to achieve robust stability, faster dynamic response, and higher efficiency in the flux-weakening region. The book covers the combined design of a PI current regulator and varying switching frequency pulse-width modulation (PWM), along with an improved linear model predictive control (MPC) strategy. Researchers and graduate students in electrical engineering, systems and control, and electric vehicles will find this book useful.

Features:

- Implements evolutionary optimization algorithms to improve PMSM performance.
- Provides coverage of PMSM control design in the flux-weakening region.
- Proposes a modern method of model predictive control to improve the dynamic performance of interior PMSM.
- Studies the dynamic performance of two kinds of PMSMs: surface-mounted and interior permanent magnet types.
- Includes several case studies and illustrative examples with MATLAB®.

This book is aimed at researchers, graduate students, and libraries in electrical engineering with specialization in systems and control and electric vehicles.

## Advances in Power Electronic Converters

Series Editor: Md. Rabiul Islam, *University of Wollongong, Australia*

Recent advances in power semiconductor devices (e.g., ultra-fast recovery, silicon carbide, gallium-nitride), soft magnetic materials (e.g., amorphous/nanocrystalline), and controllers have led to the development of new high power density efficient power converters for emerging applications (e.g., renewable energy, transportation, smart power grid, smart buildings, modern industries, and so forth). However, the design process of the application specific advanced technology-based power converters involves multiphysics problems that entail complex tradeoffs among reliability, efficiency, size and weight, and cost. Therefore, extensive multiphysics work in the field of new power electronic converter topologies, switching and control, EMI/EMC and signal integrity, protection and condition monitoring, and design and optimization of magnetic components has attracted significant interest recently to develop the next-generation power converters. This series will serve to collect recent advancements in energy conversion with advanced power converters with the emphasis on the new design, optimization, switching, and control methods and new protection and condition monitoring techniques.

### Permanent Magnet Synchronous Machines and Drives
Flux Weakening Advanced Control Techniques
*Wei Xu, Moustafa Magdi Ismail and Md. Rabiul Islam*

**For more information about this series, please visit:** www.routledge.com/ Advances-in-Power-Electronic-Converter/book-series/CRCAPEC

# Permanent Magnet Synchronous Machines and Drives

## Flux Weakening Advanced Control Techniques

Wei Xu, Moustafa Magdi Ismail and
Md. Rabiul Islam

CRC Press
Taylor & Francis Group
Boca Raton  London  New York

CRC Press is an imprint of the
Taylor & Francis Group, an **informa** business

Designed cover image: Wei Xu, Moustafa Magdi Ismail and Md. Rabiul Islam

MATLAB® is a trademark of The MathWorks, Inc. and is used with permission. The MathWorks does not warrant the accuracy of the text or exercises in this book. This book's use or discussion of MATLAB® software or related products does not constitute endorsement or sponsorship by The MathWorks of a particular pedagogical approach or particular use of the MATLAB® software.

First edition published 2024
by CRC Press
6000 Broken Sound Parkway NW, Suite 300, Boca Raton, FL 33487–2742

and by CRC Press
4 Park Square, Milton Park, Abingdon, Oxon, OX14 4RN

*CRC Press is an imprint of Taylor & Francis Group, LLC*

© 2024 Wei Xu, Moustafa Magdi Ismail and Md. Rabiul Islam

Reasonable efforts have been made to publish reliable data and information, but the author and publisher cannot assume responsibility for the validity of all materials or the consequences of their use. The authors and publishers have attempted to trace the copyright holders of all material reproduced in this publication and apologize to copyright holders if permission to publish in this form has not been obtained. If any copyright material has not been acknowledged please write and let us know so we may rectify in any future reprint.

Except as permitted under U.S. Copyright Law, no part of this book may be reprinted, reproduced, transmitted, or utilized in any form by any electronic, mechanical, or other means, now known or hereafter invented, including photocopying, microfilming, and recording, or in any information storage or retrieval system, without written permission from the publishers.

For permission to photocopy or use material electronically from this work, access www.copyright.com or contact the Copyright Clearance Center, Inc. (CCC), 222 Rosewood Drive, Danvers, MA 01923, 978–750–8400. For works that are not available on CCC please contact mpkbookspermissions@tandf.co.uk

*Trademark notice*: Product or corporate names may be trademarks or registered trademarks and are used only for identification and explanation without intent to infringe.

ISBN: 9781032330655 (hbk)
ISBN: 9781032335452 (pbk)
ISBN: 9781003320128 (ebk)

DOI: 10.1201/9781003320128

Typeset in Times
by Apex CoVantage, LLC

# Contents

# Foreword

Permanent magnet synchronous motors (PMSMs) have been widely used in the electric vehicles (EVs) due to their high-power density, large torque-to-inertia ratio, high reliability, etc. This book involves emerging control strategies of the non-salient and salient PMSMs operating in the flux-weakening region. The optimization algorithms, such as adaptive velocity particle swarm optimization (AVPSO) and genetic optimization (GA) algorithms, are also presented, including the offline optimization procedures. The advanced control strategies like improved field-oriented control (FOC) and model-predictive control (MPC) are explicitly investigated in detail.

The parameters optimization presented in this book is based on the new cost function of the error value matrix for the proportional-integral (PI) controllers. A new cost function is deduced from the Taylor series expansion form to obtain the linearized equilibrium part of the advanced control strategy. Additionally, the stability terms are considered to obtain high performance in speed tracking at the flux-weakening region. In this book, a real-time AVPSO algorithm is used in the continuous control model predictive control. An adaptive discrete linear predictive model is formulated to represent the model plant across all regions of the four-quadrant control region. Moreover, the finite control set optimization technique is presented to predict the future behavior of the linear MPC drive of a salient PMSM. The dynamic performance of the advanced control strategies is compared with that of the conventional control strategies to verify the effectiveness of the advanced strategies. Furthermore, this book also presents a systematic analysis of the prototype developmental process and relevant experiments, which will be a good reference for academicians, researchers, and industries, as well as graduate students for their study.

Finally, I would like to express my sincere congratulations to the three authors, Professor Wei Xu, Dr. Moustafa Magdi Ismail, and Dr. Rabiul Islam, for their high-level and intelligent work, especially on adaptive flux-weakening control and adaptive linear model predictive control strategies. I believe in total that this book will be an impressive milestone in the PMSM drives and related fields of applications.

Frede Blaabjerg
Professor and IEEE Fellow
Aalborg University
Aalborg, Denmark
December 2022

# Preface

Salient permanent magnet synchronous motors (PMSMs) have been widely used in the electric vehicles (EVs) sector due to their high-power density, large torque-to-inertia ratio, and high reliability. The systems of the non-salient PMSM drives are also employed in the light EVs. Unusually, the traction systems need to operate the drive in the flux-weakening region to achieve a wide speed range. Meanwhile, field-oriented control (FOC) strategy and model-predictive control (MPC) strategy are currently popular PMSM control methods. However, both methods still have the following problems when applied to PMSM field weakening control:

(1) As the speed goes up, due to the increase of the PMSM back electromotive force and the limited inverter output voltage, the field-weakening operation of PMSM will become unstable, which in turn degrades the transient and steady-state characteristics of the PMSM drive; and (2) PMSM also suffers from torque ripples in the region of flux-weakening. These ripples have multiple causes, including a time delay of the digital controller, using a pulse-width modulation (PWM) technique, and so on.

This book presents advanced field-oriented control (FOC) and model predictive control (MPC) strategies to address industrial applications based on permanent magnet synchronous motors (PMSMs). The book compiles recent advancements in these strategies and provides empirical evidence of their effectiveness. Moreover, the book emphasizes the significance of real-time control optimization and highlights the associated challenges. Furthermore, the book includes a systematic analysis of the prototype development process, which involves two 3 kW PMSMs, and presents pertinent experimental results. This analysis can serve as a valuable reference for researchers and practitioners in the field.

The **Introduction** chapter reviews the important progress in the variable frequency strategies, ripple reduction strategies, and predictive control strategies. It also introduces a historical overview of the magnet materials. Detailed performance comparisons are made among the PMSM, induction motor, and brushless direct current motor drive systems under the industrial applications.

The chapter on **Performance Analysis of PMSM Drive System Using Frequency Modulation Technique** provides background on the operation of the PMSM drives at the constant-torque region and the flux-weakening region. The maximum current and the maximum voltage are discussed as well. It also analyses the control strategy principle of the maximum torque per ampere in both regions. The torque ripple mitigation issue is interesting in this chapter for non-salient PMSM drives. The technique of the predictive stator currents ripple of the inherent inverter is employed to conclude the predicted ripple of the $q$-axis stator current of the non-salient PMSM. Thus, the predicted ripple of the induced torque can be linearly evaluated to vary the switching frequency of the inverter modulator.

The chapter on **Adaptive Flux-Weakening Control Strategy for Non-Salient Permanent Magnet Synchronous Motor Drives** deals with the optimal parameters of the FOC strategy based on a new tuning technique. This tuning technique employs the Taylor series expansion to consider the error matrix of the control loops of the advanced FOC strategy and the non-salient PMSM plant model. Meanwhile, the offline procedure of the optimization algorithm is based on the adaptive velocity particle swarm optimization (AVPSO), which can select the optimal proportional-integral (PI) parameters to achieve the minimal error matrix. Besides, the anti-windup proportional and integral (AWPI) is considered to reduce the overrun performance of the non-salient PMSM.

The chapter on **Design and Optimization of Stator Current Regulators for Surface-Mounted Permanent Magnet Synchronous Motor Drives** deals with the problems of electromagnetic torque ripple and stator current ripple at the high-speed range. This chapter aims to obtain stable speed tracking at the flux-weakening region with a low ripple peak of the induced torque and stator current. Thus, a new cost function is considered to achieve this goal. In other words, this cost function has two stability terms. One term is to force the optimization algorithm to select parameters that correspond to the high negative poles of the system. The other term is to force the optimization algorithm to specify parameters that avoid instabilities or margin stability. Meanwhile, the optimization genetic algorithm is employed to evaluate this cost function in offline work. Lastly, the FOC strategy with the new PI parameters is implemented in the advanced non-salient PMSM drive combined with the variable switching frequency algorithm.

The chapter on **Advanced Flux-Weakening Control for Interior Permanent Magnet Synchronous Drives** deals with implementing the FOC strategy on the salient pole PMSM. The reason is that salient PMSMs excel as a further component of the reluctance torque in the torque equation compared to the non-salient permanent magnet type motor. There is also a difference in the stator inductance of the air gap. Thus, the modification of the control method involves the limiter of the outer control loops to defeat the overshoot of the speed and stator currents. Furthermore, this modified limiter is based on the third-order generalized flux observer.

The chapter on **Modified First-Order Flux Observer–Based Speed Predictive Control of Interior Permanent Magnet Drives** deals with speed MPC. In this chapter, linear predictive control is based on the finite control set evaluator to select reference stator voltage components optimally. Thus, the power inverter modulator is replaced by this evaluator, which directly defines the inverter switch states. Additionally, the modified first-order flux observer is employed to support the advanced predictive control to directly evaluate the predictive motor speed in the objective function. Meanwhile, this objective function is designed directly for motor speed error and the $d$-axis required stator current error to obtain high stable speed control in the flux-weakening region.

The chapter on **Adaptive Linear Model Predictive Control for Flux-Weakening Control Based on Particle Swarm Optimization** deals with an adaptive discrete linear predictive plant model. In other words, the modification is to update the operating point or the equilibrium points of the predictive plant model by the current condition of the salient PMSM drive with the motor speed and the $d$-$q$ stator current. The modified third-order generalized integrator flux observer is also employed to support the advanced predictive control strategy, which can directly evaluate the predictive motor speed in the objective function. However, the control performance is considered in the advanced predictive speed control to solve the overshoot in motor speed and stator current.

It is worth mentioning the partial support received from the National Natural Science Foundation of China under Grant 51877093, the National Key Research and Development Program of China under Grant 2018YFE0100200, the Key Technical Innovation Program of Hubei Province under Grant 2019AAA026, the Key Research and Development Program of Sichuan Province under Grant 2021YFG0081, Technology and Innovation Commission of Shenzhen Municipality under Grant JCYJ20190809101205546, and the School of Electrical, Computer, and Telecommunications Engineering, University of Wollongong, Wollongong, NSW, Australia, in writing this book. The authors of this book also wish to acknowledge the support and inspiration of their families during the preparation of this book.

Wei Xu<br>
Moustafa Magdi Ismail<br>
Md. Rabiul Islam

# About the Authors

**Wei Xu** (M'09-SM'13) received double BE and ME degrees from Tianjin University, Tianjin, China, in 2002 and 2005, and a PhD from the Institute of Electrical Engineering, Chinese Academy of Sciences, in 2008, respectively, all in electrical engineering. His research topics mainly cover design and control of linear/rotary machines.

From 2008 to 2012, he was Postdoctoral Fellow with University of Technology Sydney, Vice Chancellor Research Fellow with Royal Melbourne Institute of Technology, and Japan Science Promotion Society Invitation Fellow with Meiji University. Since 2013, he has been a full professor with State Key Laboratory of Advanced Electromagnetic Engineering in Huazhong University of Science and Technology, China. He is a fellow of the Institute of Engineering and Technology (IET). He is the general chair of the 2021 International Symposium on Linear Drives for Industry Applications (LDIA 2021) and the 2023 IEEE International Conference on Predictive Control of Electrical Drives and Power Electronics (PRECEDE 2023), both in Wuhan, China. He has published over 140 papers and has had over 120 invention patents granted, all in the related field of electrical machines and drives. He has been the associate editor of several IEEE Transactions Journals.

**Moustafa Magdi Ismail** received his BE and ME degrees in electrical engineering from the Minia University, El-Minya, Egypt, in 2011 and 2016, respectively. He also received his PhD in 2021 at the school of Electrical and Electronics Engineering, State Key Laboratory of Advanced Electromagnetic Engineering of Huazhong University of Science and Technology, Wuhan, China. He has been awarded an honorary International Graduate Certificate for his PhD.

He is currently employed as an assistant professor at Minia University. In 2021, he started teaching at the Higher Institute of Engineering and Technology in New Minya City.

He has published scientific papers in both international conferences and high-quality SCI journals. He serves as a reviewer for *IEEE Transactions* journals (TVT, TIE, TTE, TPEL, TEC, and IAS), *IEEE-JESTIE*, and *IEEE Access*. He was selected as one of the top reviewers for *IEEE Transactions on Vehicular Technology* in 2020.

His current research interests include electric machinery, inverter systems, smart grids, modelling and control design of electric vehicles, predictive plant model, predictive control design, and optimization algorithms.

**Md. Rabiul Islam** received a PhD from the University of Technology Sydney (UTS), Sydney, Australia, in electrical engineering in 2014.

He is currently a senior lecturer with the School of Electrical, Computer, and Telecommunications Engineering (SECTE), University of Wollongong (UOW), New South Wales, Australia. He is a senior member of IEEE. His research interests are in the fields of power electronic converters, renewable energy technologies, power quality, electrical machines, electric vehicles, and smart grids. He has authored or co-authored more than 400 papers, including more than 100 *IEEE Transactions/IEEE Journal* papers. He has written or edited seven technical books. He has received several best paper awards, including two best paper recognitions from *IEEE Transactions on Energy Conversion* in 2020. He is serving as an associate editor for *IEEE Transactions on Industrial Electronics*, *IEEE Transactions on Energy Conversion*, *IEEE Power Engineering Letters*, and *IEEE Access*. As a lead guest editor, he organized the first joint IEEE Industrial Electronics Society and IEEE Power & Energy Society special section entitled Advances in High-Frequency Isolated Power Converters. He is an editor of the book series entitled Advances in Power Electronic Converters for CRC Press, Taylor & Francis Group. He has received funding from several governments and industries, including in total $5.48 million from the Australian government through the Australian Research Council (ARC) Discovery Project (DP) 2020 entitled A Next Generation Smart Solid-State Transformer for Power Grid Applications and an ARC Industrial Transformation Training Centre Project 2021 entitled ARC Training Centre in Energy Technologies for Future Grids.

# Symbols

## Introduction

| | |
|---|---|
| $d$-axis | Direct axis component |
| $q$-axis | Quadrature axis component |
| F | Voltage source frequency |
| $n_p$ | Number of pole pairs |
| $N_s$ | Synchronous speed |
| $i_{qs}$ | $q$-axis of the PMSM stator current |
| $i_{ds}$ | $d$-axis of the PMSM stator current |
| $i_{ds}^*$ | $d$-axis of the reference stator current |
| $i_{qs}^*$ | $q$-axis of the reference stator current |
| $i_A$ | Instant measured values for A-phase stator current |
| $i_B$ | Instant measured values for B-phase stator current |
| $i_C$ | Instant measured values for C-phase stator current |
| $\Theta_m$ | Counted mechanical angle of rotor position |
| $\omega_m$ | Calculated rotor velocity |
| $\omega_m^*$ | Reference rotor velocity |
| $V_{dc}$ | DC-link voltage of the DC circuit for the inverter |
| $v_{ds}^*$ | $d$-axis of the reference stator voltages |
| $v_{qs}^*$ | $q$-axis of the reference stator voltages |
| $T^*$ | Reference torque |
| $T_{e\text{-est}}$ | Estimated electromagnetic torque |
| $\psi^*$ | Reference flux vector |
| $\psi_{est}$ | Estimated stator flux vector |
| $B_T$ | Magnetic flux density |
| $v_{ds}$ | $d$-axis PMSM stator voltages |
| $v_{qs}$ | $q$-axis PMSM stator voltages |
| $\lambda$ | Stator flux of PMSM |
| $v_{\alpha s}^*$ | $\alpha$-axis reference stator voltages |
| $v_{\beta s}^*$ | $\beta$-axis reference stator voltages |
| $I_{d\text{-minCPT}}$ | $d$-axis current produced by the minimum current per torque algorithm |
| $I_{d\text{-MTPV}}$ | $d$-axis current produced by MTPV algorithm |
| $P_{max}$ | Maximum consumed power of the PMSM |
| $T_{sat}$ | Saturated reference torque |
| $I_{rated}$ | Rated current of the PMSM |
| $\lambda_m$ | Flux linkage due to the rotor magnets linking the stator part |
| $I_{s\text{-max}}$ | Maximum current passing in the inverter and the motor circuit |
| $L_{ds}$ | $d$- axis stator inductance of PMSM |
| $L_{qs}$ | $q$-axis stator inductance of PMSM |
| $z$ | Indicates the discrete domain |
| $K_p$ | Proportional gain for a PI controller of flux decreasing |
| $K_i$ | Integral gain for a PI controller of flux decreasing |
| $m_{max}$ | Maximum line-modulation ratio |

| | |
|---|---|
| $m^*_{max}$ | Maximum reference line-modulation ratio |
| $\delta_{max}$ | Maximum value of the duty cycle |
| $\delta_{min}$ | Minimum value of the duty cycle |
| $R_s$ | Stator resistance of PMSM |
| $L_s$ | Stator inductance of PMSM |
| $S$ | Indicates the Laplace domain |
| $F_{sw}$ | Sampling frequency of the inverter |
| $G_{sc}(S)$ | Serial regulator matrix |
| $G_{fc}(S)$ | Feedback regulator matrix |
| $G_P(S)$ | Plant regulator matrix |
| $k_t$ | Constant torque coefficient |
| $i_{q0}^*$ | Reference $q$-axis stator current deduced from torque controller |
| $i_{q0}^*$ | Reference $q$-axis stator current deduced from ILC algorithm |
| $u^{fw}_{d,wkfd}$ | $d$-axis reference voltage deduced from the flux-weakening control loop |
| $u^{fw}_{q,wkfd}$ | $q$-axis reference voltage deduced from the flux-weakening control loop |
| $i^*_{ds,wkfd}$ | Reference stator current of the $d$-axis deduced from the flux-weakening control loop |
| $i^*_{qs,wkfd}$ | Reference stator current of the $q$-axis deduced from the flux-weakening control loop |
| $u^{fw}_{s,wkfd}$ | Reference vector of stator voltage deduced from the flux-weakening control loop |
| $i^*_{ds,MTPA}$ | $d$-axis reference current deduced by MTPV algorithm |
| $EMF_d$ | Permanent magnets back EMF of $d$-axis |
| $EMF_q$ | Permanent magnets back EMF of $q$-axis |
| $T_d$ | Total time delay |
| $k_{PWM}$ | Inverter gain |
| $\omega_e$ | Electric PMSM velocity |
| $\eta_1, \eta_2, \eta_3, \eta_4, \eta_5, \eta_6,$ and $\eta_7$ | Coefficients function of machine parameters |
| $u_{fd}$ | $d$-axis nonlinear decoupling control term |
| $u_{fq}$ | $q$-axis nonlinear decoupling control term |
| $u_{fbd}$ | $d$-axes stabilizing fuzzy control term |
| $u_{fbq}$ | $q$-axes stabilizing fuzzy control term |
| $h_{di}$ | $d$-axis normalized weight of each if-then rule |
| $h_{qi}$ | $q$-axis normalized weight of each if-then rule |
| $m_{di}$ | $q$-axis membership function of the fuzzy logic algorithm |
| $m_{qi}$ | $q$-axis membership function of the fuzzy logic algorithm |
| $i_{de}$ | $d$-axis error of current regulator |
| $i_{qe}$ | $q$-axis error of current regulator |
| $\hat{f}_s$ | Lump of the uncertainties |
| $e_s$ | Error vector that has been estimated |
| $\hat{i}_{ds}$ | $d$-axis estimated stator current |
| $\hat{i}_{qs}$ | $q$-axis estimated stator current |
| $T_s$ | Step time |
| $\hat{f}_{df}(k)$ | $d$-axis estimated disturbances caused by parameter variations |
| $\hat{f}_{qf}(k)$ | $q$-axis estimated disturbances caused by parameter variations |

| $v_A$ | Instant measured values a-phase stator voltages |
| $v_B$ | Instant measured values b-phase stator voltages |
| $v_C$ | Instant measured values c-phase stator voltages |
| $k$ | Discrete time |
| $t$ | Time instance |
| $i^*_{qsL}$ | Reference load current |
| $C_e$ | Voltage constant of PMSM |

## Performance Analysis of PMSM Drive System Using Frequency Modulation Technique

| $V_{dqs}$ | $d$-$q$ axes stator voltage vector |
| $\lambda_{dqs}$ | Stator flux vector |
| $i_{dqs}$ | $d$-$q$ axes stator current vector |
| $i_{ds}$ | $d$-axis measured stator current |
| $i_{qs}$ | $q$-axis measured stator current |
| $L_s$ | Stator reactance of surface-mounted type permanent magnet synchronous motor |
| $\lambda^s_m$ | Rotor flux vector |
| $\Theta_r$ | Electrical angle of the rotor |
| $\omega_e$ | Electrical angular speed |
| $R_s$ | Stator resistance |
| $L_{ds}$ | $d$-axis stator reactance |
| $L_{qs}$ | $q$-axis stator reactance |
| $T_e$ | Electromagnetic developed torque |
| $n_p$ | Pair poles number |
| $T_L$ | Load torque |
| $\omega_m$ | Mechanical angular speed |
| $J$ | Inertia of the shaft |
| $B_v$ | Viscous friction coefficient |
| $\lambda_m$ | Flux linkage |
| $I_s$ | Stator current vector of the PMSM |
| $\Theta$ | Torque angle |
| $T_r$ | Reluctance torque |
| $T_m$ | Magnet excitation torque |
| $i_{ds\text{-}M}$ | Operating point of $d$-axis stator current for maximum torque per ampere algorithm |
| $i_{qs\text{-}M}$ | Operating point of $q$-axis stator current for maximum torque per ampere algorithm |
| $\vartheta_M$ | Operating point of torque angle for maximum torque per ampere algorithm |
| $\omega_{max\text{-}CT}$ | Maximum velocity of the constant-torque region at the current operating condition of the PMSM |
| $V_{s\text{-}max}$ | Saturated stator voltage of the two-level voltage source inverter |
| $I_{s\text{-}max}$ | Maximum current of the power inverter and the PMSM drive |

| | |
|---|---|
| $P_{out}$ | Output power |
| $T_{max}$ | Maximum electromagnetic torque at the operating point of $i_{ds\text{-}M}$ and $i_{qs\text{-}M}$ |
| $\omega_m$ | Mechanical velocity |
| $I_{sc}$ | Short-circuit current |
| $\omega_{FW\text{-}max}$ | Maximum velocity of the flux-weakening region at the current operating condition of the PMSM |
| $I_{wanted}$ | Desired $q$-axis current peak value |
| $\Delta T_{wanted}$ | Desired peak of the torque ripple |
| $T_s$ | Switching period |
| $d_a, d_b, d_c$ | Three-phase switching cycles of the space vector modulation algorithm |
| $T_0, T_1, T_2$ | Voltage sectors actuation time |
| $i_{an}$ | A-phase stator current |
| $v^*_{\alpha s}, v^*_{\beta s}$ | Reference voltage vector $\alpha$-$\beta$ axes components |
| $P_x, P_y$ | Highest peaks of the current ripple |
| $\Delta i_{ds}$ | Maximum predicted current ripple for $d$-axis |
| $\Delta i_a, \Delta i_b, \Delta i_c$ | Real-time maximum predicted ripple peaks of three-phase stator currents |
| $\rho$ | Voltage vector magnitude |
| $i^*_{ds}$ | $d$-axis reference current |
| $K_{P\rho}$ | Proportional of the modulation index regulator |
| $K_{I\rho}$ | Integral of the modulation index regulator |
| $\rho^*$ | Voltage vector reference magnitude |
| $e_\rho$ | Modulation index controller error |
| $\Delta i^*_{ds}$ | Difference between the limited value and the reference value of the controller |
| $E_\rho$ | Integral error of the modulation index controller |
| $i_{qs\text{-}max}$ | Maximum compensated value of the $q$-axis stator current |

## Adaptive Flux-Weakening Control Strategy for Non-Salient Permanent Magnet Synchronous Motor Drives

| | |
|---|---|
| $C_F$ | DC-link equivalent capacitor |
| $i_{dqs}$ | $d$-$q$ axes stator current vector |
| $d_{dqs}$ | $d$-$q$ axes duty-cycle vector of two-level voltage source inverter |
| $\lambda_{dqs}$ | $d$-$q$ axes stator flux linkage vector of a PMSM |
| $v_{ds\_decoupling}, v_{qs\_decoupling}$ | $d$-$q$ axes decoupling components of the inner controller of the field-oriented control |
| $I_{s\text{-}max}$ | Maximum current passing in the inverter and the motor circuit |
| $i_{qs\text{-}max}$ | Maximum stator current reference of the $q$-axis used in the limiter of the PI speed controller |
| $i^*_{ds}$ | $d$-axis component of the stator reference current for PMSM |
| $e$ | Error of the PI controller |
| $Y^*$ | Output of the saturation of the PI controller |
| $\Delta$ | Difference between the input and output for the saturation of the PI controller |

| | |
|---|---|
| $*$ | Reference superscript |
| $\Delta y$ | Anti-windup value |
| $K_{aw}$ | Anti-windup gain |
| $K_{P\text{-}d}$ | Proportional parameter of the $d$-axis inner controller |
| $K_{I\text{-}d}$ | Integral parameter of the $d$-axis inner controller |
| $K_{P\text{-}q}$ | Proportional parameter of the $q$-axis inner controller |
| $K_{I\text{-}q}$ | Integral parameter of the $q$-axis inner controller |
| $K_{aw\text{-}d}$ | Anti-windup variable of the $d$-axis inner controller |
| $K_{aw\text{-}q}$ | Anti-windup variable of the $q$-axis inner controller |
| $E_d$ | Error integrations of the $d$-axis inner controller |
| $E_q$ | Error integrations of the $q$-axis inner controller |
| $e_d$ | Error value of the $d$-axis inner controller |
| $e_q$ | Error value of the $q$-axis inner controller |
| $K_{aw\text{-}\omega}$ | Anti-windup variable of the speed outer controller |
| $K_{P\text{-}\omega}$ | Proportional parameter of the speed outer controller |
| $K_{I\text{-}\omega}$ | Integral parameter of the speed outer controller |
| $e_\omega$ | Error value of the speed outer controller |
| $E_\omega$ | Error integrations of the speed outer controller |
| $D_{dq}$ | Compensated duty-cycle vector |
| $D_{dq\text{-}max}$ | Maximum duty-cycle vector |
| $K_{aw\text{-}FW}$ | Anti-windup variable of the flux outer controller |
| $K_{P\text{-}FW}$ | Proportional parameter of the flux outer controller |
| $K_{I\text{-}FW}$ | Integral parameter of the flux outer controller |
| $E_{FW}$ | Error value of the flux outer controller |
| $E_{FW}$ | Error integrations of the flux outer controller |
| $0$ | Equilibrium point subscript of a state variable |
| $v_{qs}^{\prime*}, v_{ds}^{\prime*}$ | Reference stator voltages of the $q$-$d$ axes for the stator current inner controllers |
| $g_{best}$ | Global minimum solution for AVPSO algorithm |
| $\vec{\tilde{g}}_k$ | Evaluation vector of the $k^{th}$ particle position for the internal particle swarm algorithm |
| $g_{best}$ | Minimum value for evaluation vector |
| $\vec{\tilde{g}}_n$ | Evaluation vector of particle swarm algorithm, where $n^{th}$ indicates the particle location |
| $N$ | Swarm size of adaptive velocity particle swarm algorithm |
| $p_{global}$ | Optimal particle position for adaptive velocity particle swarm algorithm |
| $p_{best}$ | Global best location that corresponds to $k^{th}$ particle of the minimum evaluation for the internal adaptive velocity particle swarm algorithm |
| $\tilde{p}_k$ | $k^{th}$ particle location for the internal adaptive velocity particle swarm algorithm |
| $\hat{p}_n$ | Initial particle position vectors of adaptive velocity particle swarm algorithm, where $n^{th}$ indicates the particle number |
| $\hat{P}_n$ | $n^{th}$ updated particle location |
| $\tilde{P}_k$ | $k^{th}$ updated particle location |

## Design and Optimization of Stator Current Regulators for Surface-Mounted Permanent Magnet Synchronous Motor Drives

| | |
|---|---|
| $d_{ds}$ | $d$-axis duty-cycle vector of two-level voltage source inverter |
| $d_{qs}$ | $q$-axis duty-cycle vector of two-level voltage source inverter |
| $x, y$ | General state variables in a plant model |
| $0$ | Equilibrium point subscript of a state variable |
| $\delta$ | Small signal model value of a state variable |
| $\lambda_j$ | Eigenvalue of the state matrix of the state-space representation |
| $A$ | State matrix of the PMSM drive system |
| $B$ | Input matrix of the PMSM drive system |
| $u$ | Input vector of the state-space representation for the PMSM drive system |
| $x$ | States vector of the state-space representation for the PMSM drive system |
| $P$ | Penalty function |
| $\hat{p}_n$ | $n^{th}$ particle of the population |
| $V_{LOW}$ | Lower given limit of the $n^{th}$ particle |
| $V_{UP}$ | Upper given limit of the $n^{th}$ particle |
| $P_n$ | Particle probabilities |
| $f_n$ | Particle of population evaluation |
| $est$ | Indicates the estimation value |
| $T_{e\text{-TOGI}}$ | Calculated induced torque based on the third-order generalized integral flux observer |
| $\omega_c$ | Center frequency |
| $W_1, W_2, W_3, W_0$ | Integrator coefficients |
| $\rho$ | Voltage vector magnitude |

## Advanced Flux-Weakening Control for Interior Permanent Magnet Synchronous Drives

| | |
|---|---|
| $\Theta_m$ | Counted mechanical position of the PMSM rotor |
| $V_{dc}$ | DC voltage of the inherent inverter |
| $\lambda_j$ | Penalty function |
| $i_{ds}^*$ | Reference PMSM stator current of the $d$-axis component |
| $i_{qs}^*$ | Reference PMSM stator current of the $q$-axis component |
| $\omega_m^*$ | Reference PMSM velocity |
| $T_{e\text{-TOGI}}$ | Calculated induced torque based on the third-order generalized integral flux observer |
| $i_{d\text{-limit}}$ | Saturated value of the reference stator current of the $d$-axis |
| $i_{q\text{-limit1}}$ | Saturated value of the reference stator current of the $q$-axis when the motor speed is under the reference speed |
| $i_{q\text{-limit2}}$ | Saturated value of the reference stator current of the $q$-axis when the motor speed is above the reference speed |

## Modified First-Order Flux Observer–Based Speed Predictive Control of Interior Permanent Magnet Drives

| | |
|---|---|
| $\omega_{\text{max-CT}}$ | Maximum velocity of the constant-torque region at the current operating condition of the PMSM |
| $i_{\text{ds-P}}$ | Predicted stator currents of the $d$-axis component |
| $i_{\text{qs-P}}$ | Predicted stator currents of the $q$-axis component |
| $\omega_{\text{m-P}}$ | Next step value or the predicted rotor velocity |
| $W_1$ | Weighting factor of the error for the required stator current |
| $W_2$ | Weighting factor of the error for the motor speed |
| $S$ | Laplace operator |
| $\lambda_r$ | Rotor fluxes vector |
| $i_s$ | Stator currents vector |
| $v_s$ | Reference stator voltages vector |
| $L_s$ | Stator inductances vector |
| $\lambda_{r\alpha}, \lambda_{r\beta}$ | Stationary frame components rotor fluxes |
| $V^*_{s\alpha}, V^*_{s\beta}$ | Stationary frame components of the reference stator voltages |
| $\vartheta$ | Phase of the modified low-pass filter |
| $G$ | Gain of the modified low-pass filter |
| $\omega_e$ | Operating frequency of the modified low-pass filter in radian per second of the salient PMSM |
| $\omega_c$ | Cut-off frequency of the modified low-pass filter |
| $T_{\text{e-Est.}}$ | Calculated electromagnetic torque based on the modified low-pass filter |
| $r_{ce}$ | On-state active switch slope resistance |
| $r_d$ | On-state freewheeling diode slope resistance |
| $V_{ce}$ | Active switch threshold voltage |
| $Vd$ | Freewheeling diode threshold voltage |
| $T_d$ | PWM dead time |
| $T_{\text{on}}$ | Power device turn-on times |
| $T_{\text{off}}$ | Power device turn-off times |
| $v_{ds}, v_{qs}$ | $d$-$q$ axes components of each voltage vector |
| $\Theta_e$ | Velocity position of the PMSM rotor |
| $S_a$ | Arm 1 upper switch state of the inherent power two-level voltage source inverter |
| $S_b$ | Arm 2 upper switch state of the inherent power two-level voltage source inverter |
| $S_c$ | Arm 3 upper switch state of the inherent power two-level voltage source inverter |
| $C_e$ | Voltage constant of PMSM |

## Adaptive Linear Model Predictive Control for Flux-Weakening Control Based on Particle Swarm Optimization

| | |
|---|---|
| $T_{\text{e-TOGI}}$ | Calculated induced torque based on the third-order generalized integral flux observer |

| | |
|---|---|
| $\delta$ | Small signal model value of a state variable |
| $A$ | State matrix of the PMSM drive system |
| $B$ | Input matrix of the PMSM drive system |
| $x$ | States vector of the state-space representation for the PMSM drive system |
| $u$ | Input vector of the state-space representation for the PMSM drive system |
| $y$ | Output vector of the state-space representation for the PMSM drive system |
| $E$ | Load torque matrix of the state-space representation for the PMSM drive system |
| $D$ | Feedthrough matrix of the state-space representation for the PMSM drive system |
| $K$ | Sampling time |
| $A_{\mathrm{d}}$ | Discretization matrix of the state matrix |
| $B_{\mathrm{d}}$ | Discretization matrix of the input matrix |
| $E_{\mathrm{d}}$ | Discretization matrix of the load torque matrix |
| $T$ | Time domain |
| $q_0, q_1, q_2, q_3, q_4, q_5, q_6, q_7, q_8, q_9, q_{10}, q_{11}, q_{12}$ | Variables resulting from calculations of the Discretization of the $A_{\mathrm{d}}$, $B_{\mathrm{d}}$, and $E_{\mathrm{d}}$ |
| $\omega^*_{\mathrm{m}}$ | Reference PMSM velocity |
| $k$ | Sample time |
| $\omega_{\mathrm{F\text{-}ref}}$ | Persuasive reference velocity |
| $\omega_{\mathrm{max\text{-}CT}}$ | Maximum velocity of the constant-torque region at the current operating condition of the PMSM |
| $C_{\mathrm{e}}$ | Voltage constant of PMSM |
| $i_{\mathrm{d\text{-}ref}}$ | Required demagnetizing current for flux-weakening operation |
| $I_{\mathrm{S\text{-}p\text{-}max}}$ | Maximum allowable limit of the prediction stator current vector of the advanced predictive control |
| $I_{\mathrm{s\text{-}max}}$ | Maximum current passing in the inverter and the motor circuit |
| $I_{\mathrm{rated}}$ | Rated stator current PMSM |
| $I_{\mathrm{L}}$ | Rated load stator current of the current operating points of the PMSM |
| $W_{\mathrm{wm}}$ | Weighting velocity factor of the cost-function of the adaptive predictive control |
| $W_{\mathrm{id}}$ | Weighting $d$-axis stator current factor of the cost-function of the adaptive predictive control |
| $\omega_{\mathrm{m\text{-}p}}$ | Predicted motor velocity of the adaptive predictive control during the next sample time |
| $i_{\mathrm{ds\text{-}p}}$ | Predicted motor $d$-axis stator current of the adaptive predictive control during the next sample time |
| $i_{\mathrm{qs\text{-}p}}$ | Predicted motor $q$-axis stator current of the adaptive predictive control during the next sample time |
| $p_{\mathrm{global}}$ | Global minimum solution |
| $v_{\mathrm{ds\_limit}}, v_{\mathrm{qs\_limit}}$ | Lower and upper bounds of $d$-$q$ axes voltages of the real-time adaptive particle swarm velocity |

| | |
|---|---|
| $X(k)$ | State vector of the conventional predictive control strategy |
| $U(k)$ | Input vector of the conventional predictive control strategy |
| $Y(k)$ | Output vector of the conventional predictive control strategy |
| $\Delta u_{\text{d-max}}$, $\Delta u_{\text{q-max}}$ | Approximate change rates of $d$-$q$ axes stator voltage components |
| $t_s$ | Required settling time of the system |
| $\Theta_{\text{Ids-Is}}$ | Angle between the $d$-axis component current vector and the stator current magnitude vector |

# Acronyms

## Introduction

PMSM = Permanent Magnet Synchronous Motor
SPMSM = Surface-Mounted PMSM
IPMSM = Interior PMSM
BLDC = Brushless Direct Current
PM = Permanent Magnet
DC = Direct Current
AC = Alternating Current
EV = Electric Vehicle
DTC = Direct Torque Control
FOC = Field-Oriented Control
MPC = Predictive Control Model
MAX = Maximum
LPF = Low Pass Filter
MTPA = Maximum Torque per Ampere
MTPV = Maximum Torque per Minimum Reference Volt
KFW = Flux-Weakening Factor
$\Delta$ = Rate of Change
RC = Resonance controller
DRGA = Dynamic Relative Gain Array
2DOF = Two-Degree-of-Freedom
ILC = Iterative Learning Control Model
DSP = Digital Signal Processing
HVAC = Heating, Ventilation, and Air Conditioning
AWPI = Anti-Windup Proportional and Integral
AVPSO = Adaptive Velocity Particle Swarm Optimization Algorithm
PWM = Pulse-Width Modulation
VSFPWM = Varying Switching Frequency PWM
VSI = Voltage Source Inverter
GA = Genetic Algorithm
DLPM = Discrete Linear Plant Model
TOGI = Third-Order Generalized Integral
PFWC = Proposed Flux-Weakening Control
CFWC = Conventional Flux-Weakening Control
no = Nominal Values

## Performance Analysis of PMSM Drive System Using Frequency Modulation Technique

PMSM = Permanent Magnet Synchronous Motor
PM = Permanent Magnet

MTPA = Maximum Torque per Ampere
EMF = Back Electromotive Force
MMF = Magnetic Motive Force
PWM = Pulse-Width Modulation
DC = Direct Current
VSFPWM = Varying Switching Frequency PWM
DSP = Digital Signal Processor

## Adaptive Flux-Weakening Control Strategy for Non-Salient Permanent Magnet Synchronous Motor Drives

PMSM = Permanent Magnet Synchronous Motor
FOC = Field-Oriented Control
PFWC = Proposed Flux-Weakening Control
CFWC = Conventional Flux-Weakening Control
AWPI = Anti-Windup Proportional and Integral
AVPSO = Adaptive Velocity Particle Swarm Optimization Algorithm
PI = Proportional and Integral

## Design and Optimization of Stator Current Regulators for Surface-Mounted Permanent Magnet Synchronous Motor Drives

PMSM = Permanent Magnet Synchronous Motor
PI = Proportional and Integral
GA = Genetic Algorithm
FOC = Field-Oriented Control
TOGI = Third-Order Generalized Integral
FFT = Fast Fourier Transform
FMPWM = Frequency Modulation Pulse-Width Modulation
MBR= Particle Member
Chr = Chromosome

## Advanced Flux-Weakening Control for Interior Permanent Magnet Synchronous Drives

PMSM = Permanent Magnet Synchronous Motor
PI = Proportional-Integral
DSP = Digital Signal Processor
TOGI = Third-Order Generalized Integral

## Modified First-Order Flux Observer Based Speed Predictive Control of Interior Permanent Magnet Drives

PMSM = Permanent Magnet Synchronous Motor
MPC = Predictive Control Model
PWM = Pulse-Width Modulation

## Adaptive Linear Model Predictive Control for Flux-Weakening Control Based on Particle Swarm Optimization

PMSM = Permanent Magnet Synchronous Motor
FOC = Field-Oriented Control
MPC = Model-Predictive Control
SVPWM = Space Vector Pulse-Width Modulation
DLPM = Discrete Linear Plant Model
TOGI = Third-Order Generalized Integral
AVPSO = Adaptive Velocity Particle Swarm Optimization Algorithm
CPM = Conventional Plant Model
VSI = Voltage Source Inverter
PMPC = Proposed Model Predictive Control
CMPC = Conventional Model Predictive Control

# 1 Introduction

## 1.1 RESEARCH BACKGROUND AND SIGNIFICANCE OF THE PMSMS IN INDUSTRIAL APPLICATIONS AND MAIN CONSTRUCTION

The electric motor industry has recently seen a significant increase in the use of permanent magnet (PM) motors. The growth rate of PM motor development, which has exceeded 100 percent, is closely linked to the development of permanent magnet materials [1–5]. However, these growth figures for PM motors have been accumulating for many years. Several decades after the invention of electric motors, scientists began exploring the potential uses of permanent magnet synchronous motors (PMSMs). The first PM-based electric motors were used in experimental equipment. However, due to their low efficiency, these PM rods were not suitable for industrial applications. As a result, scholars have developed numerous designs of PM configurations, varying in materials, sizes, and shapes, resulting in the high magnet density that PMSMs utilize today.

### 1.1.1 History of the First Permanent Magnet Motors

Early researchers constructed a basic electric rotary machine based on permanent magnets. For example, Michael Faraday conducted one of the first experiments in the field of machine electromagnetic design. This design implemented the idea of Hans Oersted that electrical current provides an electric-magnetic field that transforms electrical energy into mechanical energy. Also, William Wollaston obtained an electromagnetic coil when carrying an electric current [6]. This Faraday device adopted a stator and rotating shaft, including PM rods with wires connected to mercury and direct-current (DC) voltage source units. Thus, the electromagnetic flux is created from the current flowing in the stator winding, and the electromagnetic torque is produced because of the intersection of the rotor shaft with the electromagnetic flux [6].

The types of motors on the market today are an extension of the Faraday device with many developments by researchers and manufacturers. For example, in 1882, the research study of pioneering Peter Barlow gave a spinning wheel or a Barlow wheel. This device generates mechanical rotational energy by dipping the wheel in mercury and applying a DC voltage to the rods, as shown in Figure 1.1 [7]. In 1837, Thomas Davenport obtained the first patent for an electric motor that was based on the improvement of earlier designs, such as the electromagnet solenoid by Joseph Henry. This motor has a rated power of 4.5 W and a rated speed and torque of around 1000 rpm and 4.5194 Nm, respectively. Meanwhile, the Davenport motor used permanent magnets, but other researchers used electromagnetic coils, which led to the growth of the industrial electric motors market until high-quality permanent magnets were created to use in PMSMs.

DOI: 10.1201/9781003320128-1

1

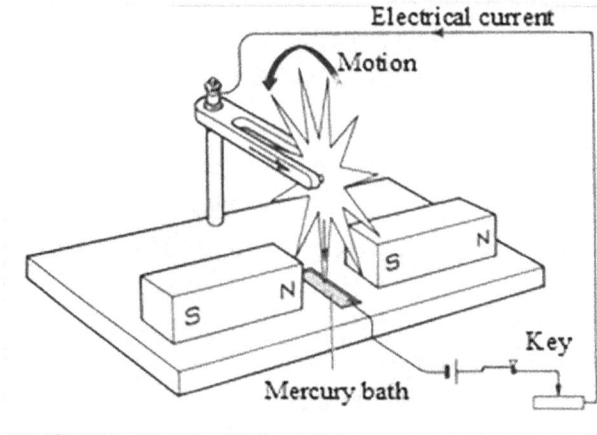

**FIGURE 1.1**    Barlow apparatus [7].

### 1.1.2 HISTORY OF ELECTROMAGNETIC MOTORS

In 1882, researchers designing electric motors preferred the electromagnet over the permanent magnet because of its cheaper price, lower maintenance cost, smaller size, and capacity to convert a larger amount of electrical energy into mechanical energy.

William Sturgeon is credited with inventing the first electromagnets in 1825. Later, in 1827, Istvan Jedlik designed the first motor with an electromagnet and commutator to expand into a modern DC motor. However, in 1834, the first practice machine of the DC electromagnets motor that had rated power and torque of 15 W and 13.5 Nm, respectively, was designed by Moritz Jacobi. In 1889, the alternating-current electromagnetic motors with their stator and rotor shaft parts were popular over the DC motors and were simpler in design than the electromagnetic DC machines [8].

The stator function is to generate rotating electromagnetic fields that the shaft intersects. Thus, electromagnetic torque is produced. This rotating electromagnetic field can be created by two or more phases of alternating current (AC) or what is called a polyphase system [9]. However, for several decades, the control problems of AC electromagnetic motors allowed the electromagnetic DC motors to be present in industrial applications, but permanent magnets are making a comeback on the horizon with further development [10].

### 1.1.3 THE RETURN OF PERMANENT MAGNET MOTORS

In the late 20th century, many scientists discovered new magnetic materials, such as cobalt, carbon, and wolfram steel, and they were used instead of the lodestone of permanent magnetics.

But these new magnetic types were not of high quality until the Alinco magnet, as it has a high quality in various industrial applications and completes the permanent magnet's journey in the machines [11].

In the 1930s, an Alnico magnet became available, which is 100 times higher in quality than any magnetic stone, as the important additives of nickel, aluminium, and cobalt are in solution with iron [12–16]. Moreover, in the 1950s, ferrite or ceramic magnets were developed and proved to be highly efficient in motors for small devices. In the 1960s, the use of permanent magnets in electric motors expanded when researchers came up with rare earth compounds of samarium and cobalt [17]. They reached neodymium, iron, and boron PMs in the 1980s, which led to high magnetic energy and great progress in PM DC motors [11, 18, 19].

For future permanent magnets, on the research prospects, a new type of permanent magnet has developed: the nanoparticle permanent magnet. This product is made of composite materials of rigid or flexible nuclear structures with dimensions less than a micrometre. This type is used in biomedicine, magnetic storage media, separation of magnetic particles, sensors, catalysts, and dyes. In the future, their use could be developed in electric motors [20, 21].

### 1.1.4 Main Construction of PMSM

The advantages of an induction motor and a synchronous motor are combined in a permanent magnet synchronous motor, so this motor deserves special attention. There are two designs of PMSM; one is sinusoidally powered and the other trapezoidal powered.

The sinusoidal PMSM is fed by three-phase sinusoidal waveforms and adopts the principle of a rotating magnetic field, as shown in Figure 1.2. Each stator phase draws an electric current from the source in a sinusoidal function at the same time, which is called a sinusoidal PMSM. The trapezoidal PMSM is also fed with a 120°

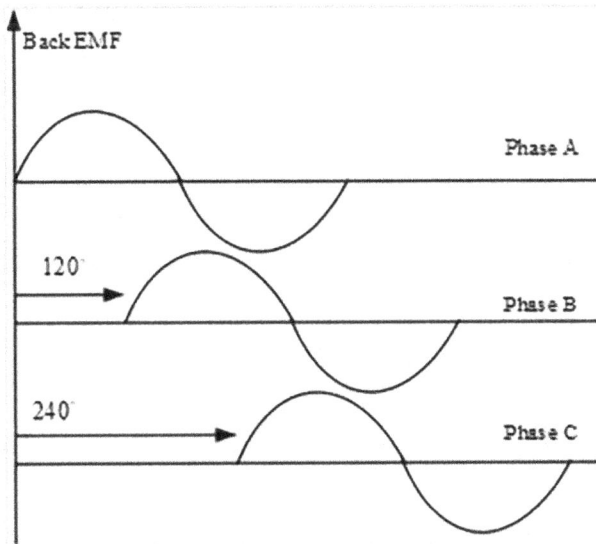

**FIGURE 1.2**   Sinusoidal excitation waveforms.

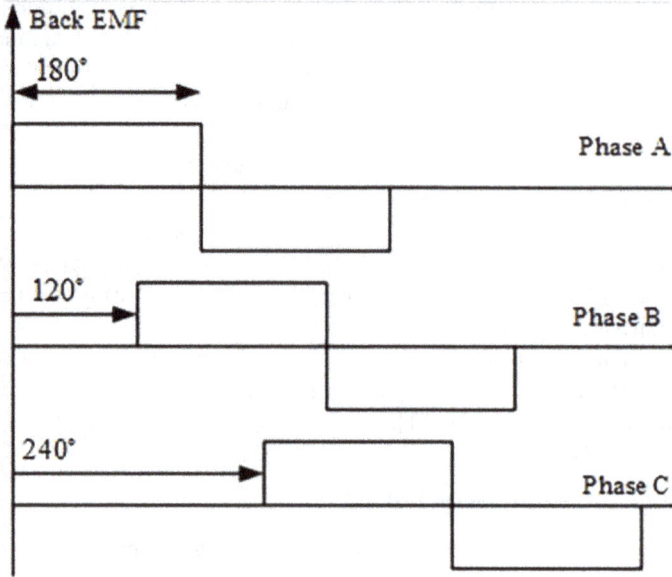

**FIGURE 1.3**  Trapezoidal excitation waveforms.

offset three-phase wave, which takes the form of a trapezoidal function, as shown in Figure 1.3 [22–26]. Based on the magnet structure, the PMSM can be classified as [27–31]

(1) Surface-mounted permanent magnet synchronous machine: For this type of PMSM, permanent magnets are usually attached to the surface of the rotor by a non-conductive material. For this type, the rotor runs at a speed no higher than the maximum rated speed written on the nameplate of the machine.

(2) Inset permanent magnet synchronous machine: For this type of PMSM, permanent magnets are usually connected inside the core of the rotor by a non-conductive material [32–36]. Also, this type is not suitable for running the rotor at a speed higher than the top speed written on the nameplate of the machine.

(3) Interior permanent magnet synchronous machine: For this type of PMSM, the magnets are buried inside the core of the rotor, which provides saliency or greater inductance for the $q$-axis than the $d$-axis inductance where $d$- and $q$- denote the abbreviation of direct and quadrature, respectively. This saliency allows operating in a high-speed range greater than the rated speed written on the nameplate of the machine. However, the magnets are difficult to volatilize while working at high speed.

Also, the smooth rotor surface is considered an advantage. Buried magnets result in mechanical durability and a smaller air gap than other design types. A comparison between external and internal PMSM is illustrated in Table 1.1 [37–41].

**TABLE 1.1**
**Comparison between External and Internal PMSM**

|  | External PMSM | Internal PMSM |
|---|---|---|
| Construction | simple | relatively complicated |
| Magnetic flux density compared to residual flux density | smaller | can be greater |
| Armature reaction flux | small | greater |
| PMs protected against armature field | no | yes |
| PM flux leakage | low | high |
| Flux-weakening capability | poor | great |
| Speed range | medium | wide |

### 1.1.5 PMSM Drive Applications Compared to the Drives of the Induction Motor and Brushless DC Motor

The benefits of using PM bars in building electrical machinery include the following [2]:

(1) There is no voltage applied to excite the machine, so the efficiency of the machine increases due to no losses in the excitation circuit.
(2) This type of electric motor has higher output power and electromagnetic torque than motors that rely on electromagnetic excitation.
(3) The air gap of the motors is dependent on PMs, which have a greater flux density than the motors dependent on the excitation circuit and, thus, better dynamic performance.
(4) The construction design is not complicated compared to the electromagnetic excitation machines. Therefore, maintenance is simple.
(5) The advantage of the PM machines is reducing maintenance costs for some types and preferences in motor life.

The use of induction motors in industrial applications grew in the early 20th century, which encouraged remarkable progress in control methods and industrial electronics [42–46]. The advantages of induction motors are that they are easy to construct, wind, and maintain and are low priced due to the electromagnet. But researchers continued to develop permanent magnets until they were used in electric motors with precision, and this was due to the low power factor of induction motors and their low efficiency.

As a result, the use of motors has become an attractive option for industrial applications, traction systems, and electric vehicles (EVs) to operate in a wide speed range in the scope of field decreasing, especially when the price of the rare earth magnet is low [47–51].

Responding to the question of which is better for traction systems and EVs, PMSM motors or induction motors, both motors are used in traction applications and offer good performance and efficiency. There are strong indications that the PMSM is better than the induction motor as it is an effective tool in producing a drive system, the total cost is lower, and the advantages lie in the mineral powders used in their manufacture.

Synchronization of the shaft speed with the frequency of the current drawn is achieved when a three-phase voltage is applied to the PMSM stator. This is because the magnetization of the rotor is permanent. This eliminates the problem of induction motor slip related to lowering the overall efficiency of the motor. Moreover, to avoid losses caused by eddy current in the induction motor, it was designed to operate at a frequency of 60 Hz, knowing that eddy current losses increase with the increase in the source frequency [52–56].

Therefore, dealing with non-rated speed requires the manufacturer to design a cooling circuit to prevent overheating. Thus, the permanent magnet motor provides a large range of energy. From this point, the PMSM provides a large range of energy with the magnet produced using metallic powder technology.

For EVs dependent on PMSM, the 50 kW PMSM weighs 13.6 kg without a power inverter for voltage and frequency configuration [57–61]. Examples of EVs that use this type of machine include the third Tesla model, the Chevy Bolt, and the Chevy Volt. The Chevy Bolt uses a 149.14 kW PMSM and a gearbox rated 7.05: 1 to reduce the speed of wheels. The PMSM is used in the third Tesla model. Little information is available to researchers, but the manufacturers may improve the performance of the motor by focusing on flow lines through a set of magnets arranged in the Halbach group.

The equation for calculating the speed of the PMSM is like the equation for calculating the speed of the induction motor and can be written as

$$N_s = 120F / (2n_p) \tag{1.1}$$

where $F$ is the voltage source frequency, $n_p$ is the number of pole pairs, and $N_s$ is the synchronous speed. However, the rotor speed of PMSM will not be less than the stator flux speed as it is in an induction motor. For example, if an induction motor was designed to be two pole and operated at a voltage source of 60 Hz, then the stator flux speed would be 3600 rpm. But because it is an asynchronous motor, the slip reduces the speed of the rotor shaft, which affects the motor's efficiency. Figure 1.4 provides the relationship between rotor velocity or slip and both torque and current. Among the disadvantages of the use of induction motors in EVs is their heavy weight; induction motors with a capacity ranging from 37.3 to 45.4 kW weigh from 317.5 to 453.6 kg. The main advantage of EVs based on induction motors is that they are relatively cheap to build.

Table 1.2 lists the performance comparison between 600 W IPMSM and 600 W induction motors. As can be seen from this table, the IPMSM gives better results than the induction motor. It is worth noting that the efficiency of the IPMSM is more than 35% higher than that of the squirrel cage induction motor [62].

While the PMSM appears to be a better alternative to the induction motor, a BLDC motor was designed to replace the brushed DC motor to avoid its design defects. The brush is absent from the PMSM design to enable it to work in the high-speed range and to increase its efficiency. Once DC voltage is applied to the stator windings, DC passes through the windings, then flux is created, and then the permanent magnet rotor is turned. Therefore, the speed is controlled by controlling the DC voltage [63–67].

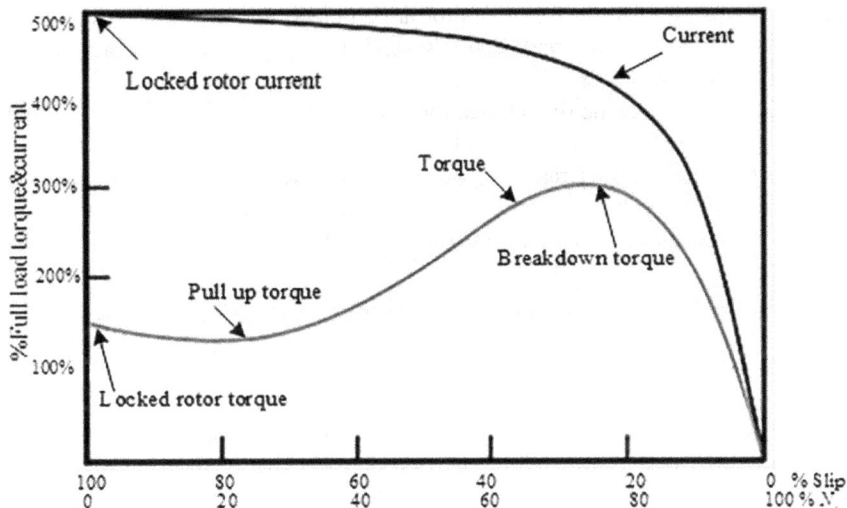

FIGURE 1.4 Performance of induction machine in terms of slip.

TABLE 1.2
Performance Comparison of Interior PMSM versus Induction Motor

| | IPMSM | Induction Motor |
|---|---|---|
| Rated stator voltage (V) | 130 | 130 |
| Rated current (A) | 3.11 | 3.43 |
| Rated power (W) | 687 | 818 |
| Rated speed (rpm) at 50 Hz | 1500 | 1434 |
| Rated torque (Nm) | 3.82 | 4 |
| Efficiency (%) | 87.3 | 73.3 |
| Power factor (W) | 98.1 | 68.8 |
| Output power (W) | 600 | 600 |

The control of a PMSM and BLDC motor is the same except for the positioning of the rotor and electric current controller. Three Hall Effect sensors are used to locate the rotor in a BLDC motor, and usually, the commutator operates in six steps, resulting in small breaks that cause torque ripple [68–71].

However, PMSM requires one Hall Effect sensor because the commutator is continuous, and thus PMSM is distinguished by the absence of high-torque ripples compared to BLDC motors [72–76]. Both motors are used in industrial applications and traction systems and can be interchanged, but the ruling standard is the load characteristics. The features that make the BLDC motor an ideal motor for continuously running EVs include its low cost and moderate efficiency. On the other hand, PMSM includes these features in addition to higher efficiency and less noise. The systems that distinguish the BLDC motor are an electronic power-steering system; heating, ventilation, and air conditioning system; hybrid EV drivetrain; and regenerative braking system.

As for PMSM, this is the first choice for working in servomechanisms in automobiles [77–81]. This is because it is resistant to wear and tear, produces less noise, and adds to its high efficiency. Brakes are one example that amplifies the power coming from the driver's feet on the pedals. Another example is the steering system because it is more energy intensive and provides many control solutions [82–86].

## 1.2 LITERATURE REVIEW FOR THE PMSM CONTROL SCHEMES

This section discusses the challenges of speed-control strategies in the regions of constant torque, weakening of flux, and reduction of current and torque ripples in a permanent magnet synchronous motor. Therefore, the drive control methods rely on more than one principle to reach the desired performance of the controlled motor, such as direct torque control (DTC), frequency converter method, field-oriented control (FOC), hysteresis current control, the principle of sliding mode control, and, finally, predictive control model (MPC) algorithms. The following sections explain more about these control strategies and the results they achieve in the dynamic performance of this motor [87–91].

### 1.2.1 VARIABLE FREQUENCY CONTROL OF PMSM

The frequency of the AC voltage source should be changed to control the speed of PMSM. Therefore, the variable frequency drive is divided into two methods of frequency control: scalar control and vector control. Figure 1.5 shows the types of variable frequency drive according to their main function. Scalar control is the

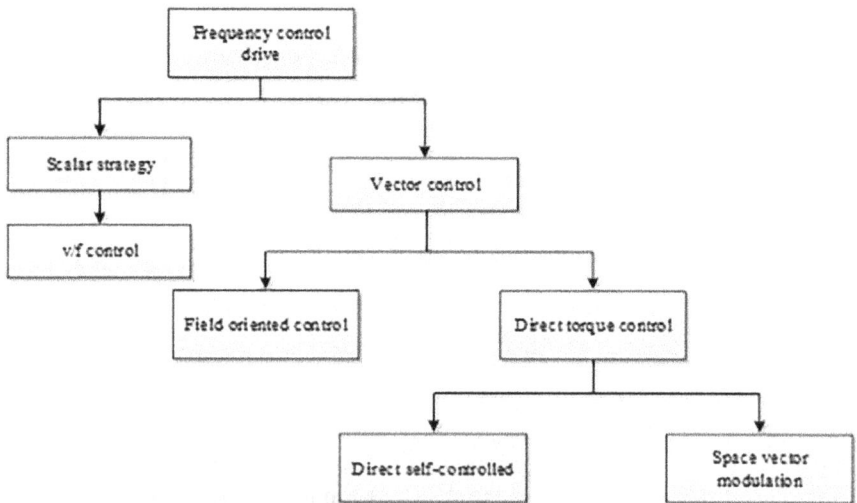

FIGURE 1.5   Overview of the frequency control algorithms.

simplest way to control the motor speed, where the frequency is set in proportion to the required speed, and the voltage is adjusted so that the ratio is constant with the frequency.

The advantages of this method are that it does not depend on machine parameters, less calculation, ease of use, and no need to control angles. However, a damper winding is used to synchronize the rotor frequency with the source frequency to increase the stability of the drive system at a certain limit of the applied frequency range, which is not suitable for PMSM [92].

Moreover, the absence of feedback reduces the dynamic performance of the systems, so a vector control algorithm is preferred [93–97]. There are two types of vector control—direct torque control and field-oriented control—both of which control the scale and angle of the flux, which gives more stability to the drive system than scalar control. In the field-oriented control, the $d$-$q$ axes of the stator current components are directly controlled. The control of the $q$-axis of the PMSM stator current ($i_{qs}$) is indirect control of the induced electromagnetic torque, and the control of the $d$-axis of the PMSM stator current ($i_{ds}$) is the indirect control of the level of flux. Therefore, it is possible to obtain the required torque at a minimum current and increase motor efficiency.

For the SPMSM (non-salient machine), the induced torque depends on the $q$-axis of the stator current. Thus, it is easy to reach the highest torque at the minimum stator current at zero $d$-axis stator current throughout the control time with the torque trajectory linear on the $d$-$q$ plane. In contrast, with the IPMSM (salient machine), it is more difficult to achieve the maximum electromagnetic torque at the minimum magnitude of stator current because the torque depends on both the $d$-$q$ axes of the stator current, which makes the torque trajectory nonlinear in the $d$-$q$ axes. The stator current calculations of $d$-$q$ axes to obtain the highest electromagnetic torque at minimum stator current are discussed in [98–102]. This point of maximum torque per ampere (MTPA) control is discussed in detail in Chapter 2.

The traditional FOC relies on an external dynamic controller to adjust the speed and an internal controller to control the stator current, and in each controller, the proportional-integral (PI) regulator is an essential component with the limiter, as shown in Figure 1.6. In this figure, PWM denotes pulse-width modulation; $i_{ds}^*$ and $i_{qs}^*$ are $d$-$q$ axes of the reference stator currents; $i_A$, $i_B$, and $i_C$ are the instant measured values for three-phase ABC current of the stator; $\Theta_m$ is the counted mechanical angle of rotor position; $\omega_m$ and $\omega_m^*$ are the calculated rotor velocity and reference velocity; $V_{dc}$ DC links voltage of the DC circuit for the inverter; and $v_{ds}^*$ and $v_{qs}^*$ are the $d$-$q$ axes of the reference stator voltages. However, these controllers depend on the machine parameters, especially the $d$-$q$ axes inductances, which change with the operating conditions due to saturation.

In 1984, Takahashi and Noguchi were the first researchers to offer direct torque control [103]. Direct self-control was introduced in 1985 by researcher Diebenbroek [104]. These two methods are characterized by their simplicity and strength in a good performance regarding the FOC method. However, one of their requirements is the estimation of torque and flux to determine the reference voltage vector, and there is no need to measure the position of the rotor, as shown in Figure 1.7 [105–109], where $T^*$ is the reference torque, $T_{e\text{-est}}$ the estimated electromagnetic torque, $\psi^*$ the

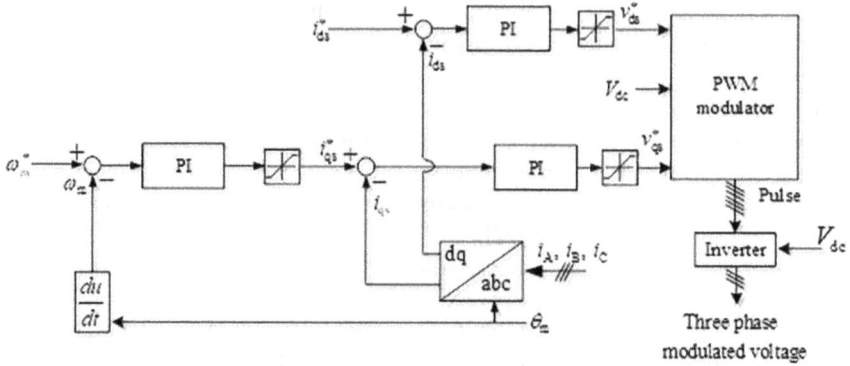

**FIGURE 1.6** Block diagram of the field-oriented control strategy.

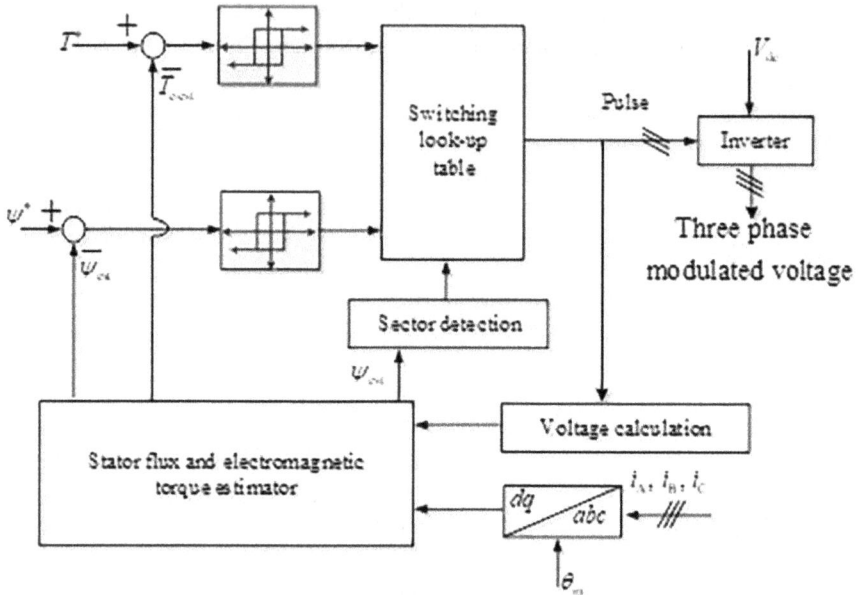

**FIGURE 1.7** Block diagram of the DTC strategy.

reference flux vector, and $\psi_{est}$ the estimated stator flux vector. The disadvantages of direct torque control appear at higher speeds, with higher stator current and torque ripple [110]. Therefore, for better direct torque control performance, researchers suggest improving the switching table, increasing the number of sectors, and using a constant switching frequency for the inverter [111, 112]. A comparison between FOC and DTC strategies is presented in Table 1.3.

There are several methods to extend the speed range (flux-weakening region) of the PMSM drive system [113–117]. To improve maximum speed range and maximum power, the novel method of PMSM motor proposed in [113] uses a three-dimensional

**TABLE 1.3**
**Comparison between FOC and DTC Strategies**

|  | FOC | DTC |
| --- | --- | --- |
| Ripple peak for induced electromagnetic torque and stator flux | lower | higher |
| Dynamic performance response | slower | faster |
| Coordinate transformations | need | no need |
| Inner current controller | need | no need |
| Depending on the machine parameters | critical | only in the resistance |

finite element to create a difference gap between the stator and the rotor in an axial direction without moving the current vector. However, the method depends heavily on calculating the no-load back-EMF was necessary through the computation of the linkage magnetic flux of stator windings and $d$-$q$ axes inductances.

These parameters were analyzed with the use of a three-dimensional finite element method according to the axial gap between the stator and the rotor. In contrast, the authors in [114] depend on the two-dimensional lookup tables function of machine torque and flux to determine $d$-$q$ axes' reference stator currents.

The authors propose a new flux-weakening control loop in the DTC strategy to determine the required reference flux operating the IPMSM in the desired reference velocity, as shown in Figure 1.8, where $v_{ds}$ and $v_{qs}$ are the PMSM stator voltages of $d$-$q$ axes, $\lambda$ is the stator flux, a MAX subscript indicates maximum value, a "base"

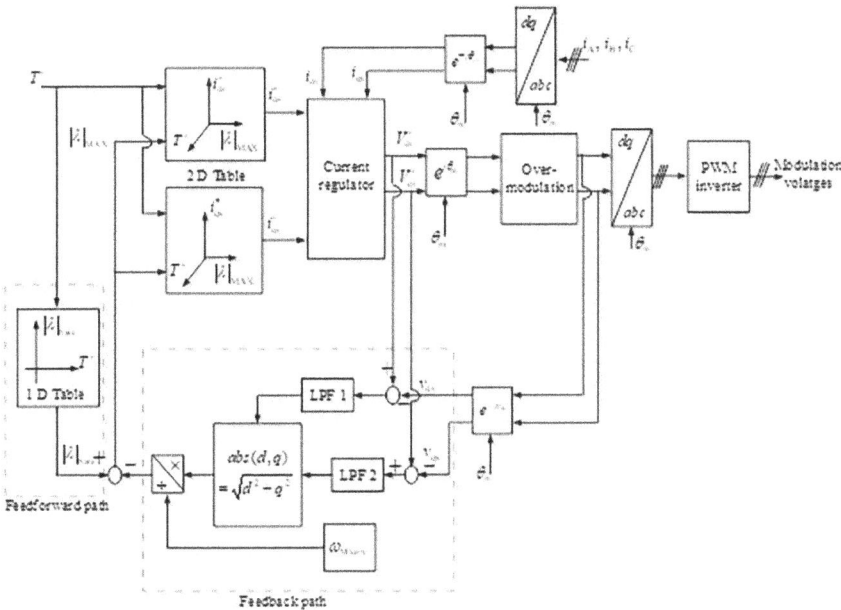

**FIGURE 1.8** Block diagram of the proposed feedback path of DTC strategy [114].

subscript indicates the nominal value, and LPF denotes a low-pass filter. The operation of the proposed feedback path for the flux-weakening goal depends on the error between the compensated voltage and the output voltage of overmodulation. The proposed strategy controls the modulation index of the inverter to implement the quasi-six-step operation. This quasi-six-step operation keeps the controllability of the synchronous-current regulators. Moreover, without the use of additional devices when testing this algorithm, it became clear that IPMSM showed excellent and durable performance with an increase of 10% in torque at the maximum speed in the region of field decreasing compared to the conventional methods. It is considered that the authors obtained these results by simply modifying the software.

In [115], the proposed method was realized based on the measurement of inverter characteristics and the resistance of the used cable. This control method uses the voltage differences of the input and output inverter to modify the compensated $d$-axis current, as shown in Figure 1.9, where $v^*_{\alpha s}$ and $v^*_{\beta s}$ are the reference voltages of the $\alpha$-$\beta$ axes, and $\Delta$ indicates the rate of change. Additionally, a PI controller uses the reference $q$-axis current to set the maximum torque per voltage. Therefore, maximizing torque is accomplished at a high-speed range. When examining this improved flux-weakening scheme on a PMSM with large inductance and operating in a high-speed region, it was concluded that the motor's performance in the current trajectory improved the performance of the maximum torque at the minimum reference voltage (MTPV). Moreover, the electromagnetic torque and the efficiency of the motor have increased. The required issues of this algorithm are that it relies on knowledge of the cable resistance, coil impedance, inverter parameters, and setting of LPF. However, it does not need complicated calculations or additional devices. The authors in [116] propose an improved DTC strategy for the IPMSM drive to work in the field-weakening region. It differs from the FOC as in DTC, the flux and torque are directly controlled, as shown in Figure 1.10.

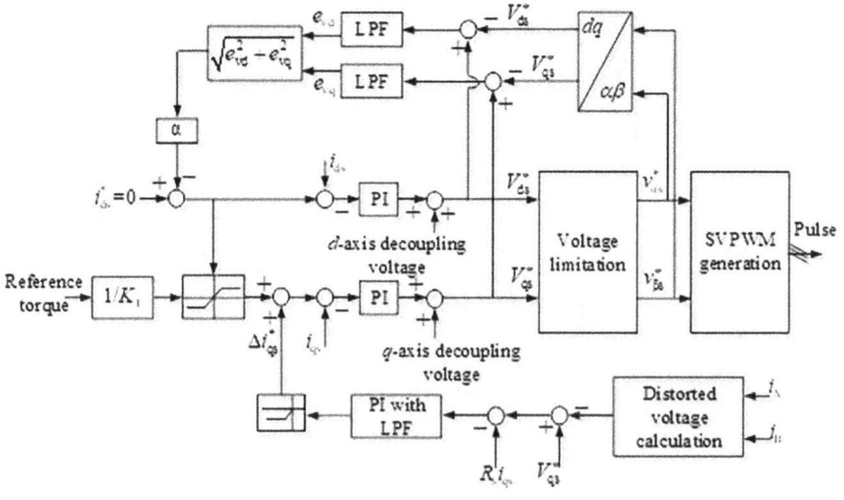

FIGURE 1.9    Block diagram of flux-weakening control of SPMSM having a large inductance [115].

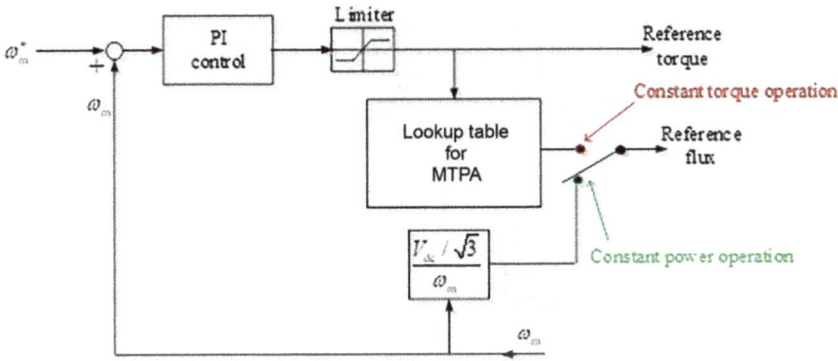

**FIGURE 1.10**   Block diagram of the outer control loop [116].

Therefore, the proposed method differs in several ways; this strategy uses a first-order estimator to estimate flux and torque. For both regions of the constant torque and constant power, the outer speed control loop, shown in Figure 1.10, calculates the references of the flux and torque. Moreover, it also uses hysteresis comparators instead of internal current control loops. This strategy gives an excellent performance to achieve the MTPA in each of the constant torque and flux-weakening regions.

The authors in [117] realized the flux-weakening control method determines the $d$-$q$ axes reference currents to drive an IPMSM at any value of speed. This method minimizes the magnitude of current to reach the reference torque, considering the addition of the voltage and current limits and battery power. For details, the magnitude of the compensated voltages of $d$-$q$ axes obtained by the current controller has been compared with the maximum voltage that the inverter can provide for generating the demagnetizing $d$-axis current. The proposed saturation equation has been used to calculate the limiting current to limit the demagnetizing required current.

This saturating equation depends on the maximum current passing through the inverter and the motor circuit, the current produced by the conventional MTPV algorithm, and the current produced by the proposed torque algorithm, as shown in Figure 1.11, where KFW is the flux-weakening factor, $I_{d\text{-minCPT}}$ is the $d$-axis current produced by the minimum current per torque algorithm, $I_{d\text{-MTPV}}$ is the $d$-axis current produced by MTPV algorithm, $P_{\max}$ is the PMSM maximum power, $T_{\text{sat}}$ is the saturated reference torque, $I_{\text{rated}}$ is the PMSM rated current, $\lambda_{m}$ is the flux linkage due to the rotor magnets linking the stator, $I_{s\text{-max}}$ is the maximum current passing through the inverter and the motor circuit, $L_{ds}$ and $L_{qs}$ are $d$-$q$ axes stator inductance of PMSM, and $z$ indicates the discrete domain.

Meanwhile, the $q$-axis reference current is responsible for generating the required reference torque. Therefore, the proposed torque algorithm has been used to limit the reference torque, considering the operating speed and the maximum limit of power and current.

The line-modulation method proposed in [118] was employed to run in the high-speed range. The negative $d$-axis reference current was determined by the regulation process of the maximum line-modulation ratio to 1 to maximize the inverter voltage

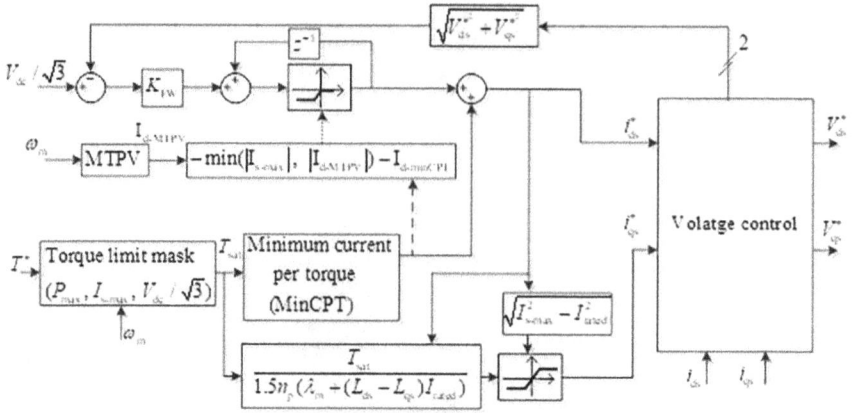

**FIGURE 1.11** Block diagram of the torque algorithm and saturation equation of current controller [117].

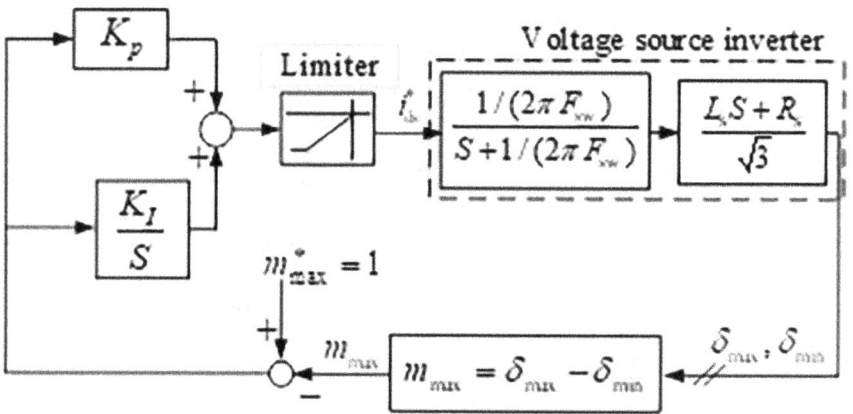

**FIGURE 1.12** Block diagram of the line-modulation control loop [118].

to run the motor in the field-decreasing region, as shown in Figure 1.12, where $K_p$ and $K_i$ are the proportional and integral gains for a PI regulator of flux decreasing, $m_{max}$ and $m^*_{max}$ are the maximum and maximum reference of a line-modulation ratio, $\delta_{max}$ and $\delta_{min}$ are the maximum and minimum values of the duty cycle, $R_s$ is the PMSM stator resistance, $L_s$ is the stator inductance of PMSM, $S$ indicates the Laplace domain, and $F_{sw}$ is the sampling frequency of the inverter.

This approach maximizes the inverter voltage to reduce copper losses. However, the performances of model-based flux-weakening controls are affected by the variation of motor parameters. Among the results added by this research is that the voltage trajectory in the $\alpha$-$\beta$ axes stationary frame takes a hexagonal shape instead of a circuit. It indicates the exploitation of the largest amount of DC voltage of the inverter and causes increasing efficiency. However, the performances of this control algorithm are affected by the variation of motor parameters.

Finally, a frequency control drive for electric motors is often turned off because the PI parameters of the PI regulators are difficult to tune. Meanwhile, when controlling the motor in a region of the flux weakening, a control loop is added to determine the reference demagnetizing current with another PI regulator, which leads to an increase in the number of PI parameters to be set. This is what will be discussed in Chapter 3, which offers a practical solution suitable for academic studies and industries.

### 1.2.2 CURRENT CONTROLLER DESIGNING FOR PERFORMANCE IMPROVEMENT OF A PMSM DRIVE

Different methods have been proposed to improve the current regulator performance of PMSM system drives [119–123]. Thus, a speed PI controller design scheme is proposed in [119] to mitigate torque ripple in PMSMs operating in one high vehicle inertia, as shown in Figure 1.13. The principle of the proposed algorithm is to modify the speed PI controller and regulate the $q$-axis reference current. Additionally, a resonance controller (RC) is adjusted to set in the external speed loop. As a result, the frequency bandwidth is increased in a specified scope. Therefore, the proposed technique effectively controls the current ripple to confirm the reduction in torque ripple. Finally, the improved speed regulator unit demonstrated a significant reduction in velocity ripple of up to 83% when compared to the classic PI controller.

A new algorithm was proposed in [120] to measure and analyze the dynamic decoupling phenomenon of a PMSM operating as one multi-input multi-output drive. The dynamic relative gain array (DRGA) has examined the performance of different decoupling current loops and optimized the design of current regulators.

Thus, a two-degree-of-freedom regulator is designed, relying on the algorithm of DRGA as shown in Figure 1.14, where $G_{sc}(S)$, $G_{fc}(S)$, and $G_P(S)$ are the regulator matrices of the serial, feedback, and plant, respectively. For details, the DRGA is a gain matrix between system drive inputs and outputs. Meanwhile, the two-degree-of-freedom (2DOF) regulator has been optimized for the feedback of the DRGA matrix.

The feature of the proposed current regulator is less sensitive to errors in machine parameters. However, this design algorithm is not concerned with lowering the

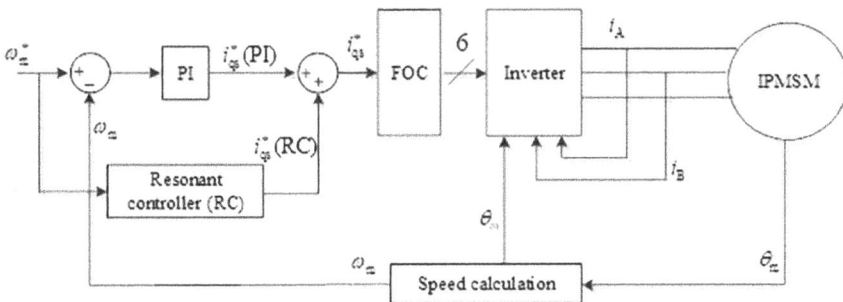

**FIGURE 1.13**   Block diagram of the resonant controller in conjunction with a PI controller control [119].

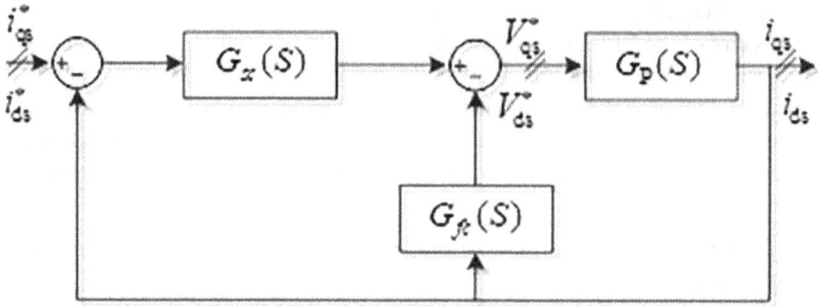

**FIGURE 1.14**   Block diagram of the structure of the 2DOF regulator [120].

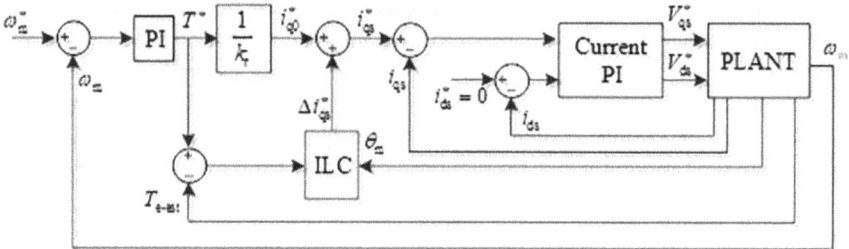

**FIGURE 1.15**   Block diagram of the PI control scheme integrated with ILC [121].

ripple peaks of current or torque. In [121], a novel technique of the current regulator is proposed based on a torque-estimation model, iterative learning control (ILC) model, and conventional PI current regulator, as shown in Figure 1.15, where $k_t$ is the constant torque coefficient, $i_{q0}{}^*$ references the $q$-axis current deduced from torque controller, and $i_{q0}{}^*$ references the $q$-axis current deduced from the ILC algorithm.

Thus, a sliding mode observer is used to estimate the flux for torque calculation, and a new shaping algorithm is employed to decrease the ripple of the estimated flux caused by the high-frequency modulation technique. Meanwhile, the ILC scheme can first record the calculated torque and reference currents to be used in the next period. The proposed approach reduces torque ripple and velocity fluctuation, according to comprehensive results [122].

For the contribution in [123], a single $q$-axis current regulator is designed to control the voltage angle to drive the PMSM in the flux-weakening range, as shown in Figure 1.16, where $u^{fw}_{d,wkfd}$ and $u^{fw}_{q,wkfd}$ are $d$-$q$ axes referencing voltage deduced from the flux-weakening control loop.

Moreover, $i^*_{ds,wkfd}$, and $i^*_{qs,wkfd}$ reference stator current of the $d$-$q$ axes deduced from the flux-weakening control loop, $u^{fw}_{s,wkfd}$ references the vector of stator voltage deduced from the flux-weakening control loop, and $i^*_{ds,MTPA}$ $d$-axis references current deduced by MTPV algorithm. The authors have demonstrated that the transfer

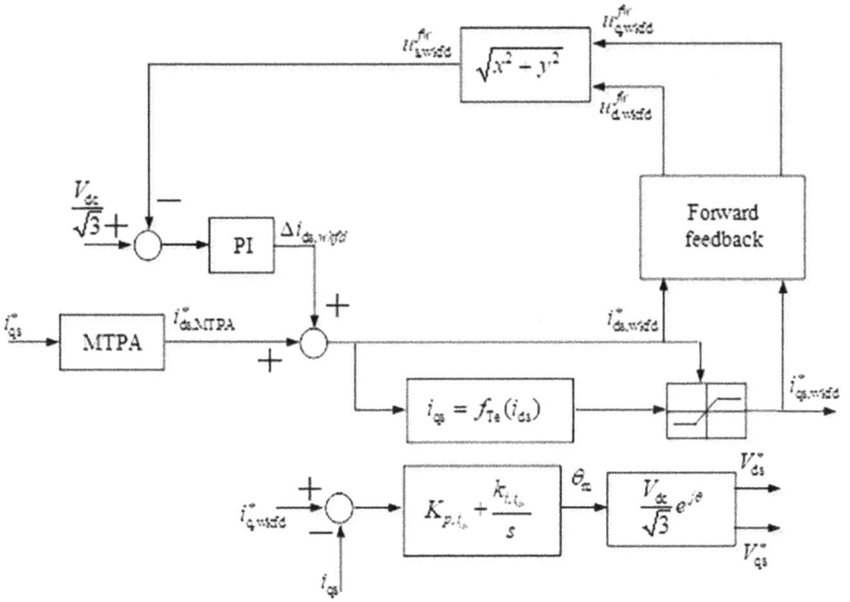

$$\sqrt{x^2 - y^2}$$

Forward feedback

$$\frac{V_{dc}}{\sqrt{3}}$$

PI

MTPA

$$i_{qs} = f_{Te}(i_{ds})$$

$$K_{p,i_q} + \frac{k_{i,i_q}}{s}$$

$$\frac{V_{dc}}{\sqrt{3}} e^{j\theta}$$

**FIGURE 1.16** Block diagram of the flux-weakening control loop to control the voltage angle [123].

function can be used to design the $q$-axis current controller, which has a voltage angle output. Meanwhile, a full analysis is made to optimize the coefficients of the proposed transfer function. Based on what was mentioned, this method has very strong stability in reaching a zero-load state and a fast-dynamic response to torque and currents when their references are changed. However, torque ripples are not considered in this study.

The method in [124] aims to reduce the scan period of the microprocessor unit in use. Hence, the proposed method is suitable for the real-time industrial drives required to operate several actuators. The scholars of this reference consider that PWM output is the ultimate in the controller unit every time. In this study, the delay time is studied and connected to the stability of the PMSM system. Researchers also reduced the system to a single input and a single output and applied the control theory to study the delay in the drive. Meanwhile, the current PI regulators and their delay time are studied to provide wide stability for the system.

Figure 1.17 proposes the suggested structure of the current loop with delay, where $EMF_d$ and $EMF_q$ are the permanent magnets back EMF of $d$-$q$ axes, $T_d$ is the total time delay, $k_{PWM}$ is the inverter gain, and $\omega_e$ is the electric PMSM velocity. Finally, the delay time is modelled to apply the current regulator to the integrated circuit co-working with the controller unit of digital signal processing (DSP) to reduce the time of the PWM loading.

The fuzzy logic optimizer in [125] is built into current PI regulators of $d$-$q$ axes for optimizing PI parameters to drive the PMSM, as shown in Figure 1.18, where $\eta_1$, $\eta_2$, $\eta_3$, $\eta_4$, $\eta_5$, $\eta_6$, and $\eta_7$ are the coefficients function of machine

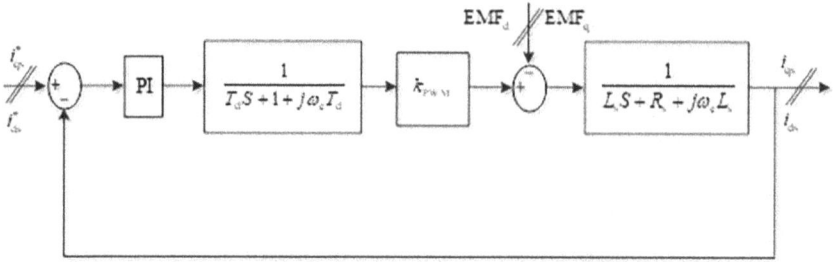

**FIGURE 1.17**    Block diagram of the time-delay control strategy [124].

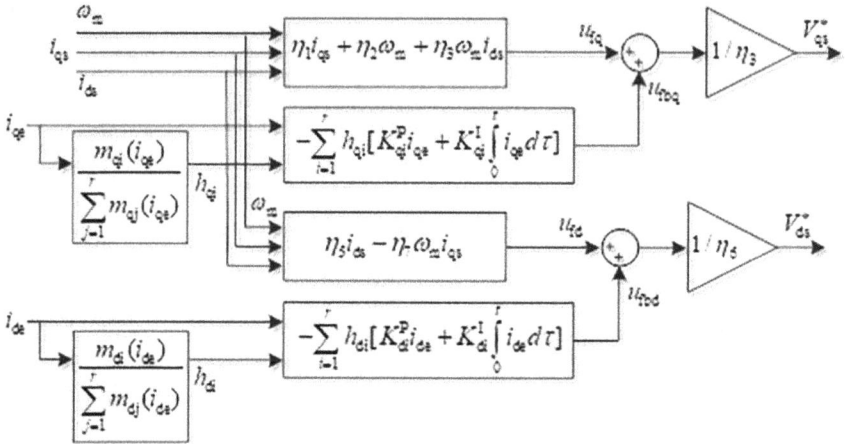

**FIGURE 1.18**    Block diagram of the fuzzy logic current regulator algorithm [125].

parameters; $u_{fd}$, and $u_{fq}$ are $d$-$q$ axes nonlinear decoupling control terms; $u_{fbd}$ and $u_{fbq}$ are $d$-$q$ axes stabilizing fuzzy control terms; $h_{di}$ and $h_{qi}$ are $d$-$q$ axes normalizing the weight of each IF-THEN rule; $m_{di}$ and $m_{qi}$ are $d$-$q$ axes of the membership function of the fuzzy logic algorithm, and $i_{de}$ and $i_{qe}$ are $d$-$q$ axes of the error of current regulators.

In this fuzzy logic work, the FOC of the machine model and PI regulators are analyzed to determine the matrix of the state error based on the PI gains and sampled signals of the speed and stator currents for the $d$-$q$ axes. Afterwards, the parameters and membership function of the fuzzy logic strategy are defined, relying on the different errors. Therefore, the proposed algorithm calculates the PI gains based on the selection of the membership functions. Lastly, stator current controllers can determine the compensated stator voltages for $d$-$q$ axes, depending on the optimized gains. As a result, the improved PI strategy based on fuzzy logic has a significant impact on the

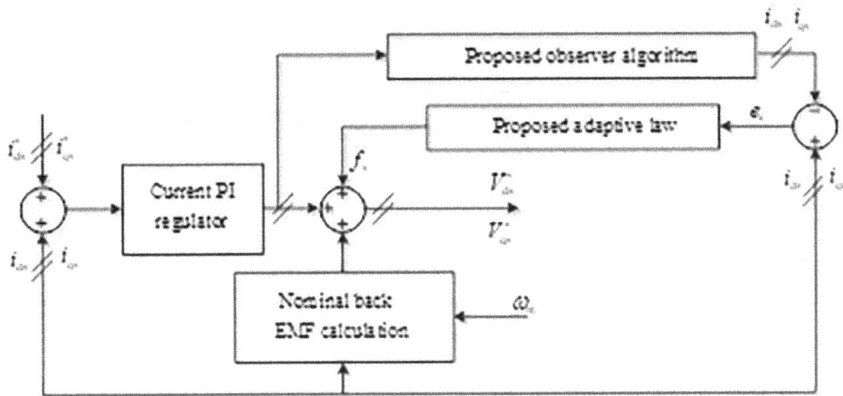

**FIGURE 1.19** Block diagram of the current regulator of the adaptive disturbance observer algorithm [126].

quality of the dynamic performance of the drive system, which can mitigate the peak ripple of both torque and current.

For a higher-performance PMSM drive system, [126] proposes a robust current control approach. Figure 1.19 shows how the suggested adaptive disturbance observer, based on the feedforward technique, can help the stator current controller at the step of compensated voltages for the $d$-$q$ axes, where $\hat{f}_s$ is a lump of the uncertainties that are caused by the variation of the parameters, $e_s$ is the error vector that has been estimated, and $\hat{i}_{ds}$, and $\hat{i}_{qs}$ are the estimated stator currents for $d$-$q$ axes, respectively. The proposed control strategy has been used to develop a simple adaptive algorithm with high bandwidth for uncertainty problems. Meanwhile, the stability of the stator current control technique has been validated using the Lyapunov function. The results show that the discrete time–proposed strategy is effective in reducing ripples in both current and torque due to the elimination of unwanted harmonics in the flux linkage. Also, the current control performance is improved to some extent.

In [127], a current regulator design scheme is proposed to improve the steady-state performance of both PMSM and converter. The proposed method, shown in Figure 1.20, controls the time delay well to estimate the disturbances in the motor caused by the parameter changes, where $T_s$ is step time; $\hat{f}_{df}(k)$ and $\hat{f}_{qf}(k)$ represent the $d$-$q$ axes' estimated disturbances caused by parameter variations; $v_A$, $v_B$, and $v_C$ are instant measured values for three-phase ABC stator voltages, and "$k$" and "no" denote discrete time and nominal values, respectively.

Meanwhile, the estimated disturbances are used to calculate the reference stator voltages of $d$-$q$ axes relying on the feedforward technique. Therefore, the frequency control strategy for electric motor drives is often substituted due to the difficulty of controlling torque and current ripple when controlling the motor in the flux-weakening region. This will be discussed in Chapter 4, which provides an improved FOC method to control the torque and current ripple.

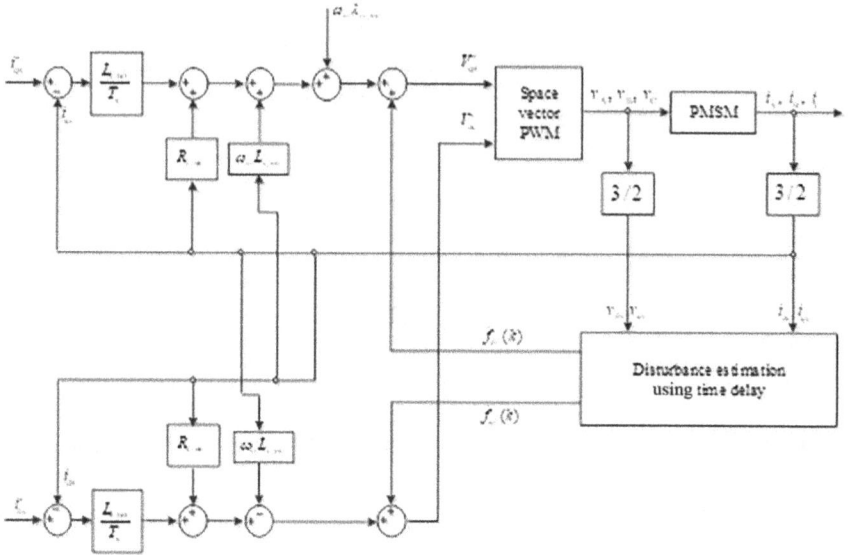

**FIGURE 1.20**   Block diagram of the current control algorithm [127].

### 1.2.3 MODEL PREDICTIVE CONTROLLER DESIGN FOR PMSM PERFORMANCE ENHANCEMENT

The model predictive control strategy is a new control algorithm that can consider all control limitations by a single prediction time. The drive control strategies of several industrial and commercial systems, such as chemical plants, electric traction systems, and oil refineries [128–132], are based on the principle of the MPC approach.

For the MPC strategy theory, the suggested cost function, based on the discrete plant model, is solved in the optimization process. At every sample time ($t$ in the time dimension), the microprocessor unit samples the motor speed and stator current of $d$-$q$ axes to predict the dynamic performance in the future based on a short time horizon $((t+T_s)$—t$)$ using a finite control set evaluation technique or a numerical optimization technique, as shown in Figure 1.21 [133–137]. In Figure 1.21, the optimization technique depends on the iterative evaluation of the measured state and reference state for the cost function. Lastly, the optimization technique backs the recommended stator voltage of reference $d$-$q$ axes as manipulated variables that correspond to a minimal value to the cost function. Thus, the best manipulating variables are employed to drive the inherent motor, now with belief that the motor performance can meet the dynamic performance requirements [138–142]. Therefore, the new sample of the plant states is measured again, and the set of orders of the optimization algorithm is repeated, starting with the new measured variable states of the system, yielding a new predicted trajectory [143–147]. It can be said that MPC is a multivariate nonlinear control strategy that uses a separate interior model to process a cost function that depicts the dynamic performance of the plant system by evaluating it using a

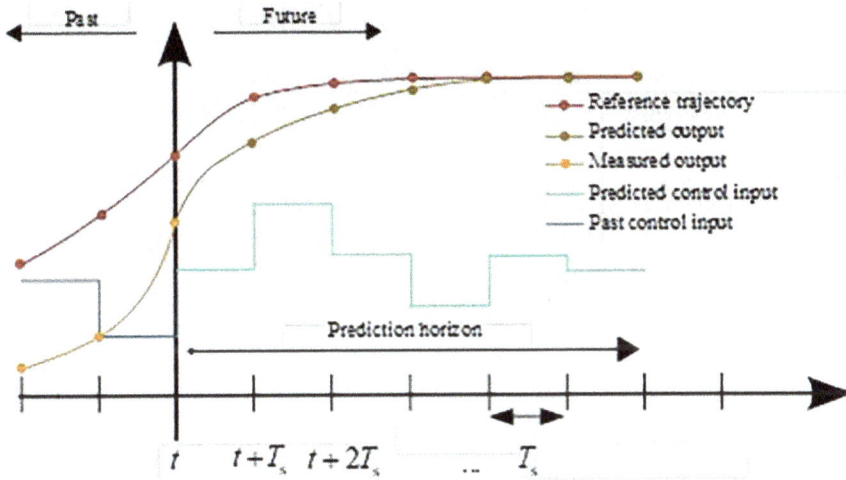

**FIGURE 1.21**   Basic model predictive control algorithm scheme.

minimizing optimization technique [148–152]. The MPC has various characteristics that make it a successful control algorithm for industrial applications:

(1) The concept of the MPC algorithm is simple. It does not rely on complex mathematical reasoning. It is a straightforward math.
(2) The productivity of the MPC algorithm comes from the fact that its parts can be tailored to the specifics of the control issue. While the MPC strategy can confirm all the challenges of industrial control systems, it has some challenges during its implementation, such as
   (a) The number of the controlled state variables can be enormous, resulting in a delay in the processing time of the digital control unit.
   (b) The cost function of the drive control system has several objectives to control the state variables. Thus, the optimization technique is needed as a high qualification to solve the cost function and find the best-manipulated variables.

The MPC strategies can predict the output variables of a plant system that change through changes in the manipulated variables. Therefore, the types of MPC controllers are determined by the approach used by the plant model, and they are linear MPC controllers, adaptive linear controllers, and nonlinear controllers. A linear MPC control is used in most nonlinear systems by approximating the system in a certain range as a linear relationship between inputs and outputs. This is because it is easy to deal with the linear MPC controller and to reach the optimal variables in the least amount of time. The adaptive MPC controller differs in that the equilibrium points are always updated to recalibrate the nonlinear plant model at another range. As for the nonlinear controller, there is no need to approximate the plant model definitively; it is content to just convert it from a continuous model to a discrete model like the two other types.

One of the fundamentals of an MPC controller is the objective function. The objective function or the fitness function is one of the most important research points in the field of MPC controllers because it is based on scientific reasoning aiming to arrive at the optimal manipulated variables at the predicted step time. Meanwhile, the fitness function links the manipulated inputs, outputs, and state variables. Additionally, each term in the objective function is multiplied by constant or variable weight factors, according to the proposed algorithm. The weight factors are used to determine the significance of the predicted output when certain conditions serve the primary objective of the MPC controller.

Based on the evaluation algorithm, MPC controllers are classified into two types: the MPC finite-set control type and the MPC continuous control type. In the finite-set control, the manipulated variables use the voltage vectors of the voltage source inverter to minimize the proposed fitness function while in the continuous control MPC, the convex or non-convex optimization algorithm is used to minimize the fitness function. However, the continuous control MPC controller excels in that it produces less ripple in torque and current than the finite-set MPC type.

The MPC algorithm is employed in the drive of PMSM in the field-decreasing scope. Reference [153] proposes a linear plant model of the gain scheduled MPC, which determines the optimal reference stator voltages by a pre-papered optimization solution for all required speed ranges, as shown in Figure 1.22, where $i^*_{qsL}$ references load current and $V_{dqs}$ are $d$-$q$ axes stator voltage vectors of the PMSM. The proposed MPC strategy in this work can pre-calculate the best global solution throughout specified cost functions stored in the digital control unit.

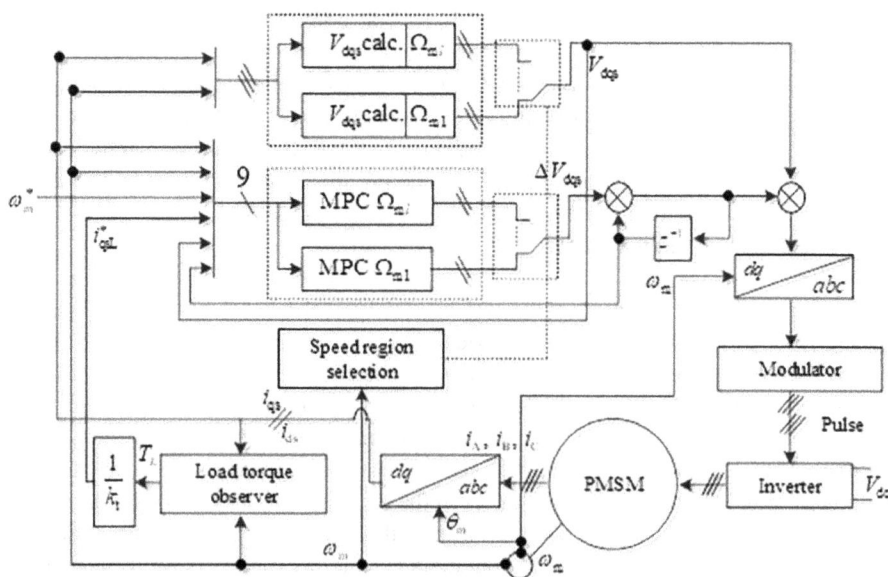

**FIGURE 1.22**   Block diagram of the explicit MPC control strategy [153].

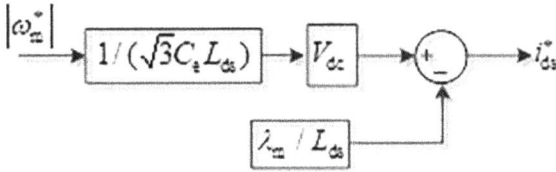

FIGURE 1.23   Block diagram of calculating demagnetizing current without a feedback loop [154].

Thus, the feature of the gain scheduled type called a tiny step time could have a significant impact on dynamic performance.

In reference [154], the researchers efficiently use the linear plant model in their MPC strategy to obtain a stable speed and current. The demagnetizing reference current for the $d$-axis is calculated based on the machine parameter of the voltage constant to operate the interior PMSM with the required speed in the flux-weakening region. However, this calculation may work inaccurately because it is based on the machine parameters, as shown in Figure 1.23, where $C_e$ is the voltage constant.

Finally, it is difficult to find a real-time optimization algorithm rather than a finite-set control algorithm for reducing torque ripple. Moreover, it is also difficult to express the plant model that is accurate in every operating condition of the motor system drive. Therefore, from this point of view, Chapter 5 provides an adaptive MPC controller and continuous control of real-time optimization techniques to solve these problems.

## 1.3  CONTRIBUTION OF THE BOOK

Model optimization algorithms and predictive methods have been used to develop control and planning strategies in electric power systems and machine drives [155–161]. The PMSM high-performance propulsion system was recently developed, solidifying its place in electric traction drive systems [162–164]. Due to the advantages of the PMSMs over BLDC and induction motors, the traction system drives based on PMSM drive systems are becoming more efficient, environmentally sustainable, and better suited to large megacity environments [165–169]. Additionally, PMSMs' industrial drive systems, such as washing units; heating, ventilation, and air conditioning (HVAC) units; and other such items have a greater initial cost.

However, the PMSM drives can provide a suitable size compared to the induction machine drives for smaller mechanical structure packages, a better torque-to-inertia ratio, higher dependability, and better control stability. The most significant parameters for these industrial drives are an instantaneous reaction to a planned update; stable control during the load disturbance period; and minimum peak ripple of torque, stator current, and velocity. As a result, they should achieve these parameters in the flux-weakening operation of the PMSM drive [170, 171].

However, there are still some problems resulting from the flux-decreasing and the current regulator saturation that leads to the complicated control characteristics of PMSMs. As a result, the stator current ripple is increased to cause an increase in torque ripple. In general, the internal gear vibration reduces the life of the motor for many reasons, such as the structure of the motor, the sampled speed fluctuations, the time delay of closed-loop control, and so on. Hence, for achieving the high performance of PMSMs, a vector control strategy is used because it can directly control the current or torque and provide a fast dynamic response. However, it suffers from adjusting the PI parameters, which deteriorates the drive performance, especially the ripples in current and torque. This problem is further exacerbated when a control loop is added to run the motor in the region of flux weakening. The number of PI regulators increases, and six to eight gains are required to adjust. Meanwhile, reliance on the control strategy on machine parameters subverts engine performance when the temperature changes. In contrast to the vector control techniques, continuous control MPC has recently been developed to achieve high performance because it is the most reliable control approach and can optimize performance online with multi-states control, low torque ripple, a faster transient period, and so on.

In general, there is still a need to comprehensively study the shortcomings of PMSM control in the region of flux weakening, such as driving the dynamic responses faster; controlling ripples of stator current and electromagnetic torque; designing and analyzing PI control parameters; and finding fitness functions to improve the system stability, reduce copper and switching losses, estimate stator flux and induced electromagnetic torque, find a linear plant model that expresses the motor more, find fitness functions to indicate the prediction of the speed and magnitude of stator current in the flux-weakening region, and develop optimization algorithms to improve performance in real time. This book attempts to overcome these shortcomings of the permanent magnet synchronous motor drive system for industrial applications and EVs. So it is essential to investigate the following issues, which are the main topics within the scope of this book:

- Proposing the anti-windup proportional and integral (AWPI) technique for the PI regulator of FOC strategy to cancel the overshoot of motor speed and stator current.
- Enhancing the performance of the FOC strategy that runs the PMSM at the flux weakening using a first fitness function, mentioned in Chapter 3, to optimize PI regulators of outer and inner regulators. For the optimization, the adaptive velocity particle swarm optimization (AVPSO) algorithm evaluates the proposed fitness function offline to select the optimal PI gains to ensure the desired performance and system stability.
- Proposing the control loop of the field decreasing case based on the reference duty cycle to reduce the dependence of the motor performance on machine parameters and run the motor in the flux-weakening region.
- Improving the torque ripple control algorithm of varying switching frequency PWM (VSFPWM) for the voltage source inverter (VSI) to reduce the peak of current and torque ripples.
- Improving a second optimization fitness function, mentioned in Chapter 4, to involve the stability issue for designing the PI gains of the current regulator of FOC strategy to reduce the peak of the stator current and electromagnetic torque ripples. For optimization, in this book, the genetic algorithm (GA) technique evaluates the proposed fitness function offline to select optimal PI

gains that reduce the ripple. Meanwhile, the VSFPWM control algorithm is integrated to work with the proposed current regulator to reduce the ripple peak.

- Improving the discrete linear plant model (DLPM) of the PMSM model to automatically adapt the operation points online through the measured states. In this DLPM, the observer of a third-order generalized integral (TOGI) is employed to estimate the stator flux to make the model better reflect the IPMSM condition.
- Modifying (Chapter 5) the control strategy proposed in Chapter 4 to fit the controlling of the motor of the interior magnet–type PMSM. Meanwhile, the limiters of the PI controllers are reconstructed to overcome the overrun of the speed and stator current signals.
- Proposing a linear model predictive control in Chapter 6 to present the third objective function in this book. The cost function depends on the speed error to improve the performance of the controlled interior PMSM in the flux-weakening region.
- Proposing a linear fitness function, as the fourth objective function in this book mentioned in Chapter 5, that the proposed linear MPC control strategy can use to optimize the reference voltages of the $d$-$q$ axes as the manipulated variables.
- Proposing a performance-control algorithm to calculate a reference mask for motor reference speed and the magnitude of the stator current as the maximum value that the linear MPC control strategy can predict in the prediction procedure of the next manipulated variables.
- Improving the real-time AVPSO algorithm that the linear MPC control strategy can use to optimize the manipulated variables as an online optimization process to improve motor performance.
- Finally, a test bench platform of two 3 kW surface-mounted and interior PMSMs is constructed to emulate the proposed drive performance. Comprehensive simulation and experiments have adequately demonstrated that each proposed control strategy in this book increases the performance of the controlled PMSM compared to previous works.

## 1.4 ORGANIZATION OF THE BOOK

The book contains eight chapters, outlined as follows:

**Chapter 1** provides a general background on the research issues, followed by a brief definition of permanent magnet (PM) motor construction, and detailed operating principles for Permanent Magnet Synchronous Motors (PMSMs). Additionally, the chapter includes abundant descriptions of the types, applications, and main features of PMSMs, which are compared to induction and Brushless DC (BLDC) motors. The chapter presents a comprehensive literature review of different dynamic models and various control techniques. Finally, the research objectives and contributions of this book are summarized.

**Chapter 2** introduces mathematical models for salient and non-salient Permanent Magnet Synchronous Motors (PMSMs). The chapter explains the control algorithm of Maximum Torque per Ampere (MTPA) in the constant torque region and demonstrates the limitations of voltage and current in the theory of flux weakening. The chapter also explains how the MTPA algorithm operates in the flux-weakening region and introduces the Vector Space Field-Oriented Pulse Width Modulation (VSFPWM) control algorithm of the Voltage Source Inverter (VSI) for the Surface Permanent Magnet Synchronous Motor (SPMSM). Finally, the chapter discusses the results of experiment and simulation case studies of the SPMSM running in the flux-weakening region to verify the effectiveness of the strategy.

**Chapter 3** introduces the proposed adaptive drive system, including the Proposed Flux Weakening Control (PFWC) strategy compared to the previous work of the Conventional Flux-Weakening Control (CFWC) strategy. Additionally, the chapter introduces the proposed algorithm for selecting the optimal Proportional-Integral (PI) gain for PI regulators and the offline application of the optimization procedure of the Adaptive Velocity-PSO (AVPSO) technique. Finally, the chapter simulates the proposed adaptive control strategy numerically and verifies it with experimental results.

**Chapter 4** proposes the adaptive Field Oriented Control (FOC) drive system and analyzes its drive model and stability, including the parameters-tuning algorithm of the proposed current regulator of FOC strategy, the flux observer based on a Third-Order Generalized Integrator (TOGI), and the offline flow chart implementation of the optimization procedure based on a Genetic Algorithm (GA) technique. Lastly, the chapter presents the simulated and experimental results in detail.

**Chapter 5** proposes the same adaptive FOC drive system discussed in Chapter 4 but reconfigured to fit the control of the interior magnet-type PMSM. The chapter reconstructs the limiters of the PI regulators to cancel the overrun in the interior PMSM performance and modifies the parameters tuning of the proposed PI regulators based on the machine parameters of the interior PMSM. Finally, the chapter presents the simulated and experimental results in detail.

**Chapter 6** analyzes the Linear Predictive Model Control (LPMC) drive of the interior PMSM. The chapter uses the Finite Control Set Technique (FCST) as an evaluator of the proposed cost function. The chapter introduces the proposed objective function to directly evaluate the error values of the speed and d-axis stator current to improve motor performance in the flux-weakening region. Moreover, the chapter obtains the prediction speed based on the flux observer of the modified Low-Pass First-Order Integrator (LPFOI). Finally, the chapter investigates and compares the simulation and experiment validations for the proposed MPC control method to the conventional MPC strategy.

**Chapter 7** analyzes the Dual Linear Plant Model (DLPM) as a proposed linear plant model. The chapter explains in detail the proposed algorithm for avoiding speed overshoot and overcurrent, introduces the proposed objective function based on the DLPM of the proposed MPC strategy, and presents the proposed real-time AVPSO and its implementation. Finally, the chapter investigates and compares the simulation and experiment validations for the proposed MPC control method to the conventional MPC strategy.

**Chapter 8** collects the main findings and contributions of this research and concludes the book. This chapter also contains future directions for further research and development of a control strategy for the PMSM drive system used in industrial and Electric Vehicle (EV) applications.

## 1.5 SUMMARY

This chapter presents a comprehensive history of permanent magnet synchronous motors (PMSMs) with a focus on the sinusoidal power source of surface-mounted and interior magnet types. In addition, a comparison of PMSMs to induction and brushless DC motors is presented to illustrate the advantages of using PMSM motors in industrial applications. Furthermore, a thorough literature review is conducted on PMSM drive systems, including traditional methods and modern control strategies.

The dynamic performance of PMSM drive systems is discussed under the guidance of conventional and improved field-oriented control strategies, with a demonstration of the superiority of the model-predictive control strategy in terms of dynamic performance in the constant torque and weak field regions. The literature review concludes that the significant parameters of the PMSM drive include less settling time when the motor speed tracks the reference speed in the flux-weakening region. In this region, the speed signal should have less overshoot and undershoot, and the ripple peak of electric current and induced torque should also be taken into account.

## BIBLIOGRAPHY

[1]   D. E. Cameron, J. H. Lang, and S. D. Umans. The origin and reduction of acoustic noise in doubly salient variable-reluctance motors. IEEE Transactions on Industry Applications, November 1992, 28(6):1250–1255.

[2]   J. F. Gieras. Permanent magnet motor technology: design and applications, CRC Press, 2002.

[3]   M. S. Islam, R. Islam, and T. Sebastian. Noise and vibration characteristics of permanent-magnet synchronous motors using electromagnetic and structural analyses. IEEE Transactions on Industry Applications, February 2014, 50(5):3214–3222.

[4]   R. Wu and G. R. Slemon. A permanent magnet motor drive without a shaft sensor. IEEE Transactions on Industry Applications, September 1991, 27(5):1005–1011.

[5]   B. Zheng, H.-W. Zhang, S.-F. Zhao, J.-L. Chen, and G.-H. Wu. The physical origin of open recoil loops in nanocrystalline permanent magnets. Applied Physics Letters, November 2008, 93(18):182503.

[6]   L. Q. Gothard and J. Rosen. Encyclopedia of physical science, Facts on File, 2010.

[7]   Martin Doppelbauer. The invention of the electric motor 1856–1893, Karlsruhe Institute of Technology (KIT), 2012.

[8]   Stanley Ryan Sifford. Multiport analysis of permanent magnet DC motors, ProQuest, 2006, p. 56.

[9]   Carroll Gantz. The vacuum cleaner: a history. McFarland, 2012, p. 40.

[10]  S. Yamamura. AC motors for high performance applications: analysis and control, CRC Press, 1986.

[11]  J. Pyrhonen, T. Jokinen, and V. Hrabovcova. Design of rotating electrical machines, John Wiley & Sons, 2013.

[12]  K. Buschow. Trends in rare earth permanent magnets. IEEE Transactions on Magnetics, March 1994, 30(2):565–570.

[13]  J. Coey. Perspective and prospects for rare earth permanent magnets. Engineering, February 2020, 6(2):119–131.

[14]  P. C. Dent. Rare earth elements and permanent magnets. Journal of Applied Physics, April 2012, 111(7):07A721.

[15]  Huber, C., C. Abert, F. Bruckner, M. Groenefeld, O. Muthsam, S. Schuschnigg, K. Sirak et al. 3D print of polymer bonded rare-earth magnets, and 3D magnetic field scanning with an end-user 3D printer. Applied Physics Letters, October 2016, 109(16):162401.

[16]  Huber C, Abert C, Bruckner F, Groenefeld M, Schuschnigg S, Teliban I, Vogler C, Wautischer G, Windl R, and Suess D. 3D printing of polymer-bonded rare-earth magnets with a variable magnetic compound fraction for a predefined stray field. Scientific Reports, August 2017, 7(1):1–8.

[17]  H. Nakamura. The current and future status of rare earth permanent magnets. Scripta Materialia, September 2018, 154:273–276.

[18]  C.-L. Xia. Permanent magnet brushless DC motor drives and controls, John Wiley & Sons, 2012.

[19]  Goldman. Modern ferrite technology, Springer Science & Business Media, 2006.

[20]  P. Bergmann and M. J. de Andrade. Nanostructured materials for engineering applications, Springer, 2011.

[21]  J. P. Liu, E. Fullerton, O. Gutfleisch, and D. J. Sellmyer. Nanoscale magnetic materials and applications, Springer, 2009.

[22]  R. M. Calfo and M. P. Krefta. Trapezoidal shaped magnet flux intensifier motor pole arrangement for improved motor torque density, Google Patents, 2005.

[23]  S. Dwari and L. Parsa. Fault-tolerant control of five-phase permanent-magnet motors with trapezoidal back EMF. IEEE Transactions on Industrial Electronics, March 2010, 58(2):476–485.

[24]  M. Hattori and N. Makino. Permanent magnet motor, Google Patents, 2005.

[25]  P. Liang, F. Chai, Y. Yu, and L. Chen. Analytical model of a spoke-type permanent magnet synchronous in-wheel motor with trapezoid magnet accounting for tooth saturation. IEEE Transactions on Industrial Electronics, April 2018, 66(2):1162–1171.

[26]  X. Zhang, C. Zhang, J. Yu, P. Du, and L. Li. Analytical model of magnetic field of a permanent magnet synchronous motor with a trapezoidal Halbach permanent magnet array. IEEE Transactions on Magnetics, March 2019, 55(7):1–5.

[27]  A. Aleksashkin and A. Mikkola. Literature review on permanent magnet generators design and dynamic behavior, Lappenranta University of Technology, 2008.

[28]  N. Bianchi and S. Bolognani. Design optimisation of electric motors by genetic algorithms. IEE Proceedings-Electric Power Applications, September 1998, 145(5): 475–483.

[29]  D. G. Dorrell, M.-F. Hsieh, M. Popescu, L. Evans, D. A. Staton, and V. Grout. A review of the design issues and techniques for radial-flux brushless surface and internal rare-earth permanent-magnet motors. IEEE Transactions on Industrial Electronics, October 2010, 58(9):3741–3757.

[30]  S. O. Edhah, J. Y. Alsawalhi, and A. A. Al-Durra. Multi-Objective optimization design of fractional slot concentrated winding permanent magnet synchronous machines. IEEE Access, November 2019, 7:162874–162882.

[31]  S. Morimoto. Trend of permanent magnet synchronous machines. IEEJ Transactions on Electrical and Electronic Engineering, March 2007, 2(2):101–108.

[32]  A. Jabbari. 2D analytical modeling of magnetic vector potential in surface mounted and surface inset permanent magnet machines. Iranian Journal of Electrical and Electronic Engineering, December 2017, 13(4):362–373.

[33]  G. Liu, X. Du, W. Zhao, and Q. Chen. Reduction of torque ripple in inset permanent magnet synchronous motor by magnets shifting. IEEE Transactions on Magnetics, October 2016, 53(2):1–13.

[34]  T. Lubin, S. Mezani, and A. Rezzoug. Two-dimensional analytical calculation of magnetic field and electromagnetic torque for surface-inset permanent-magnet motors. IEEE Transactions on Magnetics, December 2011, 48(6):2080–2091.

[35]  D.-K. Ngo, M.-F. Hsieh, and T. A. Huynh. Torque enhancement for a novel flux intensifying PMa-SynRM using surface-inset permanent magnet. IEEE Transactions on Magnetics, February 2019, 55(7):1–8.

[36]  L. Xu et al. Design and analysis of a new linear hybrid excited flux reversal motor with inset permanent magnets. IEEE Transactions on Magnetics, December 2014, 50(11):1–4.

[37]  T. Lubin, S. Mezani, and A. Rezzoug. 2-D exact analytical model for surface-mounted permanent-magnet motors with semi-closed slots. IEEE Transactions on Magnetics, November 2010, 47(2):479–492.

[38]  L. Ortombina, F. Tinazzi, and M. Zigliotto. Magnetic modeling of synchronous reluctance and internal permanent magnet motors using radial basis function networks. IEEE Transactions on Industrial Electronics, July 2017, 65(2):1140–1148.

[39]  P. Pillay and R. Krishnan. Modeling of permanent magnet motor drives. IEEE Transactions on Industrial Electronics, November 1988, 35(4):537–541.

[40]  M. Rahman, T. Little, and G. Slemon. Analytical models for interior-type permanent magnet synchronous motors. IEEE Transactions on Magnetics, September 1985, 21(5):1741–1743.

[41]  G. R. Slemon and X. Liu. Modeling and design optimization of permanent magnet motors. Electric Machines & Power systems, March 1992, 20(2):71–92.

[42]  H. Bakhtiarzadeh, A. Polat, and L. T. Ergene. editors. Design and analysis of a permanent magnet synchronous motor for elevator applications. International Conference on Optimization of Electrical and Electronic Equipment (OPTIM) & 2017 Intl Aegean Conference on Electrical Machines and Power Electronics (ACEMP), May 2017.

[43]  S. Jain, A. K. Thopukara, R. Karampuri, and V. Somasekhar. A single-stage photovoltaic system for a dual-inverter-fed open-end winding induction motor drive for pumping applications. IEEE Transactions on power Electronics, October 2014, 30(9):4809–4818.

[44]  S. Rahman et al. Design and implementation of cascaded multilevel qZSI powered single-phase induction motor for isolated grid water pump application. IEEE Transactions on Industry Applications, December 2019, 56(2):1907–1917.

[45]  M. Shreelakshmi and V. Agarwal. Trajectory optimization for loss minimization in induction motor fed elevator systems. IEEE Transactions on Power Electronics, August 2017, 33(6):5160–5170.

[46]  M. Z. Youssef. Design and performance of a cost-effective BLDC drive for water pump application. IEEE Transactions on Industrial Electronics, August 2014, 62(5):3277–3284.

[47] L. Diao, J. Tang, P. C. Loh, S. Yin, L. Wang, and Z. Liu. An efficient DSP—FPGA-based implementation of hybrid PWM for electric rail traction induction motor control. IEEE Transactions on Power Electronics, May 2017, 33(4):3276–3288.

[48] L.-J. Diao, D.-n. Sun, K. Dong, L.-T. Zhao, and Z.-G. Liu. Optimized design of discrete traction induction motor model at low-switching frequency. IEEE Transactions on Power Electronics, December 2012, 28(10):4803–4810.

[49] S. K. Sahoo and T. Bhattacharya. Rotor flux-oriented control of induction motor with synchronized sinusoidal PWM for traction application. IEEE Transactions on Power Electronics, September 2015, 31(6):4429–4439.

[50] J. Su, R. Gao, and I. Husain. Model predictive control-based field-weakening strategy for traction EV used induction motor. IEEE Transactions on Industry Applications, December 2017, 54(3):2295–2305.

[51] C. Uyulan and M. Gokasan. Modeling, simulation, and re-adhesion control of an induction motor—based railway electric traction system. Proceedings of the Institution of Mechanical Engineers, Part I: Journal of Systems and Control Engineering, January 2018, 232(1):3–11.

[52] A. R. Eastham, G. E. Dawson, V. I. John, A. Kamar, P. C. Sen, and A. K. Wallace. Voltage-controlled 60 Hz linear induction motor drives. Electric Machines and Power Systems, November 1983, 8(6):487–497.

[53] N. Kunihiro, T. Todaka, and M. Enokizono. Loss evaluation of an induction motor model core by vector magnetic characteristic analysis. IEEE Transactions on Magnetics, September 2010, 47(5):1098–1101.

[54] J. V. Leite, M. V. V. F. da Luz, N. Sadowski, and P. A. da Silva. Modelling dynamic losses under rotational magnetic flux. IEEE Transactions on Magnetics, January 2012, 48(2):895–898.

[55] G. Rakotonirina, A. Ba-Razzouk, J. Xu, A. Sevigny, and V. Rajagopalan. editors. An improved estimation of eddy-current losses distribution for squirrel cage induction motor using finite element method. Engineering Solutions for the Next Millennium. 1999 IEEE Canadian Conference on Electrical and Computer Engineering (Cat. No. 99TH8411), May 1999.

[56] M. Wciślik and K. Suchenia. editors. Core losses model of switched reluctance motor. 2015 Selected Problems of Electrical Engineering and Electronics (WZEE), September 2015.

[57] A. Athavale, K. Sasaki, B. S. Gagas, T. Kato, and R. D. Lorenz. Variable flux permanent magnet synchronous machine (VF-PMSM) design methodologies to meet electric vehicle traction requirements with reduced losses. IEEE Transactions on Industry Applications, May 2017, 53(5):4318–4326.

[58] D. Fodorean, L. Idoumghar, M. Brévilliers, P. Minciunescu, and C. Irimia. Hybrid differential evolution algorithm employed for the optimum design of a high-speed PMSM used for EV propulsion. IEEE Transactions on Industrial Electronics, May 2017, 64(12):9824–9833.

[59] A. Hezzi, Y. Bensalem, S. B. Elghali, and M. N. Abdelkrim. editors. Sliding mode observer based sensorless control of five phase PMSM in electric vehicle. 2019 19th International Conference on Sciences and Techniques of Automatic Control and Computer Engineering (STA), March 2019.

[60] F. Momen, K. M. Rahman, Y. Son, and P. Savagian. Electric motor design of general motors' Chevrolet Bolt electric vehicle. SAE International Journal of Alternative Powertrains, July 2016, 5(2):286–293.

[61] X. Xiao, Y. Zhang, J. Wang, and H. Du. New adaptive sliding-mode observer design for sensorless control of PMSM in electric vehicle drive system. International Journal on Smart Sensing and Intelligent Systems, March 2016, 9(1):377–396.

[62]  J. Puranen. Induction motor versus permanent magnet synchronous motor in motion control applications: a comparative study. 2006.

[63]  C. Hwang, P. Li, C. Liu, and C. Chen. Design and analysis of a brushless DC motor for applications in robotics. IET Electric Power Applications, August 2012, 6(7):385–389.

[64]  K. Liu, M. Yin, W. Hua, Z. Ma, M. Lin, and Y. Kong. Design and analysis of Halbach ironless flywheel BLDC motor/generators. IEEE Transactions on Magnetics, May 2018, 54(11):1–5.

[65]  S. S. Nair, S. Nalakath, and S. J. Dhinagar, editors. Design and analysis of axial flux permanent magnet BLDC motor for automotive applications. IEEE International Electric Machines & Drives Conference (IEMDC), May 2011.

[66]  P. S. Shin, H.-D. Kim, G.-B. Chung, H. S. Yoon, G.-S. Park, and C. S. Koh. Shape optimization of a large-scale BLDC motor using an adaptive RSM utilizing design sensitivity analysis. IEEE Transactions on Magnetics, March 2007, 43(4):1653–1656.

[67]  S. Wang and J. Kang. Shape optimization of BLDC motor using 3-D finite element method. IEEE Transactions on Magnetics, July 2000, 36(4):1119–1123.

[68]  J. Bae and D.-H. Lee. Position control of a rail guided mover using a low-cost BLDC motor. IEEE Transactions on Industry Applications, February 2018, 54(3): 2392–2399.

[69]  H. Immaneni. Mathematical modelling and position control of brushless dc (BLDC) motor. International Journal of Research and Applications (IJERA), May 2013, 3(3):1050–1057.

[70]  C. Navaneethakkannan and M. Sudha. Analysis and implementation of ANFIS-based rotor position controller for BLDC motors. Journal of Power Electronics, March 2016, 16(2):564–571.

[71]  M. J. Prabu, P. Poongodi, and K. Premkumar. Fuzzy supervised online coactive neuro-fuzzy inference system-based rotor position control of brushless DC motor. IET Power Electronics, September 2016, 9(11):2229–2239.

[72]  R. Ortega, L. Praly, A. Astolfi, J. Lee, and K. Nam. Estimation of rotor position and speed of permanent magnet synchronous motors with guaranteed stability. IEEE Transactions on Control Systems Technology, April 2010, 19(3):601–614.

[73]  G. Wang, J. Kuang, N. Zhao, G. Zhang, and D. Xu. Rotor position estimation of PMSM in low-speed region and standstill using zero-voltage vector injection. IEEE Transactions on Power Electronics, October 2017, 33(9):7948–7958.

[74]  G. Zhang, G. Wang, D. Xu, and N. Zhao. ADALINE-network-based PLL for position sensorless interior permanent magnet synchronous motor drives. IEEE Transactions on Power Electronics, April 2015, 31(2):1450–1460.

[75]  Y. Zhao, W. Qiao, and L. Wu. Improved rotor position and speed estimators for sensorless control of interior permanent-magnet synchronous machines. IEEE Journal of Emerging and Selected Topics in Power Electronics, January 2014, 2(3): 627–639.

[76]  Y. Zhao, Z. Zhang, W. Qiao, and L. Wu. An extended flux model-based rotor position estimator for sensorless control of salient-pole permanent-magnet synchronous machines. IEEE Transactions on Power Electronics, September 2014, 30(8):4412–4422.

[77]  P. C. Berri, M. D. L. Dalla Vedova, P. Maggiore, and M. Scanavino. editors. Permanent magnet synchronous motor (PMSM) for aerospace servomechanisms: proposal of a lumped model for prognostics. 2018 2nd European Conference on Electrical Engineering and Computer Science (EECS), December 2018.

[78]  P. Giangrande, V. Madonna, G. Sala, A. Kladas, C. Gerada, and M. Galea. editors. Design and testing of PMSM for aerospace EMA applications. IECON 2018–44th Annual Conference of the IEEE Industrial Electronics Society, October 2018.

[79]   T. Heikkilä. Permanent magnet synchronous motor for industrial inverter applications-
       analysis and design, November 2002.

[80]   S. Wang, J. Na, and X. Ren. RISE-based asymptotic prescribed performance track-
       ing control of nonlinear servo mechanisms. IEEE Transactions on Systems, Man, and
       Cybernetics: Systems, November 2017, 48(12):2359–2370.

[81]   S. Wang, X. Ren, J. Na, and T. Zeng. Extended-state-observer-based funnel control for
       nonlinear servomechanisms with prescribed tracking performance. IEEE Transactions
       on Automation Science and Engineering, November 2016, 14(1):98–108.

[82]   O. Babayomi, A. Balogun, and C. Osheku. editors. Loss Minimizing Control of PMSM
       for Electric Power Steering. 2015 17th UKSim-AMSS International Conference on
       Modelling and Simulation (UKSim), March 2015.

[83]   B. Basler, T. Greiner, and P. Heidrich. editors. Fault-tolerant strategies for double three-
       phase PMSM used in Electronic Power Steering systems. 2015 IEEE Transportation
       Electrification Conference and Expo (ITEC), June 2015.

[84]   K. Scicluna, C. S. Staines, and R. Raute. editors. Sensorless position control of a
       PMSM for steer-by-wire applications. International Conference on Control, Decision
       and Information Technologies (CoDIT), April 2016.

[85]   J. Xuewu and L. Yingchao. editors. Field weakening control of PMSM used in an elec-
       tric power steering system. 2011 International Conference on Electric Information and
       Control Engineering, April 2011.

[86]   X. Zang and F. Qi. editors. Steering system fault tolerance simulation of PMSM con-
       trol. Proceedings of 2014 IEEE Chinese Guidance, Navigation and Control Conference,
       August 2014.

[87]   A. K. Junejo, W. Xu, C. Mu, M. M. Ismail and Y. Liu. Adaptive speed control of PMSM
       drive system based a new sliding-mode reaching law. IEEE Transactions on Power
       Electronics, November 2020, 35(11):12110–12121. doi: 10.1109/TPEL.2020.2986893.

[88]   M. M. Ismail, W. Xu, X. Wang, A. K. Junejo, Y. Liu and M. Dong. Analysis and optimi-
       zation of torque ripple reduction strategy of surface-mounted permanent-magnet motors
       in flux-weakening region based on genetic algorithm. IEEE Transactions on Industry
       Applications, July-August 2021, 57(4):4091–4106. doi: 10.1109/TIA.2021.3074609.

[89]   Y. Tang, W. Xu, D. Dong, Y. Liu and M. M. Ismail. Low-complexity multistep sequential
       model predictive current control for three-level inverter-fed linear induction machines.
       IEEE Transactions on Industrial Electronics, 2022. doi: 10.1109/TIE.2022.3192688.

[90]   M. G. Hussien, Y. Liu, W. Xu, and M. M. Ismail. Voltage regulation-based sensorless
       position observer with high-frequency signal injection topology for BDFIGs in ship
       power microgrid systems. International Journal of Electrical Power & Energy Systems,
       2022, 140:108091.

[91]   M. M. Ismail, W. Xu, J. Ge, Y. Tang, A. K. Junejo and M. G. Hussien. Adaptive linear
       predictive model of an improved predictive control of permanent magnet synchronous
       motor over different speed regions. IEEE Transactions on Power Electronics, Decem-
       ber 2022, 37(12):15338–15355. doi:10.1109/TPEL.2022.3194839.

[92]   P. C. Perera, F. Blaabjerg, J. K. Pedersen, and P. Thogersen. A sensorless, stable V/f
       control method for permanent-magnet synchronous motor drives. IEEE Transactions
       on Industry Applications, May 2003, 39(3):783–791.

[93]   S.-C. Agarlita, C.-E. Coman, G.-D. Andreescu, and I. Boldea. Stable V/f control system
       with controlled power factor angle for permanent magnet synchronous motor drives.
       IET Electric Power Applications, April 2013, 7(4):278–286.

[94]   R. S. Colby and D. W. Novotny. An efficiency-optimizing permanent-magnet syn-
       chronous motor drive. IEEE Transactions on Industry Applications, May 1988, 24(3):
       462–469.

[95]   W.-J. Kim and S.-H. Kim. editors. A Sensorless V/f control technique based on MTPA operation for PMSMS. IEEE Energy Conversion Congress and Exposition (ECCE), September 2018, 1716–1721.

[96]   Y. Nakamura, T. Kudo, F. Ishibashi, and S. Hibino. High efficiency drives due to power factor control of a permanent magnet synchronous motor. IEEE Transactions on Power Electronics, March 1995, 10(2):247–253.

[97]   S. Paitandi and M. Sengupta. Analysis, design and implementation of sensorless V/f control in a surface mounted PMSM without damper winding. Sādhanā, August 2017, 42(8):1317–1333.

[98]   H. Chaoui, M. Khayamy, and O. Okoye. MTPA based operation point speed tracking for PMSM drives without explicit current regulation. Electric Power Systems Research, October 2017, 151:125–135.

[99]   R. Heidari. Model predictive combined vector and direct torque control of SM-PMSM with MTPA and constant stator flux magnitude analysis in the stator flux reference frame. IET Electric Power Applications, August 2020, 14(12): 2283–2292.

[100]  T. Inoue, Y. Inoue, S. Morimoto, and M. Sanada. Mathematical model for MTPA control of permanent-magnet synchronous motor in stator flux linkage synchronous frame. IEEE Transactions on Industry Applications, March 2015, 51(5):3620–3628.

[101]  A. Rabiei, T. Thiringer, M. Alatalo, and E. A. Grunditz. Improved maximum-torque-per-ampere algorithm accounting for core saturation, cross-coupling effect, and temperature for a PMSM intended for vehicular applications. IEEE Transactions on Transportation Electrification, February 2016, 2(2):150–159.

[102]  A. Shinohara, Y. Inoue, S. Morimoto, and M. Sanada. Influence of stator flux estimation on reference flux for MTPA operation in PWM-based DTC PMSM drives. International Journal of Power Electronics, January 2016, 8(1):23–37.

[103]  I. Takahashi and T. Noguchi. A new quick-response and high-efficiency control strategy of an induction motor. IEEE Transactions on Industry applications, September 1986, (5):820–827.

[104]  M. Depenbrock. Direct self-control (DSC) of inverter fed induction machine. 1987 IEEE Power Electronics Specialists Conference, June 1987.

[105]  S. Mathapati and J. Bocker. Analytical and offline approach to select optimal hysteresis bands of DTC for PMSM. IEEE Transactions on Industrial Electronics, February 2012, 60(3):885–895.

[106]  H. Mesloub, R. Boumaaraf, M. T. Benchouia, A. Goléa, N. Goléa, and K. Srairi. Comparative study of conventional DTC and DTC_SVM based control of PMSM motor—simulation and experimental results. Mathematics and Computers in Simulation, January 2020, 167:296–307.

[107]  X. Wang, Z. Wang, M. Cheng, and Y. Hu. Remedial strategies of T-NPC three-level asymmetric six-phase PMSM drives based on SVM-DTC. IEEE Transactions on Industrial Electronics, March 2017, 64(9):6841–6853.

[108]  Y. Wang, L. Geng, W. Hao, and W. Xiao. Control method for optimal dynamic performance of DTC-based PMSM drives. IEEE Transactions on Energy Conversion, January 2018, 33(3):1285–1296.

[109]  A. Yassin, Amir, S. Wahsh and M. Badr. Cuckoo Search Based DTC of PMSM. International Journal of Power Electronics and Drive Systems, September 2018, 9(3):1106–1115.

[110]  J. Kaukonen. Salient pole synchronous machine modelling in an industrial direct torque-controlled drive application. 1999.

[111]  X. Shaobang, L. Yinsheng, and H. Xiaoxin. editors. Efficiency improvement on PMSM twelve sectors DTC system. Chinese Control and Decision Conference (CCDC), May 2011.

[112]   Y. Wang and J. Zhu. editors. Modelling and implementation of an improved DSVM scheme for PMSM DTC. International Conference on Electrical Machines and Systems, October 2008.

[113]   K.-C. Kim. A novel magnetic flux weakening method of permanent magnet synchronous motor for electric vehicles. IEEE Transactions on Magnetics, October 2012, 48(11):4042–4045.

[114]   T.-S. Kwon, G.-Y. Choi, M.-S. Kwak, and S.-K. Sul. Novel flux-weakening control of an IPMSM for quasi-six-step operation. IEEE Transactions on Industry Applications, November 2008, 44(6):1722–1731.

[115]   H. Liu, Z. Zhu, E. Mohamed, Y. Fu, and X. Qi. Flux-weakening control of nonsalient pole PMSM having large winding inductance, accounting for resistive voltage drop and inverter nonlinearities. IEEE Transactions on Power Electronics, June 2011, 27(2): 942–952.

[116]   M. F. Rahman, L. Zhong, and L. Khiang Wee. A direct torque-controlled interior permanent magnet synchronous motor drive incorporating field weakening. IEEE Transactions on Industry Applications, November 1998, 34(6):1246–1253.

[117]   L. Sepulchre, M. Fadel, M. Pietrzak-David, and G. Porte. MTPV flux-weakening strategy for PMSM high speed drive. IEEE Transactions on Industry Applications, July 2018, 54(6):6081–6089.

[118]   W. Wang, J. Zhang, and M. Cheng. Line-modulation-based flux-weakening control for permanent-magnet synchronous machines. IET Power Electronics, January 2018, 11(5):930–936.

[119]   H. Chuan, S. M. Fazeli, Z. Wu, and R. Burke. Mitigating the torque ripple in electric traction using proportional integral resonant controller. IEEE Transactions on Vehicular Technology, July 2020, 69(10):10820–10831.

[120]   K. Cui, C. Wang, L. Gou, and Z. An. Analysis and design of current regulators for PMSM drives based on DRGA. IEEE Transactions on Transportation Electrification, May 2020, 6(2):659–667.

[121]   J.-X. Xu, S. K. Panda, Y.-J. Pan, T. H. Lee, and B. Lam. A modular control scheme for PMSM speed control with pulsating torque minimization. IEEE Transactions on Industrial electronics, June 2004, 51(3):526–536.

[122]   W. Xu, M. M. Ismail, Y. Liu, and M. R. Islam. Parameter optimization of adaptive flux-weakening strategy for permanent-magnet synchronous motor drives based on particle swarm algorithm. IEEE Transactions on Power Electronics, April 2019, 34(12):12128–12140.

[123]   Z. Zhang, C. Wang, M. Zhou, and X. You. Flux-weakening in PMSM drives analysis of voltage angle control and the single current controller design. IEEE Journal of Emerging and Selected Topics in Power Electronics, May 2019, 7(1):437–445.

[124]   Z. Guangzhen, Z. Feng, W. Yongxing, W. Xuhui, and C. Wei. editors. Analysis and optimization of current regulator time delay in permanent magnet synchronous motor drive system. 2013 International Conference on Electrical Machines and Systems (ICEMS), October 2013.

[125]   J.-W. Jung, Y.-S. Choi, V. Leu, and H. Choi. Fuzzy PI-type current controllers for permanent magnet synchronous motors. IET Electric Power Applications, February 2011, 5(1):143–152.

[126]   Y. A.-R. I. Mohamed. Design and implementation of a robust current-control scheme for a PMSM vector drive with a simple adaptive disturbance observer. IEEE Transactions on Industrial Electronics, July 2007, 54(4):1981–1988.

[127]   K.-H. Kim and M.-J. Youn. A simple and robust digital current control technique of a PM synchronous motor using time delay control approach. IEEE Transactions on Power Electronics, January 2001, 16(1):72–82.

[128]   F. Allgöwer and A. Zheng. Nonlinear model predictive control, Birkhäuser, December 2012.

[129]   T. Geyer. Model predictive control of high-power converters and industrial drives, John Wiley & Sons, November 2016.

[130]   E. N. Hartley, P. A. Trodden, A. G. Richards, and J. M. Maciejowski. Model predictive control system design and implementation for spacecraft rendezvous. Control Engineering Practice, July 2012, 20(7):695–713.

[131]   P. Kazmierkowski. Model predictive control of high-power converters and industrial drives. IEEE Industrial Electronics Magazine, September 2018, 12(3):55–56.

[132]   L. WanAB®g. Model predictive control system design and implementation using MATL, Springer Science & Business Media, February 2009.

[133]   E. Fernandez-Camacho and C. Bordons-Alba. Model predictive control in the process industry, Springer, 1995.

[134]   Y. Jiang, C. Wan, J. Wang, Y. Song, and Z. Y. Dong. Stochastic receding horizon control of active distribution networks with distributed renewables. IEEE Transactions on Power Systems, November 2018, 34(2):1325–1341.

[135]   J. M. Maestre and R. R. Negenborn. Distributed model predictive control made easy, Springer, 2014.

[136]   T. Qu, H. Chen, D. Cao, H. Guo, and B. Gao. Switching-based stochastic model predictive control approach for modeling driver steering skill. IEEE Transactions on Intelligent Transportation Systems, July 2014, 16(1):365–375.

[137]   K. Tian, J. Wang, B. Wu, Z. Cheng, and N. R. Zargari. A virtual space vector modulation technique for the reduction of common-mode voltages in both magnitude and third-order component. IEEE Transactions on Power Electronics, March 2015, 31(1): 839–848.

[138]   D. Bernardini and A. Bemporad. Stabilizing model predictive control of stochastic constrained linear systems. IEEE Transactions on Automatic Control, November 2011, 57(6):1468–1480.

[139]   T. P. do Nascimento, G. F. Basso, C. E. Dórea, and L. M. G. Gonçalves. Perception-driven motion control based on stochastic nonlinear model predictive controllers. IEEE/ASME Transactions on Mechatronics, May 2019, 24(4):1751–1762.

[140]   N. Guo, B. Lenzo, X. Zhang, Y. Zou, R. Zhai, and T. Zhang. A real-time nonlinear model predictive controller for yaw motion optimization of distributed drive electric vehicles. IEEE Transactions on Vehicular Technology, March 2020, 69(5):4935–4946.

[141]   F. Oldewurtel, C. N. Jones, A. Parisio, and M. Morari. Stochastic model predictive control for building climate control. IEEE Transactions on Control Systems Technology, August 2013, 22(3):1198–1205.

[142]   J. Sun, G. Xing, X. Liu, X. Fu, and C. Zhang. A novel torque coordination control strategy of a single-shaft parallel hybrid electric vehicle based on model predictive control. Mathematical Problems in Engineering, June 2015, 2015.

[143]   T. Homer and P. Mhaskar. Output-feedback Lyapunov-based predictive control of stochastic nonlinear systems. IEEE Transactions on Automatic Control, July 2017, 63(2):571–577.

[144]   M. Lorenzen, F. Dabbene, R. Tempo, and F. Allgöwer. Constraint-tightening and stability in stochastic model predictive control. IEEE Transactions on Automatic Control, November 2016, 62(7):3165–3177.

[145]   D. Moser, R. Schmied, H. Waschl, and L. del Re. Flexible spacing adaptive cruise control using stochastic model predictive control. IEEE Transactions on Control Systems Technology, February 2017, 26(1):114–127.

[146]   K. Sampathirao, P. Sopasakis, A. Bemporad, and P. P. Patrinos. GPU-accelerated stochastic predictive control of drinking water networks. IEEE Transactions on Control Systems Technology, March 2017, 26(2):551–562.

[147]  X. Zeng and J. Wang. A parallel hybrid electric vehicle energy management strategy using stochastic model predictive control with road grade preview. IEEE Transactions on Control Systems Technology, March 2015, 23(6):2416–2423.

[148]  Ismail, Moustafa Magdi, Wei Xu, Jian Ge, Yirong Tang, Abdul Khalique Junejo, and Mohamed G. Hussien. Adaptive linear predictive model of an improved predictive control of permanent magnet synchronous motor over different speed regions. IEEE Transactions on Power Electronics, 2022, 37(12):15338–15355.

[149]  Bordons and C. Montero. Basic principles of MPC for power converters: Bridging the gap between theory and practice. IEEE Industrial Electronics Magazine, September 2015, 9(3):31–43.

[150]  E. Garcia, D. M. Prett, and M. Morari. Model predictive control: theory and practice—a survey. Automatica, May 1989, 25(3):335–348.

[151]  S. Naidu and C. G. Rieger. Advanced control strategies for HVAC&R systems—an overview: part II: soft and fusion control. HVAC&R Research, April 2011, 17(2):144–158.

[152]  M. Nikolaou. Model predictive controllers: a critical synbook of theory and industrial needs-VII future needs-B is more MPC theory needed? Advances in Chemical Engineering, 2001, 26:198.

[153]  Z. Mynar, L. Vesely, and P. Vaclavek. PMSM model predictive control with field-weakening implementation. IEEE Transactions on Industrial Electronics, April 2016, 63(8):5156–5166.

[154]  J. Liu, C. Gong, Z. Han, and H. Yu. IPMSM model predictive control in flux-weakening operation using an improved algorithm. IEEE Transactions on Industrial Electronics, March 2018, 65(12):9378–9387.

[155]  Habib, Habib Ur Rahman, Asad Waqar, Sharoze Sohail, Abdul Khalique Junejo, Mahmoud F. Elmorshedy, Sheheryar Khan, Yun-Su Kim, and Moustafa Magdi Ismail. Optimal placement and sizing problem for power loss minimization and voltage profile improvement of distribution networks under seasonal loads using Harris Hawks Optimizer. International Transactions on Electrical Energy Systems 2022.

[156]  Su, Haonan, and Cheolkon Jung. Perceptual enhancement of low light images based on two-step noise suppression. IEEE Access 6, 2018: 7005–7018.

[157]  S. A. Hamad, W. Xu, Y. Liu, M. M. Ali, M. Magdi Ismail and J. Rodriguez. Improved MPCC with duty cycle modulation strategy for linear induction machines based on linear metro. 2021 IEEE International Conference on Predictive Control of Electrical Drives and Power Electronics (PRECEDE), 2021, pp. 35–40. doi: 10.1109/PRECEDE51386.2021.9680996.

[158]  A. A. Elbaset, M. M. Ismail and A. H. K. Alaboudy. Particle swarm optimization for layout design of utility interconnected wind parks. 2018 IEEE Power & Energy Society Innovative Smart Grid Technologies Conference (ISGT), 2018, pp. 1–6. doi: 10.1109/ISGT.2018.8403367.

[159]  M. M. Ismail, W. Xu, Y. Liu and M. Dong. Improved torque ripple reduction method for surface-mounted permanent magnet synchronous motor in flux-weakening region. 2019 22nd International Conference on Electrical Machines and Systems (ICEMS), pp. 1–6. doi: 10.1109/ICEMS.2019.8921942.

[160]  M. M. Ismail et al. Performance enhancement of salient permanent-magnet motors over wide speed range based on finite-set model predictive control. 2021 IEEE International Conference on Predictive Control of Electrical Drives and Power Electronics (PRECEDE), 2021, pp. 75–80. doi: 10.1109/PRECEDE51386.2021.9680931.

[161]  M. M. Ismail, W. Xu, Y. Liu, A. K. Junejo and M. G. Hussien. Improved controller optimization of flux-weakening strategy for salient permanent-magnet synchronous motor based on genetic algorithm. 2021 IEEE Energy Conversion Congress and Exposition (ECCE), 2021, pp. 4927–4932. doi: 10.1109/ECCE47101.2021.9595729.

[162] K. Liao, W. Xu, L. Bai, Y. Gong, M. M. Ismail and I. Boldea. Improved position sensorless piston stroke control method for linear oscillatory machine via an hybrid terminal sliding-mode observer. IEEE Transactions on Power Electronics, December 2022, 37(12):14186–14197. doi: 10.1109/TPEL.2022.3188828.

[163] Junejo, Abdul Khalique, Wei Xu, Ashfaque Ahmed Hashmani, Fayez FM El-Sousy, Habib Ur Rahman Habib, Yirong Tang, Muhammad Shahab, Muhammad Usman Keerio, and Moustafa Magdi Ismail. Novel Fast Terminal Reaching Law Based Composite Speed Control of PMSM Drive System. IEEE Access 10 (2022): 82202–82213.

[164] M. M. Ali, W. Xu, M. F. Elmorshedy, S. A. Hamad, A. K. Junejo and M. Ismail. Improved drive performance of linear induction machine based on direct thrust control and sliding mode control with extended state observer applied for linear metro. 2021 13th International Symposium on Linear Drives for Industry Applications (LDIA), 2021, pp. 1–6. doi: 10.1109/LDIA49489.2021.9505952.

[165] S. Deilami, A. S. Masoum, P. S. Moses, and M. A. Masoum. Real-time coordination of plug-in electric vehicle charging in smart grids to minimize power losses and improve voltage profile. IEEE Transactions on Smart Grid, August 2011, 2(3):456–467.

[166] S.-I. Kim, S. Park, T. Park, J. Cho, W. Kim, and S. Lim. Investigation and experimental verification of a novel spoke-type ferrite-magnet motor for electric-vehicle traction drive applications. IEEE Transactions on Industrial Electronics, February 2014, 61(10):5763–5770.

[167] X. Liu, H. Chen, J. Zhao, and A. Belahcen. Research on the performances and parameters of interior PMSM used for electric vehicles. IEEE Transactions on Industrial Electronics, February 2016, 63(6):3533–3545.

[168] E. Sortomme, M. M. Hindi, S. J. MacPherson, and S. Venkata. Coordinated charging of plug-in hybrid electric vehicles to minimize distribution system losses. IEEE Transactions on Smart Grid, December 2010, 2(1):198–205.

[169] W. Tang and Y. J. A. Zhang. A model predictive control approach for low-complexity electric vehicle charging scheduling: Optimality and scalability. IEEE Transactions on Power Systems, June 2016, 32(2):1050–1063.

[170] K. Putri, M. Hombitzer, D. Franck, and K. Hameyer. Comparison of the characteristics of cost-oriented designed high-speed low-power interior PMSM. IEEE Transactions on Industry Applications, June 2017, 53(6):5262–5271.

[171] S.-K. Sul. Control of electric machine drive systems, John Wiley & Sons, April 2011.

# 2 Performance Analysis of PMSM Drive System Using Frequency Modulation Technique

## 2.1 INTRODUCTION: BACKGROUND OF DYNAMIC MODEL AND CHAPTER MOTIVATION

In general, controllable models of plants should be identified through differential equations, based on the chosen control strategy. Therefore, this chapter establishes dynamic models for both salient and non-salient permanent magnet synchronous motors (PMSMs). PMSMs consist of permanent magnets (PMs) in the rotor shaft, installed in the stator frame of a three-phase induction machine while maintaining the stator winding geometry [1–5]. The main parameters that determine the performance of a permanent magnet synchronous machine are [6]:

- Permanent magnet location in the rotor
- Rotor configuration
- Number of stator poles
- Dampers for the rotor

The following sections describe the mathematical models of both non-salient and salient PMSMs [7–11].

### 2.1.1 MATHEMATICAL MODEL OF A NON-SALIENT PMSM

The *d-q* axes stator voltage vector of the PMSM are defined by

$$V_{\text{dqs}} = R_{\text{s}} + \frac{d\lambda_{\text{dqs}}}{dt} \tag{2.1}$$

where $V_{\text{dqs}}$ is the *d-q* axes stator voltage, and $\lambda_{\text{dqs}}$ is the stator flux vector, which can be given as

$$\lambda_{\text{dqs}} = L_{\text{s}} i_{\text{dqs}} + \lambda_{\text{m}}^{s} \tag{2.2}$$

DOI: 10.1201/9781003320128-2

39

where $i_{dqs}$ is the *d-q* axes stator current vector including $i_{ds}$ and $i_{qs}$ of the *d-q* axes stator current. $L_s$ is the stator reactance of surface-mounted type, and $\lambda^s_m$ is the rotor flux vector that can be defined as

$$\lambda^s_m = \lambda_m e^{j\theta_r} \tag{2.3}$$

where $\theta_r$ is the electrical angle of the rotor. Assuming the flux produced by the permanent magnets is low frequency, the stator flux derivative is defined as

$$\frac{d\lambda_{dqs}}{dt} = L_s \frac{di_{dqs}}{dt} + j\omega_e \lambda_m e^{j\theta_r} \tag{2.4}$$

where $\omega_e$ is the electrical angular speed. Meanwhile, substituting (2.4) in (2.1) and rearranging it can get the following:

$$L_s \frac{di_{dqs}}{dt} = V_{dqs} - (R_s + j\omega_e L_s)i_{dqs} - j\omega_e \lambda_m \tag{2.5}$$

where $R_s$ is the stator resistance. Finally, the *d-q* axes model can be concluded from splitting up (2.5) as

$$v_{ds} = R_s i_{ds} + L_s \frac{di_{ds}}{dt} - \omega_e L_s i_{qs} \tag{2.6}$$

$$v_{qs} = R_s i_{qs} + L_s \frac{di_{qs}}{dt} - \omega_e L_s i_{ds} + \omega_e \lambda_m \tag{2.7}$$

## 2.1.2 MATHEMATICAL MODEL OF A SALIENT PMSM

For interior PMSM, the salient PMSM does not have the same stator inductance around the air gap. The *d-q* axes and the coordinates of the *α-β* axes of interior PMSM are clarified in Figure 2.1; $v_{ds}$ and $v_{qs}$ can be written based on (2.6) and (2.7) as [12]

$$v_{ds} = R_s i_{ds} + L_{ds} \frac{di_{ds}}{dt} - \omega_e L_{qs} i_{qs} \tag{2.8}$$

$$v_{qs} = R_s i_{qs} + L_{qs} \frac{di_{qs}}{dt} - \omega_e L_{ds} i_{ds} + \omega_e \lambda_m \tag{2.9}$$

where $L_{ds}$ and $L_{qs}$ are the *d-q* axes stator reactance. Furthermore, (2.8) and (2.9) can be written in state-space presentation as

$$\frac{d}{dt}\begin{bmatrix} i_{ds} \\ i_{qs} \end{bmatrix} = \begin{bmatrix} -R_s / L_{ds} & \omega_e L_{qs} / L_{ds} \\ \omega_e L_{ds} / L_{qs} & -R_s / L_{qs} \end{bmatrix}\begin{bmatrix} i_{ds} \\ i_{qs} \end{bmatrix} + \begin{bmatrix} 1 / L_{ds} & 0 \\ 0 & 1 / L_{qs} \end{bmatrix}\begin{bmatrix} v_{ds} \\ v_{qs} - \omega_e \lambda_m \end{bmatrix} \tag{2.10}$$

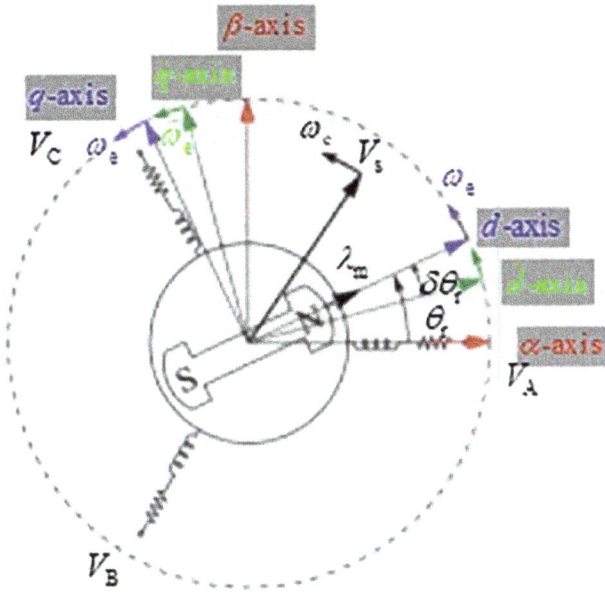

**FIGURE 2.1**   Coordinates of an interior PMSM.

Figure 2.2 represents the electrical circuit of the system in (2.10). Meanwhile, the electromagnetic developed torque ($T_e$) can be written as

$$T_e = 1.5n_p (\lambda_m i_{qs} + (L_{ds} - L_{qs})i_{qs}i_{ds})$$   (2.11)

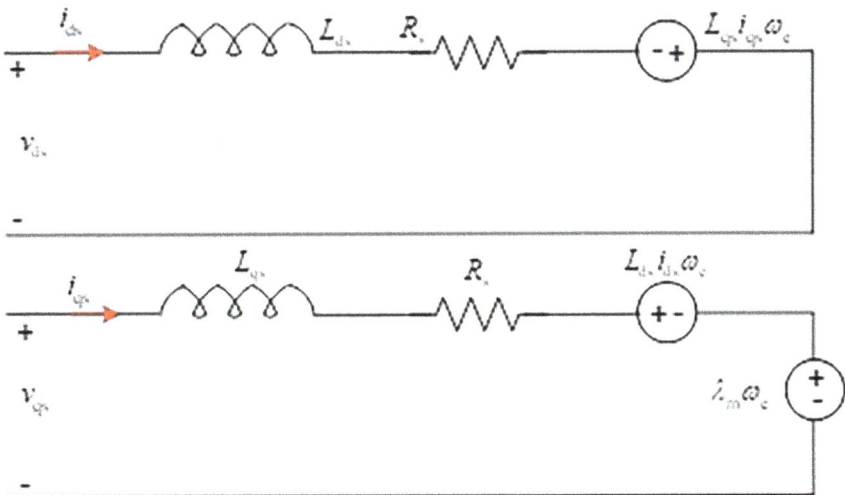

**FIGURE 2.2**   Electrical equivalent circuits for an interior PMSM.

where $n_p$ is the pair poles number. For the dynamic equation, (2.12) can express the shaft speed of PMSM as

$$\frac{d\omega_m}{dt} = \frac{1}{J}(-\omega_m B_v - T_L + T_e)  \tag{2.12}$$

where $T_L$ is the load torque, $\omega_m$ is the mechanical angular speed, $J$ is the inertia of the shaft, and $B_v$ is the viscous friction coefficient.

### 2.1.3 MOTIVATION

The mathematical model of PMSM should be studied to develop a plant model of predictive control; it could be a linear or nonlinear model. Meanwhile, the transfer function of output states to the input of PMSM can be determined in terms of the mathematical model to correctly adjust the parameters of the field-oriented control (FOC) strategy based on the proportional-integral (PI) gains of the PI regulator. Additionally, the error values of the difference between the reference state and the measured state can be modelled depending on the mathematical model of PMSM to optimally design the parameters of a control strategy.

Working in the flux-weakening region is necessary for many industrial applications such as traction drives. Thus, it is important to study the difference between the constant-torque region and the flux-decreasing region. The voltage and stator current limitations are also necessary to avoid saturation during control at the flux-weakening region, and the MTPA principle is discussed to increase operating efficiency. Lastly, the ripple of the stator current and torque can be reduced based on the variable switching frequency technique. This chapter discusses the ripple control algorithm based on the variable switching frequency technique to verify its effectiveness in the constant torque and flux-weakening regions.

## 2.2 MAXIMUM TORQUE PER AMPERE PRINCIPLE OF PMSM DRIVE

The operating mode of the PMSM automatically shifts between the constant torque scope and the flux-decreasing range, based on the reference speed instruction. Therefore, an important issue is to improve the drive systems to operate in a high-speed range, which means in a flux-weakening region. Furthermore, a parameter is required to increase the overall efficiency of driving systems in a constant-torque region, which can be applied by reducing the magnitude of $d$-$q$ axes stator currents for each loading torque required. These pairs of $d$-$q$ axes currents are necessary to achieve the MTPA under both regions without overriding the current- and voltage-limiting restrictions. The following sections examine how the MTPA algorithm operates in regions of constant torque and flux weakening.

### 2.2.1 MTPA CONTROL IN THE CONSTANT TORQUE RANGE

The principle of maximum torque per ampere is to optimize the stator current vector for the sake of reducing the $d$-$q$ axes of the stator current to produce the required

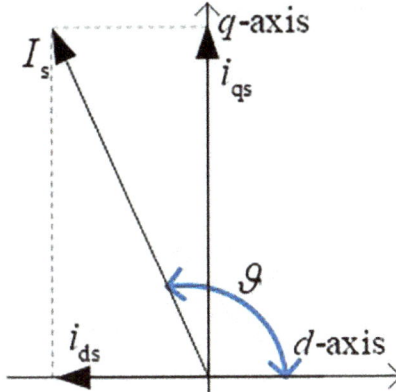

**FIGURE 2.3**   The angle of the developed torque.

electromagnetic torque. In the case of a surface-mounted PMSM torque, the reluctance torque is faded due to a lack of salience ($L_{ds} = L_{qs}$).

The electromagnetic torque can be written as

$$T_e = 1.5 n_p \lambda_m i_{qs} \tag{2.13}$$

As can be seen from (2.13), the maximum torque of the surface-mounted PMSM is achieved by maintaining the current of the $q$-axis stator with the stator current vector ($I_s$), and the $d$-axis current is equal to zero ($i_{ds} = 0$). For the torque of interior PMSM, reluctance torque is present because it has a greater salience ($L_{qs} > L_{ds}$). Thus, the electromagnetic torque can depend on the $d$-axis stator current as well as the $q$-axis stator current. The $d$-$q$ axes stator current can be concluded from Figure 2.3 as [13–17]

$$i_{qs} = I_s \sin(\vartheta) \tag{2.14}$$

$$i_{ds} = I_s \cos(\vartheta) \tag{2.15}$$

where $\vartheta$ and $I_s$ are the torque angle and stator current vector of interior PMSM, respectively. The torque angle is the angle between the positive direction of $i_{ds}$ and $I_s$, which ranges from $0°$ to $180°$. Hence, the resulting electromagnetic torque of interior PMSM can be written as

$$T_e = 1.5 n_p (\lambda_m I_s \sin(\vartheta) + (L_{ds} - L_{qs}) I_s \cos(\vartheta) I_s \sin(\vartheta)) \tag{2.16}$$

Because $sin(\vartheta)$ $cos(\vartheta)$ equals 0.5 $sin(2\vartheta)$, (2.16) can be rewritten as

$$T_e = 1.5 n_p (\lambda_m I_s \sin(\vartheta) + 0.5 (L_{ds} - L_{qs}) I_s^2 \sin(2\vartheta)) \tag{2.17}$$

the reluctance torque ($T_r$) and excitation torque of the magnet ($T_m$) can be expressed from (2.17) as

$$T_e = T_m + T_r \tag{2.18}$$

where

$$T_m = 1.5 n_p \lambda_m I_s \sin(\vartheta) \tag{2.19}$$

$$T_r = 0.75 n_p (L_{ds} - L_{qs}) I_s^2 \sin(2\vartheta) \tag{2.20}$$

Figure 2.4 shows the sinusoidal relationship between the torque angle and both the excitation torque of the magnet and the reluctance torque produced by the simulation results by substituting the torque equations (2.19) and (2.20) for the parameters given in Table 2.1.

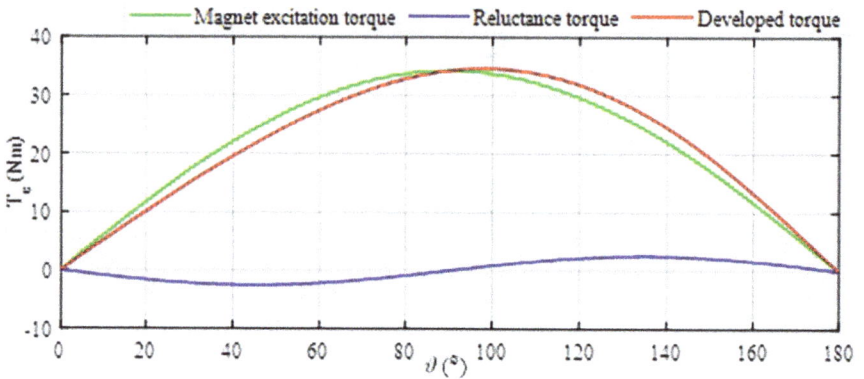

**FIGURE 2.4**  The magnet excitation torque and the reluctance torque versus the torque angle.

**TABLE 2.1  Main Parameters of the Interior PMSM Motor**

| Parameters | Symbol | Unit | Quantity |
|---|---|---|---|
| Flux linkage | $\lambda_m$ | Wb | 0.38 |
| $d$-axis component of the stator inductance | $L_{ds}$ | H | 0.00191 |
| $q$-axis component of the stator inductance | $L_{qs}$ | H | 0.00473 |
| Number pole pairs | $n_p$ | 3 | 3 |
| Space vector of the stator current | $I_s$ | A | 20 |

The interior PMSM torque is maximized at a torque angle that results in setting the torque derivative in (2.18) to zero. This angle can be derived as

$$(\cos(\vartheta)I_s)\lambda_m + ((\cos(\vartheta)I_s)^2 - ((\sin(\vartheta)I_s)^2)(L_{ds} - L_{qs}) = 0 \qquad (2.21)$$

Concerning $I_s = \sqrt{i_{ds} + i_{qs}}$, two solutions of $i_{ds}$ giving the maximum developed torque dependent on $I_s$ can be derived from (2.21) as

$$i_{ds} = \begin{cases} \dfrac{\lambda_m + \sqrt{8I_s^2(L_{qs} - L_{ds})^2 + \lambda_m^2}}{4(L_{qs} - L_{ds})} < 0 \\[4mm] \dfrac{\lambda_m - \sqrt{8I_s^2(L_{qs} - L_{ds})^2 + \lambda_m^2}}{4(L_{qs} - L_{ds})} > 0 \end{cases} \qquad (2.22)$$

Since $L_{ds}$ is smaller than $L_{qs}$, the torque angle should be larger than 90°. Thus, the current of $i_{ds}$ has a negative magnitude on the $d$-axis. Therefore, the torque angle and stator current of the $d$-$q$ axes that produce maximum torque can be derived as

$$i_{ds\text{-}M} = \frac{\lambda_m - \sqrt{8I_s^2(L_{qs} - L_{ds})^2 + \lambda_m^2}}{4(L_{qs} - L_{ds})} \qquad (2.23)$$

$$i_{qs\text{-}M} = \sqrt{I_s^2 - i_{ds-M}^2} \qquad (2.24)$$

$$\vartheta_M = \tan^{-1}(\frac{i_{qs\text{-}M}}{i_{ds\text{-}M}}) \qquad (2.25)$$

where $i_{ds\text{-}M}$, $i_{qs\text{-}M}$, and $\vartheta_M$ are the stator current for $d$-$q$ axes and the torque angle for the operating point of MTPA. It is evident from (2.18) to (2.25) that the maximum torque is limited by the supply current. Hence, the limiting voltage limitation was not considered in the constant-torque region.

There is a specific pair of $d$-axis and $q$-axis currents for each required loading torque, resulting in MTPA under this condition. Therefore, the torque angle ($\vartheta_M$) of the MTPA strategy is determined not only by the interior PMSM parameters but also by the stator current values of the $d$-$q$ axis. Finally, several torque angles for MTPA control are obtained due to different supply currents of the limit current circuits that cross the MTPA trajectory, as shown in Figure 2.5.

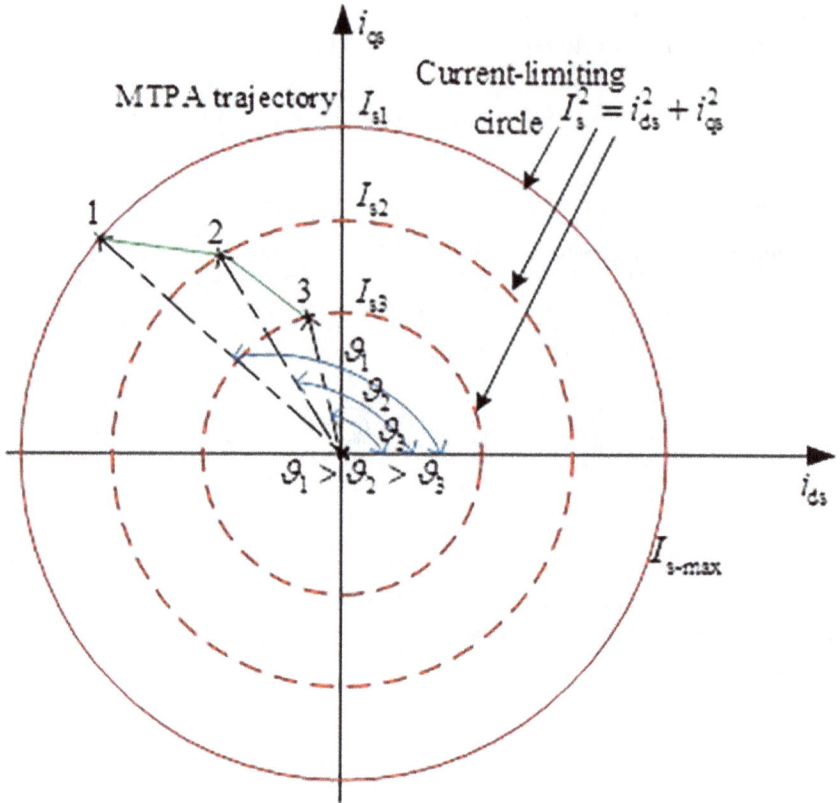

**FIGURE 2.5**   The current-limiting circle and MTPA trajectory for interior PMSM motors.

### 2.2.2 PMSM CONTROL AND OPERATING LIMITS WITHIN THE FLUX-WEAKENING RANGE

In the case of electric vehicle traction and four-wheel electric scooters, the PMSM drive systems typically require a wide constant speed range. However, the base for the field decreasing is that the magnetic field of the motor at the high-velocity level has been reduced because the back electromotive force (EMF) rises. This, in turn, has an impact on the current control loop, resulting in its not producing the desired effect and performing in a degraded manner [18–21]. Hence, the power converter of the drive system cannot control PMSM at a high rate of speed.

If the back EMF exceeds the motor voltage level, the PMSM will not be able to draw current and, hence, will not be able to develop torque, and the rotor speed of the motor will only be increased if the air gap flux is decreased. Since the rotating magnetic field induced by PMs can only be reduced indirectly by demagnetizing the magnetic motive force (MMF) from the armature, an increased velocity range can be obtained by the flux-weakening algorithms. For running in the field-decreasing region, the MMF is generated to demagnetize the stator and coil currents to counteract the apparent MMF produced by the PMs installed on the rotor. As a result, the air gap flux is implicitly reduced, thus increasing the motor speed [22, 23].

If the flux-weakening control is ignored, the back EMF will continue to increase with speed. Therefore, the output voltage of the voltage source inverter should be increased to preserve the necessary phase currents. In the scope of constant torque, as seen in Figure 2.6 and Figure 2.7, the motor can be accelerated by the maximum torque until its voltage of the motor approaches the voltage limit value at the operating point of ($\omega_{\text{CT-max}}$), which is known as the maximum speed in the constant-torque region. This base speed is the maximum speed of a PMSM controlled by the strategy of MTPA [24–28]. The base speed can be defined as

$$\omega_{\text{CT-max}} = \frac{V_{\text{s-max}}}{\sqrt{(\lambda_m + i_{\text{ds-M}} L_{\text{ds}})^2 + (i_{\text{qs-M}} L_{\text{qs}})^2}} \tag{2.26}$$

Moreover, the electrical PMSM velocity and the stator voltage space vector ($V_s$) for any operating point having a nonlinear relationship can be represented as

$$\omega_e = \frac{V_s}{\sqrt{(i_{\text{qs}} L_{\text{qs}})^2 + (\lambda_m + i_{\text{ds}} L_{\text{ds}})^2}} \tag{2.27}$$

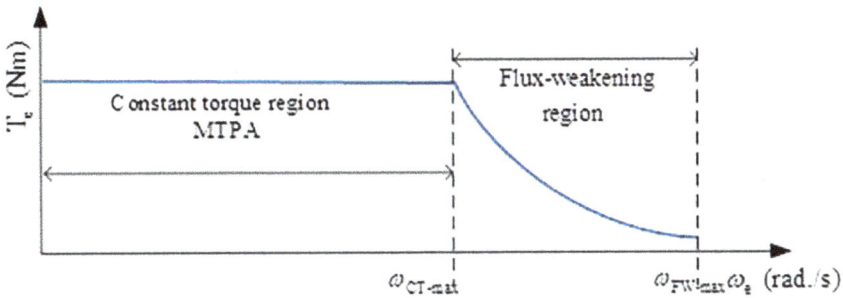

**FIGURE 2.6** The electromagnetic torque in regions of constant torque and flux weakening.

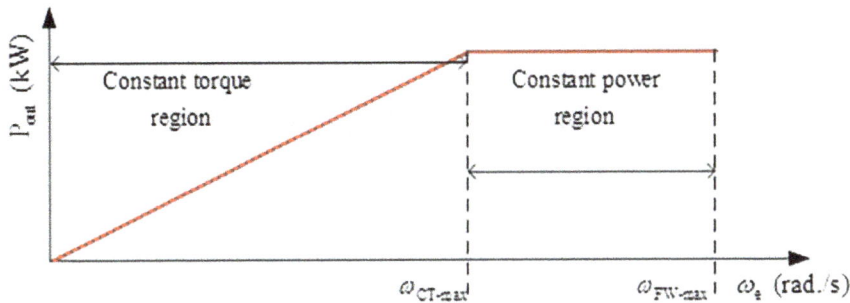

**FIGURE 2.7** The output power in regions of constant torque and flux weakening.

where $V_s$ can be stated as

$$V_s = \sqrt{(\omega_e(\lambda_m + L_{ds}i_{ds}))^2 + (L_{qs}i_{qs}\omega_e)^2} \tag{2.28}$$

For the technique used in the inverter, the space vector pulse-width modulation (PWM) technique switches the voltage source inverter (VSI). The voltage vector falls under the flux-weakening control algorithms on the hexagon boundaries, which saturates the voltage that determines the highest voltage ($V_{s\text{-}max} = 0.577V_{dc}$) [29]. Both the converter's power rating and the motor's thermal rating are used to calculate the maximum current $I_{s\text{-}max}$. This represents the peak of the maximum phase current that can be plotted as a current limit circuit in the $d$-$q$ frame as shown in Figures 2.8 and Figure 2.9. As a result, the motor's voltage and current limits can be demonstrated as

$$v_{ds}^{*2} + v_{qs}^{*2} = V_s \leq V_{s\text{-}max}^2 \tag{2.29}$$

$$i_{ds}^{*2} + i_{qs}^{*2} = I_s \leq I_{s\text{-}max}^2 \tag{2.30}$$

where $I_s$ is the stator current vector. The surface-mounted PMSM voltage limitation is a circle in the $d$-$q$ frame because $L_{ds} = L_{qs}$, as seen in Figure 2.8, which shows voltage and current limit circles for surface-mounted PMSM [30–32]. However, the voltage constraint is an ellipse for interior PMSM in the $d$-$q$ frame, as seen in Figure 2.9,

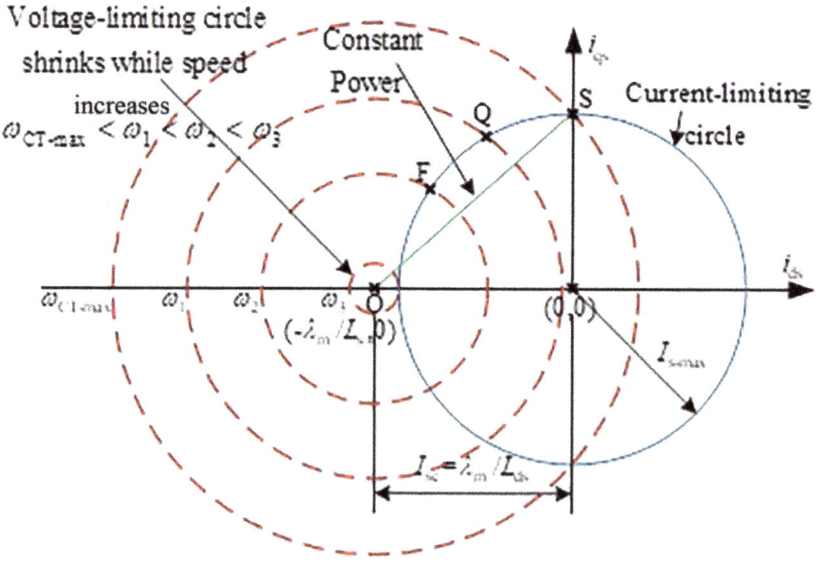

**FIGURE 2.8** MTPA determination by the voltage and current limit circles in the $i_{ds}$-$i_{qs}$ plane.

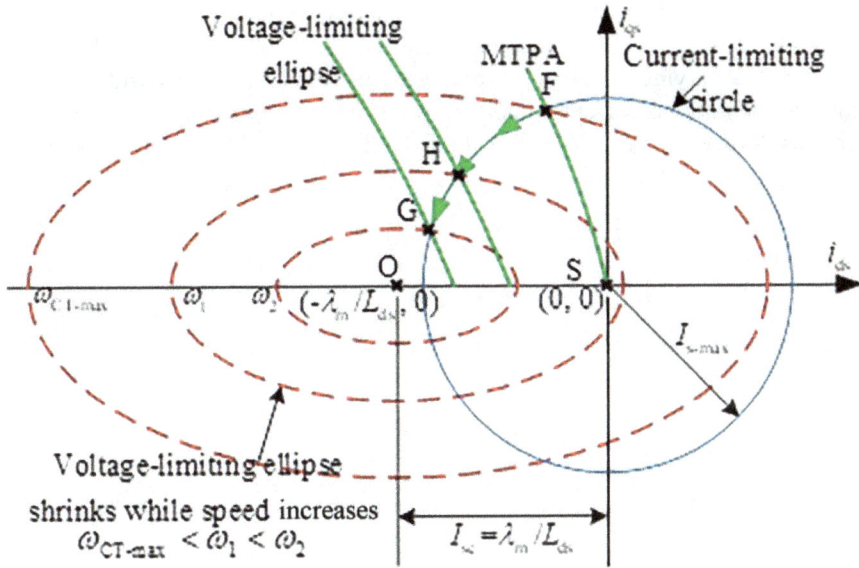

**FIGURE 2.9** Voltage-limiting ellipse and current circle for MTPA control algorithm in the $i_{ds}$-$i_{qs}$ plane.

which presents the voltage ellipse limit and current circle limit for interior PMSM, where the terminal voltage magnitude can be described from (2.28) as

$$V_s = \sqrt{\omega_e^2 (L_{qs}^2 i_{qs}^2 + L_{ds}^2 (i_{ds} + (\lambda_m / L_{ds}))^2)}, \text{ at } V_s \le V_{s\text{-max}} \tag{2.31}$$

The constant-power control technique is a commonly used method for driving sur-face-mounted PMSM in the range of flux weakening due to its simplicity and having no additional hardware requirement.

In this procedure, the current vector takes the fixed power direction that is the $S$-$O$ line to hold the power steady, as seen in Figure 2.8. Therefore, the output power ($P_{out}$) can be determined as

$$P_{out} = (30 / \pi)\omega_{CT\text{-max}} T_{max} \tag{2.32}$$

where $T_{max}$ is the maximum electromagnetic torque at the operating point of $i_{ds\text{-M}}$ and $i_{qs\text{-M}}$, as seen in Figure 2.8, which shows the electromagnetic torque in the ranges of the constant torque and flux weakening. Moreover, the output power in the regions of the constant torque and flux weakening, as shown in Figure 2.9, indicates the me-chanical velocity ($\omega_m$), which can be written as

$$T_e\omega_m = T_{max}\omega_{CT\text{-max}} = \text{Constant} \tag{2.33}$$

For surface-mounted PMSM, Figure 2.8 explains the circles of a motor voltage limit with a radius of $V_{s\text{-max}}/(\omega_e L_s)$ and centered at $(-\lambda_m/L_s, 0)$. This diagram depicts the circles with a varying diameter voltage level; the circle intersects with the circle with a constant radius current limit. The operating point shifts from $S$ to $Q$ as speed increases. As a result, the $q$-axis current point of $T_{max}$ is the intersection of the circles, which can be described as

$$i_{qs\text{-M}} = (L_s/2\lambda_m)\sqrt{(2\lambda_m I_{s\text{-max}}/L_s)^2 - ((\lambda_m/L_s)^2 - (V_{s\text{-max}}/(\omega_e L_s))^2 + I_{s\text{-max}}^2)} \quad (2.34)$$

The $d$-axis reference stator current in terms of $\omega_{\text{CT-max}}$ can be described as

$$i_{ds\text{-M}} = -\sqrt{I_{s\text{-max}}^2 - i_{qs\text{-M}}^2} \quad (2.35)$$

For interior PMSM, Figure 2.9 describes the ellipses of the voltage level centered at $(-\lambda_m/L_s, 0)$ and intersecting the current limiting circle. If the velocity increases, as the center of the voltage-limiting ellipse stays fixed, the ellipse gets smaller and smaller.

The $i_{ds\text{-M}}$ and $i_{qs\text{-M}}$ of $T_{max}$ in the flux-weakening region can be defined as follows [33]:

$$i_{ds\text{-M}} = \frac{1}{L_{qs}^2 - L_{ds}^2}(L_{ds}\lambda_m - ((L_{ds}\lambda_m)^2 + (L_{qs}^2 - L_{ds}^2)(\lambda_m^2 + (L_{ds}I_s)^2 - (V_f/\omega_e)^2)^{0.5})) \quad (2.36)$$

$$i_{qs\text{-M}} = sign(I_s)\sqrt{I_s^2 - i_{ds}^2} \quad (2.37)$$

where

$$\begin{cases} \text{if } I_s \geq 0, \text{ then } sign(I_s)=1 \\ \\ \text{otherwise } sign(I_s) = -1 \end{cases}$$

where the variable of $V_f$ can be determined as

$$V_f = V_{s\text{-max}} - R_s I_{s\text{-max}} \quad (2.38)$$

As the rotor velocity increases, the operating points of the current vector should move along the boundaries of the current circuit defined in the figure (i.e., from $F$ to $H$, from $H$ to $G$, and so on).

Failure to determine the reference current of the $d$-axis component means saturation of the control current and loss of control power. Therefore, the flux level should be changed depending on the parameters of the machine and the torque of the load to avoid saturation of the internal current controllers. It is also beneficial to set the starting point of flux weakening because the late start of flux attenuation can lead to an unwanted reduction of torque output, and an early start leads to the deterioration of acceleration control. The short-circuit current ($I_{sc}$) is defined when the radius of the voltage-limiting ellipse is the smallest and typically specified by $\lambda_m/L_{ds}$, as shown in Figure 2.9. Thus, in theory, the rotor permanent magnet flux should be omitted if the power converter draws a high-current equivalent to or greater than the short-circuit current ($I_{s\text{-max}} \geq \lambda_m/L_{ds}$). Thus, the maximum velocity ($\omega_{\text{FW-max}}$) in the flux-weakening region becomes limitless in the lossless PMSM. Moreover, all operating points have voltage limiting only and adopt a voltage-limiting ellipse instead of a current-limiting circuit to a maximal electromagnetic torque. The maximum velocity in the flux-weakening scope ($\omega_{\text{FW-max}}$) can be derived as

$$V_{s\text{-max}}^2 = \omega_e^2 (\lambda_m - L_{ds} I_{s\text{-max}})^2 \tag{2.39}$$

$\omega_e = \omega_{\text{FW-max}}$ at $i_{ds} = -I_{s\text{-max}}$ and $i_{qs} = 0$.
Therefore, $\omega_{\text{FW-max}}$ can be concluded as

$$\omega_{\text{FW-max}} = \frac{V_{s\text{-max}} / L_{ds}}{(\lambda_m / L_{ds} - I_{s\text{-max}})} \tag{2.40}$$

However, if $I_s = I_{s\text{-max}}$ of a large power inverter, the ($\lambda_m/L_{ds\text{-}} I_{s\text{-max}}$) can be zero; thus, the theoretical velocity is unlimited. In contrast, if the power converter draws less than the short-circuit current ($I_{s\text{-max}} < \lambda_m/L_{ds}$), the maximum velocity ($\omega_{\text{FW-max}}$) is limited. Hence, the maximum torque can be located at the intersection of the current circuit of that velocity and the voltage ellipse. Yet permanent magnets are powerful enough to be fully flux bonded to the stator in most construction cases.

## 2.3 FREQUENCY MODULATION TECHNIQUE FOR TORQUE RIPPLE REDUCTION

There is a growing demand for the use of surface-mounted PMSMs in light vehicle drive systems and commercial applications [34–38]. Because of the energy efficiency of surface-mounted PMSMs, the motors could provide higher magnetic flux density. However, when the motor speed is operating in the region of flux weakening, pulse-width modulation with a variable switching frequency control algorithm is not sufficient to mitigate the torque ripple. Therefore, this chapter considers an algorithm for torque ripple control called varying switching frequency pulse-width modulation.

The schematic representation of the control system used in this book is provided in Figure 2.10. In this study, a digital signal processor TMS320F28335 is used to implement the control technique as well as to control the motor and the load motor through the voltage supply inverters. The inverter used in this study is a two-level three-phase inverter that can handle a maximum current of 20 A. The load motor is coupled with the controlled motor and is controlled by the control card based on closed-loop torque mode to work as a load. For clarification, if the controlling motor is a surface-mounted PMSM, then the loading motor is an interior PMSM and vice versa [39–42].

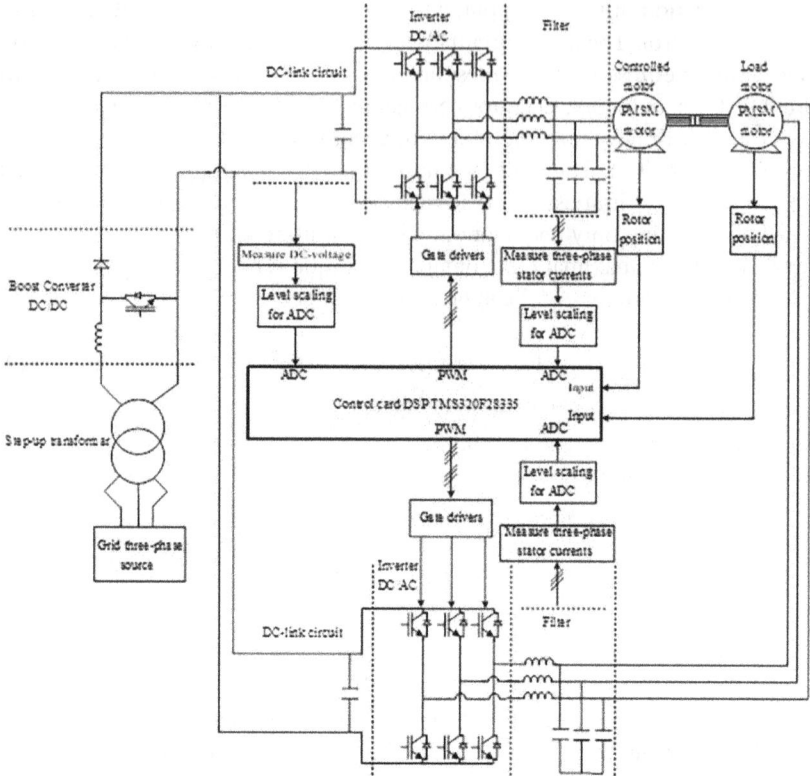

**FIGURE 2.10**   Overview of the drive system.

   Meanwhile, the counting mechanical position, the measured direct current (DC) link voltage, and the measured phase currents are necessary for the control algorithms [43–47]. The peak value of torque ripple mitigation depends on the variable switching frequency PWM (VSFPWM) for the three-phase PWM space vector inverter. The varying switching frequency PWM technique relies on predicting the ripple peak of current to control the current ripple and, hence, the torque ripple. This is because the electromagnetic torque is created by the winding currents. Therefore, any ripple in the current would cause some ripple in the electromagnetic torque [48–50].

   For building frequency switching control, it is possible to describe a connection between both the predicted peak of the current and the torque. The most widely used torque math formula for surface-mounted PMSM is as follows:

$$T_e = 1.5n_p \{\lambda_m (i_{ds}, i_{qs})\, i_{qs} + [L_{ds}(i_{ds}, i_{qs})\text{-}L_{qs}(i_{ds}, i_{qs})]\, i_{qs} i_{ds}\} \qquad (2.41)$$

As shown in (2.41), the $q$-axis of the stator winding is proportional to the torque (2.13). This can be refined despite magnetic saturation, spatial harmonics, field-oriented control angle error, and so on, if $L_{ds} = L_{qs}$. Overall, this clarity also ensures that torque ripple is controlled by regulating the ripple current. Furthermore, the method in this unit relies on (2.13) to approximate the torque equation rather than explaining the relationship between the surface mounted PMSM's various work states. In terms of the algorithm, the estimation of the torque ripple peak ($T_e$) is shown by calculating the maximum predicted peak of the $q$-axis current ripple ($i_{qs}$): i.e.,

$$\Delta T_e = 1.5 n_p \lambda_m \Delta i_{qs} \tag{2.42}$$

The authors in [51] and [52] recommend a linear relationship between the updated frequency ($F_s$) and the maximum peak ripple of the $q$-axis current ($i_{qs}$) for varying switch frequency.

$$F_s = F_{sw} \frac{\Delta i_{qs}}{\Delta I_{wanted}} \tag{2.43}$$

where $I_{wanted}$ is the necessary $q$-axis current peak value. Under the same assumptions, this calculation mainly reduces the average switching frequency in comparison to the constant switching frequency PWM method, which could reduce switching losses in all switching cycles [51–53]. As a result of (2.42) and (2.43), the updated frequency can be calculated as:

$$F_s = F_{sw} \frac{\Delta T_e}{\Delta T_{wanted}} \tag{2.44}$$

where $\Delta T_{wanted}$ is the desired peak of the torque ripple. The maximum peak of the AC stator current ripples can be decided by the DC voltage and the specified modulation ratio when determining $i_{qs}$. This algorithm aims to determine the three-phase switching cycles in Figure 2.11, which can be represented as [54]

$$d_a = (T_2 + 0.5 T_0)/T_s \tag{2.45}$$

$$d_b = T_0 / (0.5 T_s) \tag{2.46}$$

$$d_c = (T_1 + T_2 + 0.5 T_0)/T_s \tag{2.47}$$

where $T_s$ is the switching period; $d_a$, $d_b$, and $d_c$ are three-phase switching cycles; and $T_0$, $T_1$, and $T_2$ are the voltage sectors' actuation times. In Figure 2.11, the slope of the predicted maximum peaks for the first phase current ($i_{an}$) is determined in Table 2.2 and, where $\Theta$ is the voltage vector angle, and $\bar{d}_i$ is the quasi-duty cycle, can be written as

$$\bar{d}_i = 2 d_i - 1, i \in \{a, b, c\} \tag{2.48}$$

$$\theta = \tan^{-1} \left( \frac{v_{\beta s}^*}{v_{\alpha s}^*} \right) \tag{2.49}$$

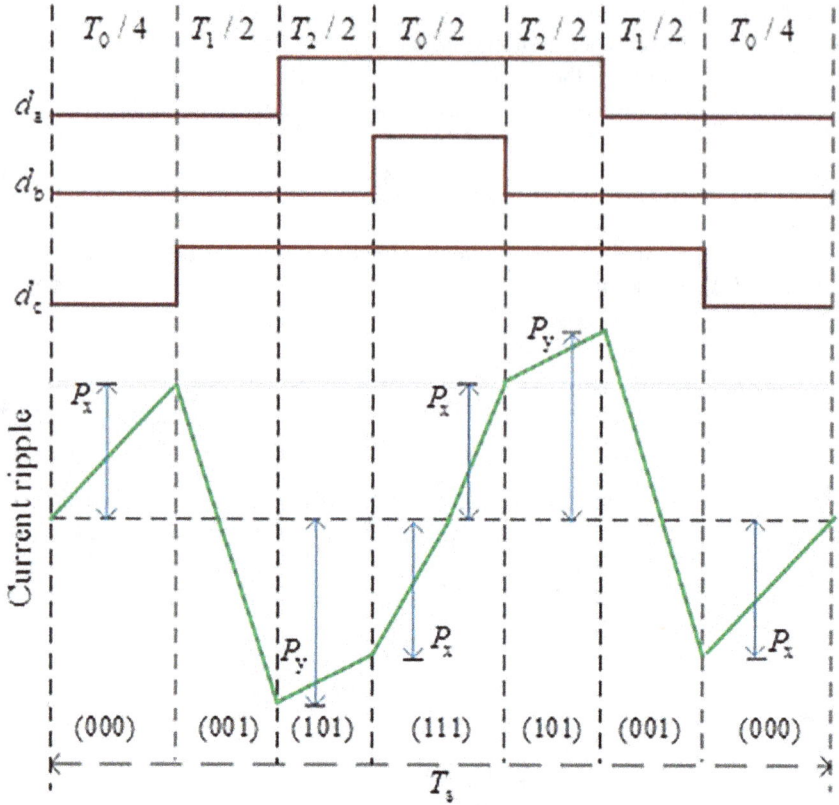

**FIGURE 2.11** Single-phase predicted current ripple locus through one switching cycle [51].

where $v^*_{\alpha s}$ and $v^*_{\beta s}$ are the reference voltage vector $\alpha$-$\beta$ axes components [52]. Finally, as shown in Figure 2.11, the highest peaks of the current ripple are the maximums of $P_x$ and $P_y$. As a result, the park conversion can determine the maximum $q$-axis ripple from the peak current ripple in the ABC coordinate:

$$\begin{bmatrix} \Delta i_{ds} \\ \Delta i_{qs} \end{bmatrix} = 0.667 \begin{bmatrix} \cos(\theta) & \cos(\theta-\frac{2\pi}{3}) & \cos(\theta-\frac{4\pi}{3}) \\ -\sin(\theta) & -\sin(\theta-\frac{2\pi}{3}) & -\sin(\theta-\frac{4\pi}{3}) \end{bmatrix} \begin{bmatrix} \Delta i_a \\ \Delta i_b \\ \Delta i_c \end{bmatrix} \quad (2.50)$$

where $\Delta i_{ds}$ is the maximum predicted current ripple for the $d$-axis and $\Delta i_a$, $\Delta i_b$, and $\Delta i_c$ are real-time maximum predicted ripple peaks of three-phase stator currents. Finally, the major drawbacks of the PWM control algorithm can be identified as follows:

- This algorithm depends on the stator inductance as a machine parameter.
- This algorithm depends on the table of the predicted current ripple slope.

**TABLE 2.2**

**Current Ripple Slope Predicted with Different Voltage Sectors [51].**

| Angle | Vector | Slope |
|---|---|---|
| Zero voltage sector | 000 or 111 | $\dfrac{di_a}{dt} = \dfrac{V_{dc}}{3L_s}\left(\dfrac{\overline{d}_b + \overline{d}_c}{2} - \overline{d}_a\right)$ |
| $0 \le \theta < 60$ | 100 | $\dfrac{di_a}{dt} = \dfrac{2V_{dc}}{3L_s}\left(1 + \dfrac{\overline{d}_b + \overline{d}_c}{4} - \dfrac{\overline{d}_a}{2}\right)$ |
| $60 \le \theta < 120$ | 110 | $\dfrac{di_a}{dt} = \dfrac{2V_{dc}}{3L_s}\left(\dfrac{\overline{d}_b + \overline{d}_c}{4} - \dfrac{\overline{d}_a}{2} + 0.5\right)$ |
| $120 \le \theta < 180$ | 010 | $\dfrac{di_a}{dt} = \dfrac{V_{dc}}{3L_s}\left(\dfrac{\overline{d}_b + \overline{d}_c}{2} - 1 - \overline{d}_a\right)$ |
| $180 \le \theta < 240$ | 011 | $\dfrac{di_a}{dt} = \dfrac{2V_{dc}}{3L_s}\left(\dfrac{\overline{d}_b + \overline{d}_c}{4} - 1 - \dfrac{\overline{d}_a}{2}\right)$ |
| $240 \le \theta < 300$ | 001 | $\dfrac{di_a}{dt} = \dfrac{V_{dc}}{3L_s}\left(\dfrac{\overline{d}_b + \overline{d}_c}{2} - 1 - \overline{d}_a\right)$ |
| $300 \le \theta < 360$ | 101 | $\dfrac{di_a}{dt} = \dfrac{2V_{dc}}{3L_s}\left(\dfrac{\overline{d}_b + \overline{d}_c}{4} + 0.5 - \dfrac{\overline{d}_a}{2}\right)$ |

## 2.4 VARIABLE SWITCHING FREQUENCY IMPLEMENTATION FOR SURFACE-MOUNTED PMSM DRIVE

In the previous sections of this chapter, the details of the surface-mounted PMSM for the flux-weakening operation and torque ripple mitigation process are explained. This section introduces the field-oriented control strategy that drives the surface-mounted PMSM in the high-speed range. The predictive current ripple of the $q$-axis is used to change the switching frequency of the inverter mentioned in the previous section.

Hence, the maximum torque ripple value can be controlled. To avoid the level of the current regulators when operating in the flux-weakening area, the outer regulation loop of the voltage vector magnitude of the VSI normally adjusts the onset of flux weakening. Finally, simulation and experimental results have demonstrated the effectiveness of the control strategy. Therefore, the following sections describe the flux-weakening control loop and relevant simulation and experimental results.

### 2.4.1 Flux-Weakening Control Loop Based on Voltage Vector Feedback

Concentrating here on the traditional vector control scheme, the main idea of controlling the flux-level algorithm is the application of the additional outer control loop of voltage vector magnitude ($\rho$), as seen in Figure 2.12. If the voltage vector magnitude is greater than 1, the motor acts in the field-decreasing area. Otherwise, the motor is working in the region of constant torque. Meanwhile, the reference current ($i^*_{ds}$) of the $d$-axis when the motor is operating in the flux-weakening area is negatively supplied by this outer loop to control the mean value of $\rho$ in 1.

However, the $\rho$ can be defined as

$$\rho = \sqrt{3}\frac{\sqrt{v^{*2}_{\alpha s}+v^{*2}_{\beta s}}}{V_{dc}} \tag{2.51}$$

The flux-weakening control loop in Figure 2.12 can be expressed as

$$e_\rho = \frac{i^*_{ds}}{K_{P\rho}} - \frac{K_{I\rho}}{K_{P\rho}}E_\rho + \int\left(\frac{K_{I\rho}}{K_{P\rho}}\right)^2 \Delta i^*_{ds} - \rho^* + \rho \tag{2.52}$$

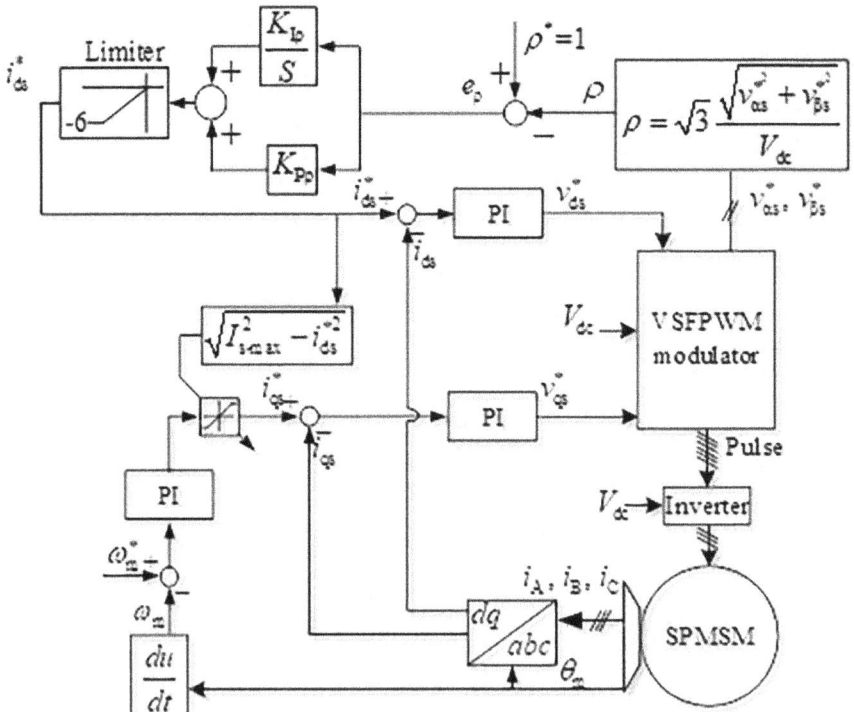

FIGURE 2.12   The proposed flux-weakening control loop.

where $K_{P_\rho}$ and $K_{I_\rho}$ are the parameters for the proportional and integral of the modulation index regulator, $\rho^*$ is the voltage vector reference magnitude, $e_\rho$ is the modulation index controller error, $\Delta i^*_{ds}$ is the difference between the limited value and the reference value of the controller, and $E_\rho$ is the integral error of the modulation index controller. The compensated value of the $d$-axis current determines the $q$-axis current peak $(i_{qs\text{-}max})$, providing the electromagnetic torque that can be described as

$$i_{qs\text{-}max} = \sqrt{I^2_{s\text{-}max} - i^{*\,2}_{ds}} \qquad (2.53)$$

## 2.4.2 SIMULATION RESULTS AND DISCUSSION

The performance of the control strategy is validated in a MATLAB software environment [55]. Table 2.3 lists the parameters and requirements of the surface-mounted PMSM motor to use for simulation. VSI's PWM has a constant switching frequency of 2.5 kHz and a controlled DC voltage of 400 V. The offline random theory is used to adjust the proportional and integral (PI) parameters, which means increasing and decreasing the parameters until the motor speed behavior achieves acceptable performance.

**TABLE 2.3**
**Simulation and Experimental Parameters of Surface-Mounted PMSM.**

| Parameters | Symbol | Unit | Quantity |
|---|---|---|---|
| Rated line voltage | $v_N$ | V | 270 |
| Rated current | $I_N$ | A | 6.8 |
| Rated power | $P_r$ | kW | 3 |
| Rated speed | $N_m$ | rpm | 2000 |
| Rated torque | $T_N$ | Nm | 14.3 |
| Stator resistance | $R_s$ | $\Omega$ | 0.8 |
| Stator inductance | $L_s$ | H | $5 \times 10^{-3}$ |
| Permanent magnet flux linkage | $\lambda_m$ | Wb | 0.35 |
| Moment of inertia | $J$ | Kg.m$^2$ | $3.78 \times 10^4$ |
| Viscous coefficient | $B_v$ | - | $7.403 \times 10^5$ |
| Number pole pairs | $n_p$ | - | 3 |
| Constant switching frequency of the inverter | $F_n$ | kHz | 2.5 |
| Maximum current of the inverter | $I_{s\text{-}max}$ | A | 20 |
| Time step | $T_s$ | s | $5 \times 10^{-4}$ |

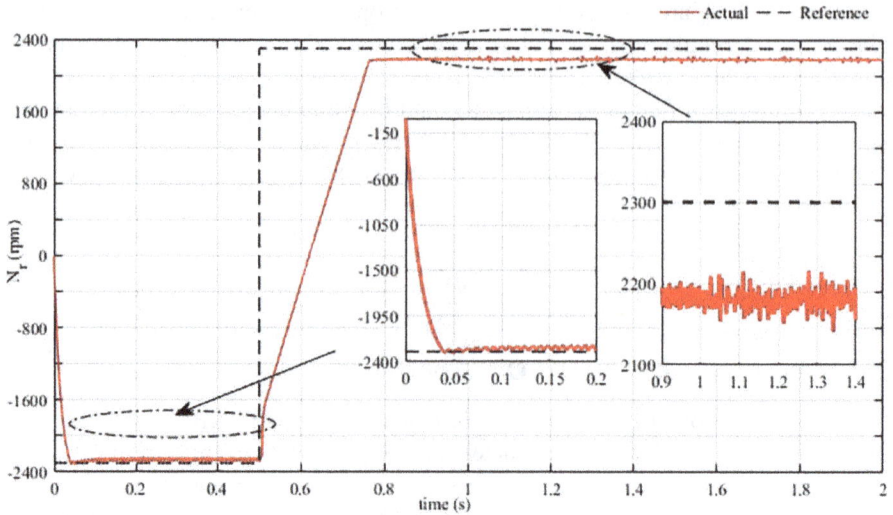

**FIGURE 2.13** Motor speed.

The motor is speeded up to -2300 rpm with an 8 Nm load torque, and then the motor angle is changed every 0.5 s. At 1 s, the VSFPWM strategy is enabled. As shown in Figure 2.13, the speed is about 2200 rpm with a steady-state error of about 5%. The error value of the steady-state period indicates that the external control loop fails to operate to pass the motor from the constant-torque region to the area of the flux weakening. This failure occurs for the following reasons:

- The PI-tuning parameters of the field control strategy are not suitable for operating in the flux-weakening region, especially the parameters of the flux controller and the inner controller of the stator currents for $d$-$q$ axes. Therefore, the driving system reaches the maximum speed in the constant-torque region at this operating condition of the required torque. In other words, the demagnetization current is not provided by the PI controller of the flux weakening in the negative trajectory of the $d$-axis.
- The flux-weakening PI controller lacks a filter to boost the negative average of the stator current for the $d$-axis if an optimum PI parameter is dropped.
- Therefore, the design failure of this PI controller is reflected in the mean value of the stator current of the $d$-axis and the voltage-vector magnitude of the VSI.

As shown in Figure 2.14, the varying switching frequency operation is enabled at 1 s and has a range of 2.5 to 10 kHz.

As shown in Figure 2.15, the predictive peak ripple of the electromagnetic torque is limited abruptly at a rate of about 75.4%. Figure 2.16 shows the stator current of the $d$-axis, which has an average value of about -0.2 A. This magnetizing current is not sufficient, thus keeping the motor speed in a constant-torque region.

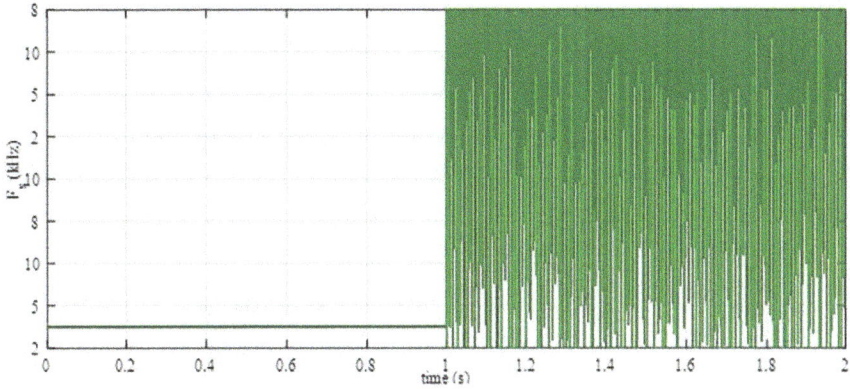

**FIGURE 2.14**   Varying switching frequency.

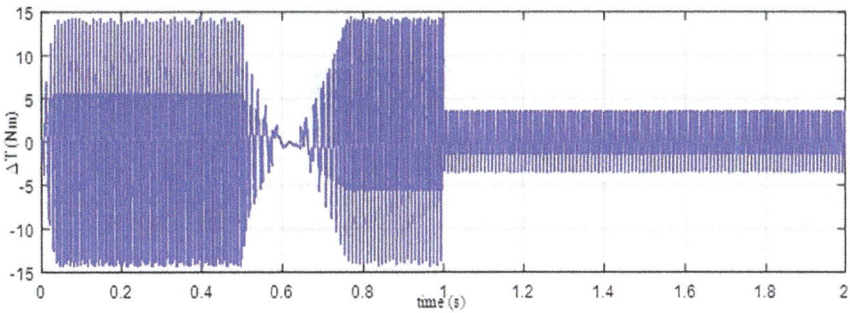

**FIGURE 2.15**   Torque ripple prediction.

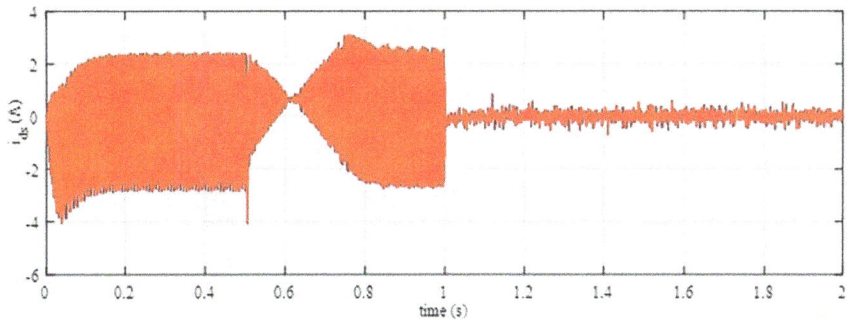

**FIGURE 2.16**   The *d*-axis stator current.

The peak of the *d*-axis current centered on the VSFPWM algorithm is limited from 2.6 to 0.6 A to limit the torque ripple. This figure confirms the efficiency of the VSFPWM control algorithm in the region of constant torque. Meanwhile, the peak of the *q*-axis current ripple is slightly reduced from 6 to 5.8 A, as shown in Figure 2.17, due to the high ripple in the reference of the *q*-axis current caused by the constant difference between the reference speed and the actual speed. However, the ripple is

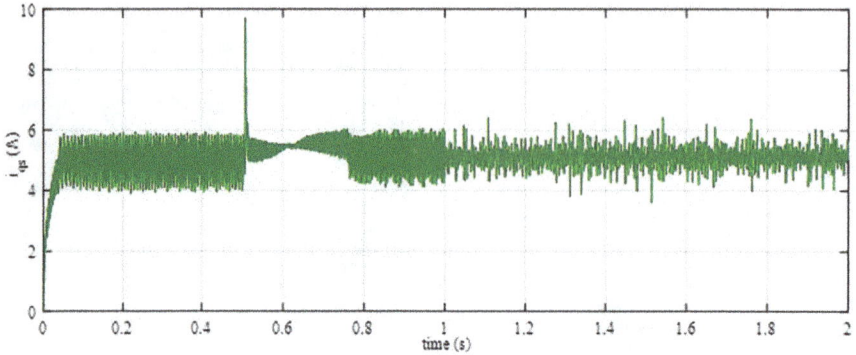

**FIGURE 2.17** The $q$-axis stator current.

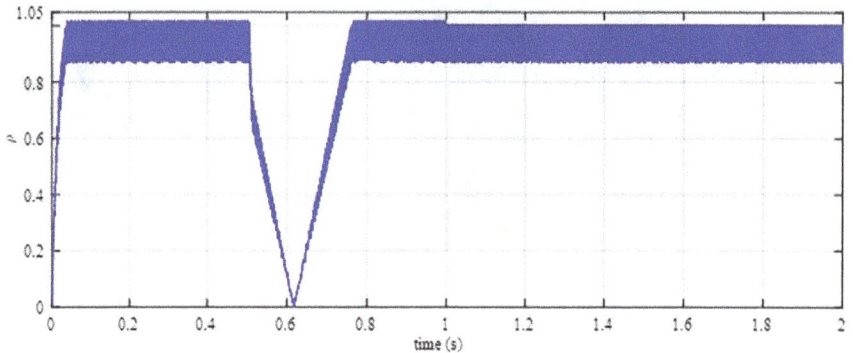

**FIGURE 2.18** Voltage vector magnitude.

at some moments greater than that before the VSFPWM algorithm is enabled due to the current controllers operating at the maximum inverter voltage under the adjusted PI gains. As a result, their errors increase at the site of the loss of their controllability.

Figure 2.18 shows the voltage vector magnitude, which has an average value of about 0.8728 for the reverse rotational direction, and that of the rotational direction of forwarding, about 0.9575. This figure confirms that this PI controller design fails to support the flux because it cannot use more DC-link voltage to drive the motor to run in the high-speed range.

A lower amplitude of the stator phase current is achieved after enabling the torque ripple control algorithm, as shown in Figure 2.19. The amplitude values of stator currents are reduced under the VSFPWM algorithm, from around 6.3 to 5.5 A. Several more order harmonics are seen in the stator phase currents before activating the ripple, achieving an efficiency as shown in Figure 2.20. Thanks to the VSFPWM control algorithm, it reduces the unwanted harmonics, as shown in Figure 2.21.

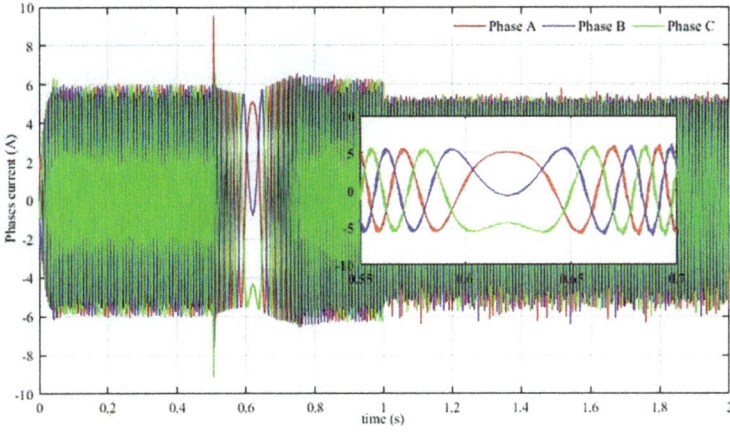

**FIGURE 2.19** Three-phases stator currents.

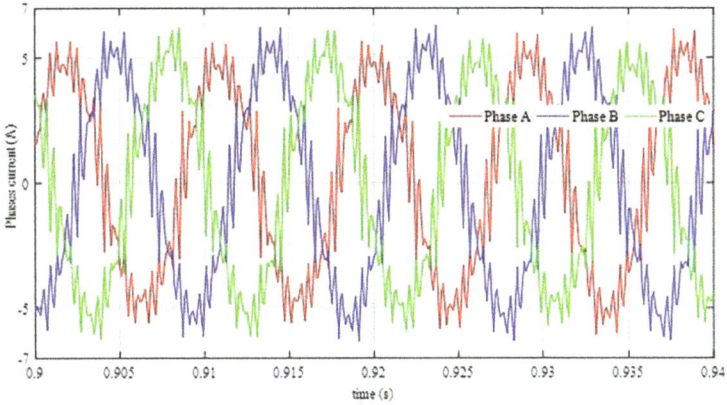

**FIGURE 2.20** Three phases of stator currents before enabling the torque ripple control algorithm.

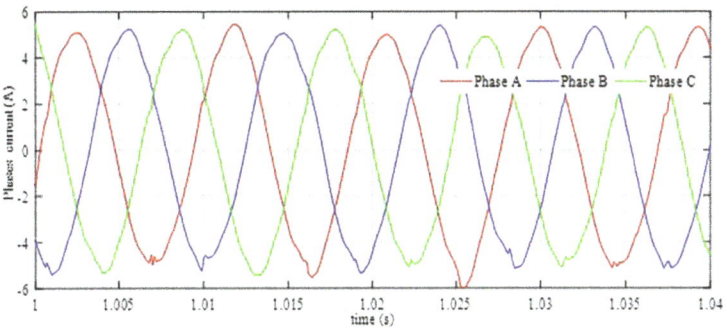

**FIGURE 2.21** Three phases of stator currents after enabling the torque ripple control algorithm.

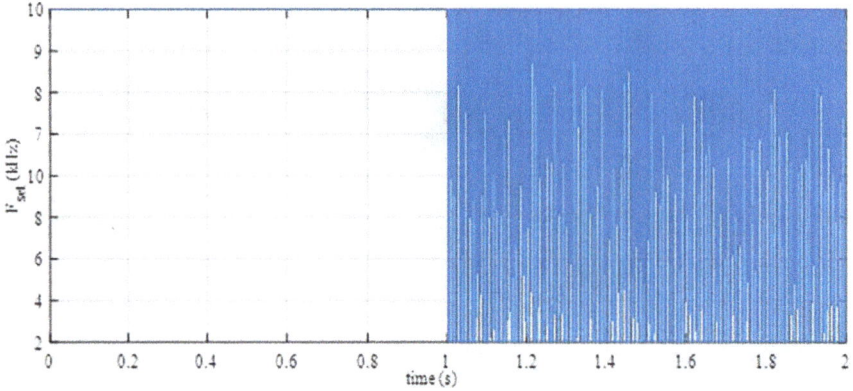

**FIGURE 2.22**    Varying switching frequency in the case of 10 kHz for a constant frequency.

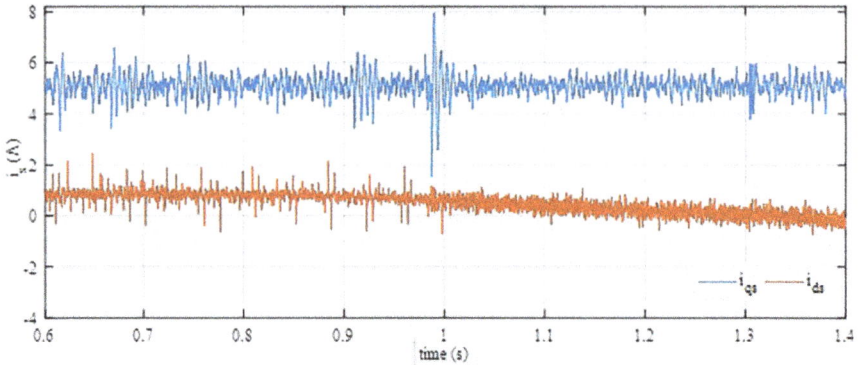

**FIGURE 2.23**    The $d$-$q$ axes stator currents in the case of 10 kHz for a constant frequency.

Figures 2.22 and 2.23 show the motor dynamic performance after adjusting the constant frequency PWM by 10 kHz to better prove the effectiveness of varying the frequency. The motor is derived to 2300 rpm as a reference speed under an 8 Nm load torque. The inverter works under 10 kHz as a constant switching frequency.

The ripple control algorithm is enabled at 1 s, as shown in Figure 2.22. As a result, the proposed control algorithm is slightly reducing the peak of the $q$-axis stator current, as shown in Figure 2.23. However, the reduction in the peak is not as much at a higher frequency of 10 kHz because the control strategy operates at parameters that are not optimal. Meanwhile, the algorithm reduces the ripple peak of the $d$-axis current, as shown in Figure 2.23.

### 2.4.3 EXPERIMENTAL RESULTS AND DISCUSSION

In this section, one experimental scene for the surface-mounted PMSM is applied in the laboratory using the control unit of the floating-point digital signal processor to

**TABLE 2.4**

**Simulation and Experimental Parameters of Interior PMSM**

| Parameters | Symbol | Unit | Quantity |
|---|---|---|---|
| Rated line voltage | $v_N$ | V | 270 |
| Rated power | $P_r$ | kW | 3 |
| Rated speed | $N_m$ | rpm | 2000 |
| Rated torque | $T_N$ | Nm | 14.3 |
| Rated current | $I_N$ | A | 6 |
| Stator resistance | $R_s$ | Ω | 1.14 |
| $d$-axis inductance | $L_{ds}$ | H | $1.91 \times 10^{-3}$ |
| $q$-axis inductance | $L_{qs}$ | H | $4.73 \times 10^{-3}$ |
| Permanent magnet flux linkage | $\lambda_m$ | Wb | 0.38 |
| Viscous coefficient | $B_v$ | - | $7.403 \times 10^{-5}$ |
| Moment of inertia | $J$ | Kg.m$^2$ | $3.78 \times 10^{-4}$ |
| Number of pole pairs | $n_p$ | - | 3 |
| Motor voltage constant | $C_e$ | - | 3.2035 |
| Maximum inverter current | $I_{s\text{-max}}$ | A | 20 |
| Constant switching frequency of the inverter | $F_n$ | kHz | 2.5 |
| Time step | $T_s$ | s | $5 \times 10^{-4}$ |

validate the effectiveness of the proposed method. Table 2.4 lists the parameters of the surface-mounted PMSM and inverter used in the test. Figure 2.24 also depicts the laboratory working model. As a load condition, a 3-kW interior PMSM is combined with the controlled motor in this figure. The control strategy's sampling frequency is 2 kHz.

Figure 2.24 illustrates the demonstrated experimental platform. This prototype is designed based on the surface-mounted PMSM and interior PMSM. From this figure, the digital controller called digital signal processor (DSP) TMS320F28335 drives the controlled motor and load motor through the voltage supply inverters. This study used a two-level, three-phase inverter that can handle a maximum current of 20 A; the maximum switching frequency of the inverters used is 10 kHz. The load motor is coupled with the controlled motor and is controlled by the control card based on closed-loop torque mode to work as a load. For clarification, if the controlling motor is a surface-mounted PMSM, then the loading motor is an interior PMSM and vice versa. For the implementation of the digital control card of DSP, all the proposed control algorithms in this study have been converted in discrete mode to apply as a programming C code. The counting mechanical position, the DC-link voltage, and the phase currents are necessary for the control algorithm. A 4096 pulse-per-cycle encoder is used to calculate the mechanical position, and a hall current sensor is used to measure current.

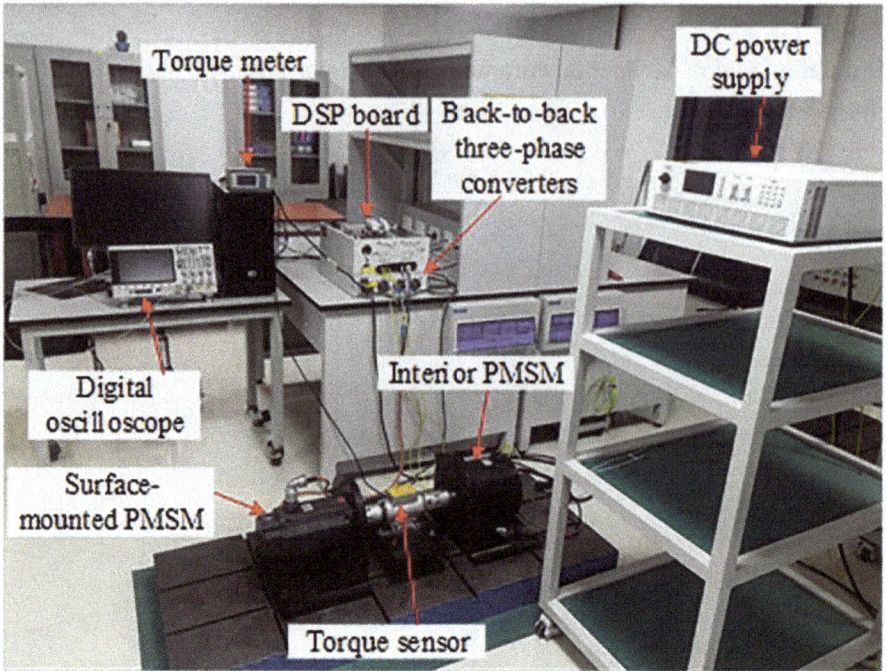

**FIGURE 2.24**  A photograph of the experimental test platform.

The torque sensor is used to show the measured electromagnetic torque in the torque meter display. Finally, a digital oscilloscope is used to display and save the necessary signals, using a Tektronix oscilloscope with a 1× passive voltage probe and a 2 MHz current bandwidth probe.

Figures 2.25 to 2.33 depict the case of abruptly modifying the load torque and motor speed. The velocity operates at a reference speed of 2100 rpm with an initial load of 5 Nm. The motor speed and load torque will be cut in half at 6.6 s and 10.9 s, respectively. At 4.4 s, the peak-ripple mitigation torque control scheme is enabled, following the completion of the transient period.

Figure 2.25 depicts the acceleration with a starting-over shoot of approximately 7.2%. When stepping down the velocity, the peak speed is 7.9%, and the overshoot is about 7% upon sudden reduction of torque. As shown in Figure 2.26, the ripple peak of the $q$-axis current and the torque in the case of 2100 rpm and 5 Nm are slightly higher than 1050 rpm and 2.5 Nm. In the case of the speed reduction, the mean value of the $q$-axis current is reduced by about 3.2 A, whereas if the torque step changes, the mean value of the $q$-axis current is reduced by about 1.4 A. The current for the $d$-axis is presented in Figure 2.27, having an average value of about -0.17 A. In addition, the ripple peak of the $d$-axis current when the torque ripple control algorithm is reduced from 2.4 to 0.9 A. As shown in Figure 2.28, this shift reduces the size of the voltage vector from 0.92 to 0.48.

The varying switching frequency in kHz is displayed in Figure 2.29. In this figure, the switching frequency is modified between 2.5 and 10 kHz. As shown in

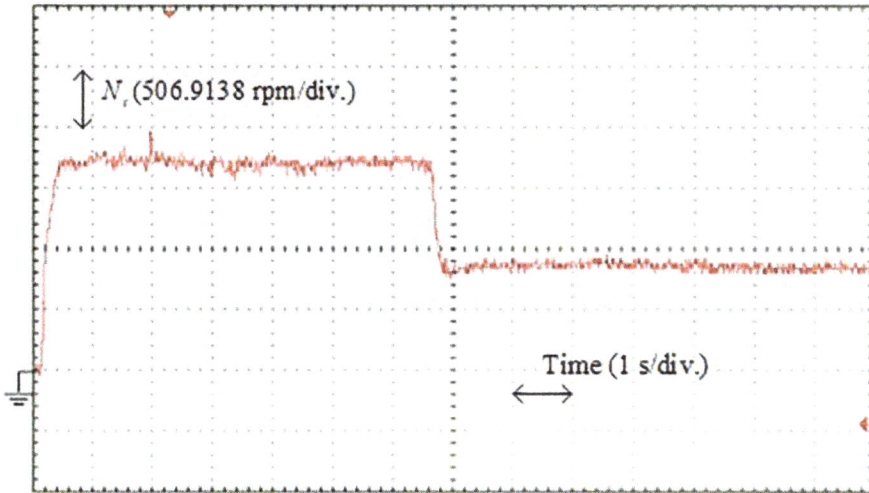

**FIGURE 2.25**  Experimental measured speed.

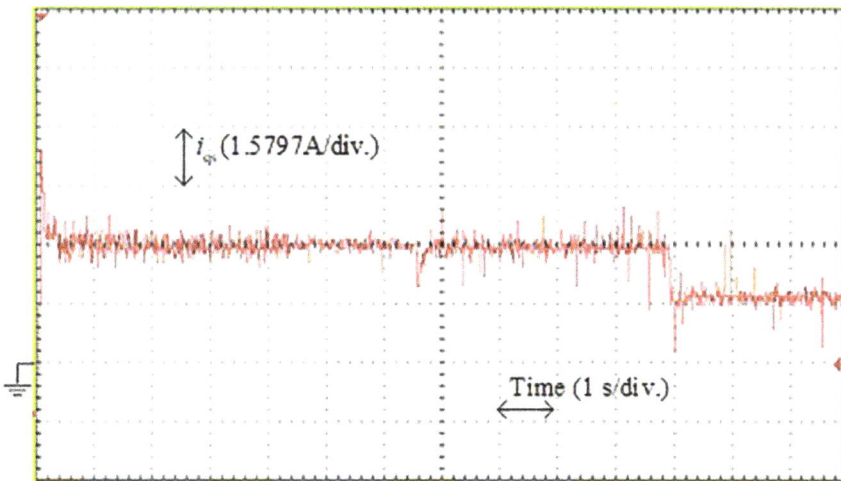

**FIGURE 2.26**  The $q$-axis measured stator current.

Figure 2.30, this change affects the torque predictive ripple. In this figure, the torque ripple reduction rate is 75% at 4.4 s and 60.3% at 6.6 s. Figure 2.31 shows the A and B phases of the stator currents. It can be seen from Figure 2.32 and Figure 2.33 that stator currents have fewer harmonics under the torque ripple reduction algorithm than constant switching frequency control. Meanwhile, the copper losses are minimized while the rotational speed is lowered, as indicated by the amplitude of the constant current phases at transition times of 4.4, 6.6, and 10.9 seconds. The current amplitudes are diminished from approximately 4.3 to 3.4 A at 4 s and from 3.4 to 1.91 at 10.9 s.

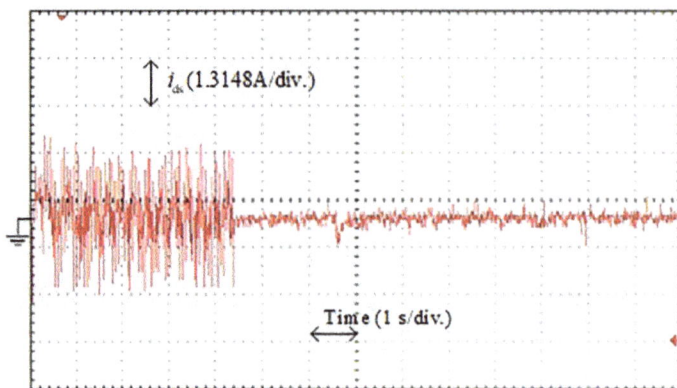

**FIGURE 2.27**   The $d$-axis measured stator current.

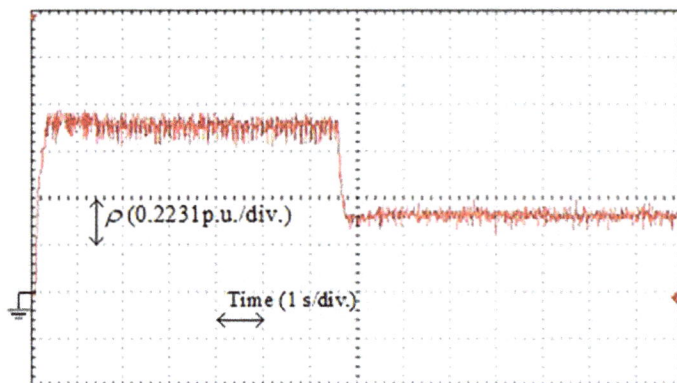

**FIGURE 2.28**   Voltage vector magnitude.

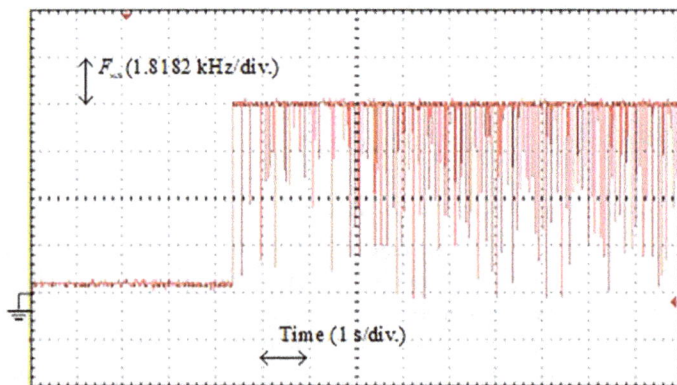

**FIGURE 2.29**   Varying switching frequency.

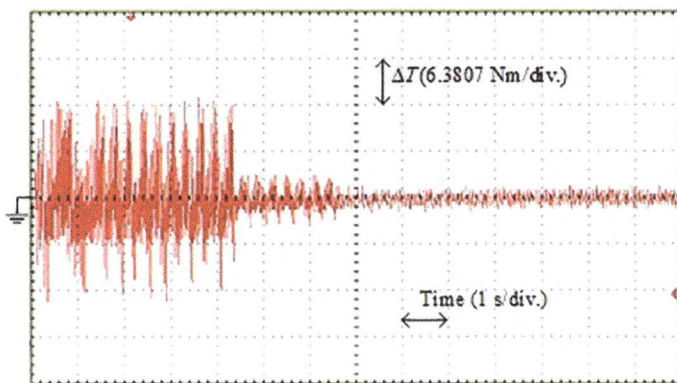

**FIGURE 2.30**   Torque ripple prediction.

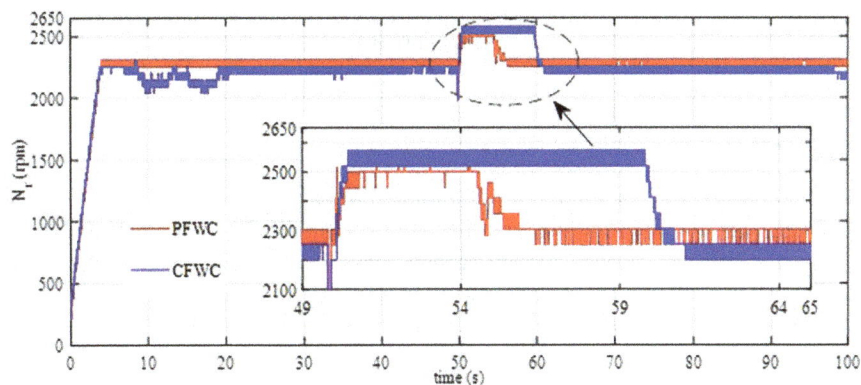

**FIGURE 2.31**   Motor speed under proposed method and the conventional method.

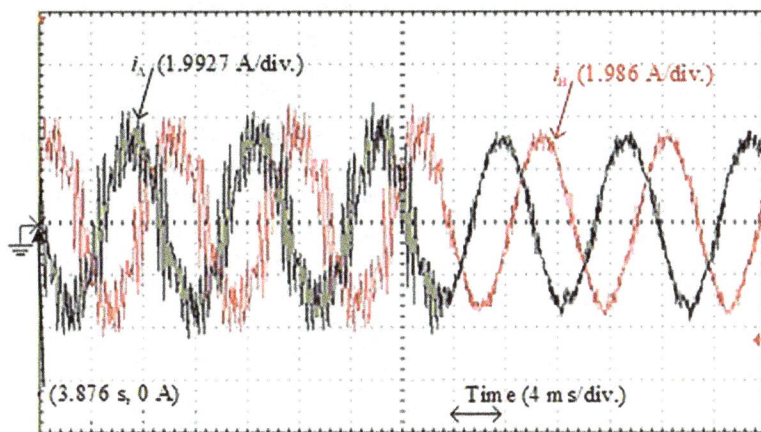

**FIGURE 2.32**   Stator currents of $i_A$ and $i_B$ phases.

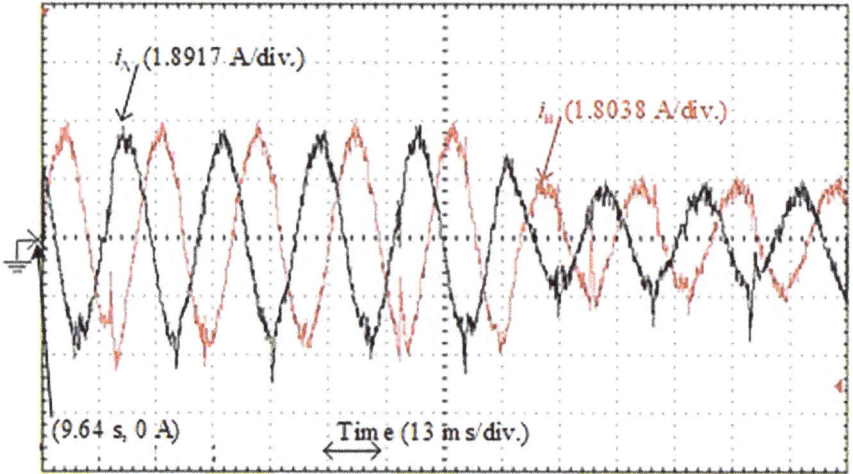

**FIGURE 2.33**   The $i_A$ and $i_B$ phases of stator currents at the moment of sudden motor speed reduction.

## 2.5  SUMMARY

This chapter provides a detailed description of PMSM classification, the theoretical explanation of salient and non-salient PMSM mathematical models, the MTPA theory, and the concept of flux-weakening operation. The type of PMSM depends on the placement of the permanent magnets in the rotor shaft, which results in varied dynamic performance among them. Salient PMSMs exhibit additional reluctance torque compared to other types.

The space current vector is inferred by the torque derivative, considering the machine parameters to validate the MTPA algorithm. The MTPA control is illustrated in the flux-weakening region, in terms of the voltage ellipse and current circle limiting. Furthermore, a proposal is made for reducing current and torque ripples based on frequency modulation. The main recommendations derived from the implementation of the VSFPWM algorithm can be summarized as follows:

(1)  The dynamic performance of the control strategy underlines the importance of setting the optimum parameters for the PI regulators of the control strategy to avoid overshoot and steady-state error. Therefore, careful tuning of the control strategy is necessary to achieve good dynamic performance.

(2)  The demagnetization current of the flux-weakening process should be increased enough, in the negative trajectory of the d-axis, to operate at the reference speed with the required load torque. Based on this definition, the negative d-axis current of the proposed loop of voltage vector magnitude is responsible for the speed drop in the constant-torque region due to the weakness of the d-axis current.

(3)  It is suggested that the PI gains of the inner controller should be adjusted to reduce the error between the reference and actual q-axis current. This error

is a necessary parameter that increases peak torque ripple, and the VSF-PWM technique may not be effective in the area of flux weakening due to the high steady-state error of the motor speed, which further increases peak ripple. However, VSFPWM shows an excellent reduction in ripple in the constant-torque region.

The methods for adjusting the PI controller parameters to enhance dynamic performance of the motor are discussed in these chapters to address all the problems identified in the study presented in this chapter.

## BIBLIOGRAPHY

[1] P. Yi, Z. Sun, and X. Wang. Research on PMSM harmonic coupling models based on magnetic co-energy. IET Electric Power Applications, April 2019, 13(4):571–579.

[2] B. N. Cassimere and S. D. Sudhoff. Population-based design of surface-mounted permanent-magnet synchronous machines. IEEE Transaction on Energy Conversion, May 2009, 24(2):338–346.

[3] R. Islam, I. Husain, A. Fardoun, and K. McLaughlin. editors. Permanent magnet synchronous motor magnet designs with skewing for torque ripple and cogging torque reduction. IEEE Industry Applications Annual Meeting, 23–27 September 2007.

[4] G. Zhao and W. Hua. Comparative study between a novel multi-tooth and a V-shaped flux-switching permanent magnet machines. IEEE Transaction on Magnetics, March 2019, 55(7):1–8.

[5] X. Sun, Z. Shi, G. Lei, Y. Guo, and J. Zhu. Analysis and design optimization of a permanent magnet synchronous motor for a campus patrol electric vehicle. IEEE Transaction on Vehicular Technology, September 2019, 68(11):10535–10544.

[6] G. R. Slemon. Electric machines and drives, Addison-Wesley, 1992.

[7] K. Kim, S. Lim, D. Koo, and J. Lee. The shape design of permanent magnet for permanent magnet synchronous motor considering partial demagnetization. IEEE Transaction on Magnetics, September 2006, 42(10):3485–3487.

[8] L. Liu, W. Liu, and D. A. Cartes. Permanent magnet synchronous motor parameter identification using particle swarm optimization. International Journal of Computational Intelligence Research, September 2008, 4:211.

[9] Y. L. Karnavas and C. D. Korkas. editors. Optimization methods evaluation for the design of radial flux surface PMSM. International Conference on Electrical Machines (ICEM), 2–5 September 2014.

[10] Lakshmikanth, S. Noise and Vibration Reduction in PMSM-A Review. International Journal of Electrical and Computer Engineering, June 2012, 2(3):405.

[11] K. Yamano, S. Morimoto, M. Sanada, and Y. Inoue. Design of surface permanent magnet synchronous motor using design assist system for PMSM. IEEJ Journal of Industry Applications, June 2017, 6(6):409–415.

[12] H. Lennart. Control of variable-speed drives. Applied Signal Processing and Control, Department of Electronics, Malardalen University, 2002.

[13] C. Calleja, A. López-de-Heredia, H. Gaztañaga, L. Aldasoro, and T. Nieva. Validation of a modified direct-self-control strategy for PMSM in railway-traction applications. IEEE Transaction on Industrial Electronics, May 2016, 63(8):5143–5155.

[14] M. Preindl and S. Bolognani. Optimal state reference computation with constrained MTPA criterion for PM motor drives. IEEE Transaction on Power Electronics, September 2015, 30(8):4524–4535.

[15] M. Khayamy and H. Chaoui. Current sensorless MTPA operation of interior PMSM drives for vehicular applications. IEEE Transaction on Vehicular Technology, April 2018, 67(8):6872–6881.

[16] Y. Lee and J. Ha. Control method of monoinverter dual parallel drive system with interior permanent magnet synchronous machines. IEEE Transaction on Power Electronics, December 2016, 31(10):7077–7086.

[17] G. Feng, C. Lai, and N. C. Kar. An analytical solution to optimal stator current design for PMSM torque ripple minimization with minimal machine losses. IEEE Transaction on Industrial Electronics, April 2017, 64(10):7655–7665.

[18] R. Ni, G. Wang, X. Gui, and D. Xu. Investigation of d- and q-axis inductances influenced by slot-pole combinations based on axial flux permanent-magnet machines. IEEE Transaction on Industrial Electronics, November 2014, 61(9):4539–4551.

[19] K. Baoquan, L. Chunyan, and C. Shukang. Flux-weakening-characteristic analysis of a new permanent-magnet synchronous motor used for electric vehicles. IEEE Transaction on Plasma Science, December 2011, 39(1):511–515.

[20] A. Sarikhani and O. A. Mohammed. Demagnetization control for reliable flux weakening control in PM synchronous machine. IEEE Transaction on Energy Conversion, October 2012, 27(4):1046–1055.

[21] Y. Inoue, S. Morimoto, and M. Sanada. Comparative study of PMSM drive systems based on current control and direct torque control in flux-weakening control region. IEEE Transaction on Industry Applications, November 2012, 48(6):2382–2389.

[22] G. Schoonhoven and M. N. Uddin. MTPA- and FW-based robust nonlinear speed control of IPMSM drive using Lyapunov stability criterion. IEEE Transaction on Industry Applications, 2016, 52(5):4365–4374.

[23] S. Bolognani, S. Calligaro, and R. Petrella. Adaptive flux-weakening controller for interior permanent magnet synchronous motor drives. IEEE Journal of Emerging and Selected Topics in Power Electronics, January 2014, 2(2):236–248.

[24] H. Ge, Y. Miao, B. Bilgin, B. Nahid-Mobarakeh, and A. Emadi. Speed range extended maximum torque per ampere control for PM drives considering inverter and motor nonlinearities. IEEE Transaction on Power Electronics, November 2017, 32(9):7151–7159.

[25] T. Inoue, Y. Inoue, S. Morimoto, and M. Sanada. Maximum torque per ampere control of a direct torque-controlled PMSM in a stator flux linkage synchronous frame. IEEE Transaction on Industry Applications, February 2016, 52(3):2360–2367.

[26] A. Shinohara, Y. Inoue, S. Morimoto, and M. Sanada. Maximum torque per ampere control in stator flux linkage synchronous frame for DTC-based PMSM drives without using q-axis inductance. IEEE Transaction on Industry Applications, March 2017, 53(4):3663–3671.

[27] A. Ahmed, Y. Sozer, and M. Hamdan. Maximum torque per ampere control for buried magnet PMSM based on DC-link power measurement. IEEE Transaction on Power Electronics, March 2017, 32(2):1299–1311.

[28] C. Lai, G. Feng, J. Tjong, and N. C. Kar. Direct calculation of maximum-torque-per-ampere angle for interior PMSM control using measured speed harmonic. IEEE Transaction on Power Electronics, January 2018, 33(11):9744–9752.

[29] C. Miguel-Espinar, D. Heredero-Peris, G. Gross, M. Llonch-Masachs, and D. M. I. Miracle. Maximum torque per voltage flux-weakening strategy with speed limiter for PMSM drives. IEEE Transaction on Industrial Electronics, September 2020, 68(10):9254–9264.

[30] M. Sanada, S. Morimoto, and Y. Takeda. Interior permanent magnet linear synchronous motor for high-performance drives. IEEE Transaction on Industry Applications, August 1997, 33(4):966–972.

[31] F. Chai, K. Zhao, Z. Li, and L. Gan. Flux weakening performance of permanent magnet synchronous motor with a conical rotor. IEEE Transaction on Magnetics, May 2017, 53(11):1–6.

[32]    K. Zhou, M. Ai, D. Sun, N. Jin, and X. Wu. Field weakening operation control strategies of PMSM based on feedback linearization. Energies, January 2019, 12(23):4526.

[33]    K. Jang-Mok and S. Seung-Ki. Speed control of interior permanent magnet synchronous motor drive for the flux weakening operation. IEEE Transaction on Industry Applications, February 1997, 33(1):43–48.

[34]    J. Liu, H. Li, and Y. Deng. Torque ripple minimization of PMSM based on robust ILC via adaptive sliding mode control. IEEE Transaction on Power Electronics, June 2018, 33(4):3655–3671.

[35]    K. Wang, Z. Q. Zhu, and G. Ombach. Torque enhancement of surface-mounted permanent magnet machine using third-order harmonic. IEEE Transaction on Magnetics, March 2014, 50(3):104–113.

[36]    C. Xia, B. Ji, and Y. Yan. Smooth speed control for low-speed high-torque permanent-magnet synchronous motor using proportional—integral—resonant controller. IEEE Transaction on Industrial Electronics, April 2015, 62(4):2123–2134.

[37]    D. Flieller, N. K. Nguyen, P. Wira, G. Sturtzer, D. O. Abdeslam, and J. Mercklé. A self-learning solution for torque ripple reduction for nonsinusoidal permanent-magnet motor drives based on artificial neural networks. IEEE Transaction on Industrial Electronics, February 2014, 61(2):655–666.

[38]    A. Mora, O. Á, J. Juliet, and R. Cárdenas. Model predictive torque control for torque ripple compensation in variable-speed PMSMs. IEEE Transaction on Industrial Electronics, July 2016, 63(7):4584–4592.

[39]    D. Dujic, M. Jones, E. Levi, J. Prieto, and F. Barrero. Switching ripple characteristics of space vector PWM schemes for five-phase two-level voltage source inverters—part 1: flux harmonic distortion factors. IEEE Transaction on Industrial Electronics, August 2011, 58(7):2789–2798.

[40]    G. Carrasco and C. A. Silva. Space vector PWM method for five-phase two-level VSI with minimum harmonic injection in the overmodulation region. IEEE Transaction on Industrial Electronics, May 2013, 60(5):2042–2053.

[41]    W. Fei. Sine-triangle versus space-vector modulation for three-level PWM voltage-source inverters. IEEE Transaction on Industry Applications, August 2002, 38(2):500–506.

[42]    O. López, D. Dujic, M. Jones, F. D. Freijedo, J. Doval-Gandoy, and E. Levi. Multidimensional two-level multiphase space vector PWM algorithm and its comparison with multifrequency space vector PWM method. IEEE Transaction on Industrial Electronics, April 2011, 58(2):465–475.

[43]    M. Chen and D. Sun. A unified space vector pulse width modulation for dual two-level inverter system. IEEE Transaction on Power Electronics, June 2017, 32(2):889–893.

[44]    S. Srinivas and K. R. Sekhar. Theoretical and experimental analysis for current in a dual-inverter-fed open-end winding induction motor drive with reduced switching PWM. IEEE Transaction on Industrial Electronics, July 2013, 60(10):4318–4328.

[45]    V. T. Somasekhar, S. Srinivas, and K. K. Kumar. Effect of zero-vector placement in a dual-inverter fed open-end winding induction-motor drive with a decoupled space-vector PWM strategy. IEEE Transaction on Industrial Electronics, May 2008, 55(6):2497–2505.

[46]    T. Chen. Dual-modulator compensation technique for parallel inverters using space-vector modulation. IEEE Transaction on Industrial Electronics, May 2009, 56(8):3004–3012.

[47]    R. Baranwal, K. Basu, and N. Mohan. Carrier-based implementation of SVPWM for dual two-level VSI and dual matrix converter with zero common-mode voltage. IEEE Transaction on Power Electronics, April 2015, 30(3):1471–1487.

[48]    D. Jiang and F. Wang. A general current ripple prediction method for the multiphase voltage source converter. IEEE Transaction on Power Electronics, October 2014, 29(6):2643–2648.

[49]    D. Jiang and F. Wang. Current-ripple prediction for three-phase PWM converters. IEEE Transaction on Industry Applications, June 2014, 50(1):531–538.

[50]  C. Zhu, Z. Zeng, and R. Zhao. Comprehensive analysis and reduction of torque ripples in three-phase four-switch inverter-fed PMSM drives using space vector pulse-width modulation. IEEE Transaction on Power Electronics, July 2017, 32(7):5411–5424.

[51]  D. Jiang and F. Wang. Variable switching frequency PWM for three-phase converters based on current ripple prediction. IEEE Transaction on power Electronics, January 2013, 28(11):4951–4961.

[52]  D. Jiang and F. Wang. Current-ripple prediction for three-phase PWM converters. IEEE Transaction on Industry Applications, June 2014, 50(1):531–538.

[53]  D. Jiang and F. Wang, A general current ripple prediction method for the multiphase voltage source converter. IEEE Transaction on Power Electronics, October 2014, 29(6):2643–2648.

[54]  C. Zhu, Z. Zeng, and R. Zhao. Comprehensive analysis and reduction of torque ripples in three-phase four-switch inverter-fed PMSM drives using space vector pulse-width modulation. IEEE Transaction on Power Electronics, July 2017, 32(7):5411–5424.

[55]  C.-M. Ong. Dynamic simulation of electric machinery: using MATLAB/SIMULINK, Prentice Hall, November 1998.

# 3 Adaptive Flux-Weakening Control Strategy for Non-Salient Permanent Magnet Synchronous Motor Drives

## 3.1 INTRODUCTION

The primary objective of traction system drives is to maintain a stable speed while operating within a high range of speed. However, the back electromotive force effect caused by flux weakening is considered a setback in high-speed regions, limiting the compensated voltages [1]. Chapter 2 addressed the challenge of adjusting PI controller parameters to achieve speed stability of a permanent magnet synchronous motor (PMSM) in the flux-weakening region and the strength of the required magnetizing current. Thus, a cascade field-oriented control system based on proportional-integral regulators must be developed, and this chapter considers significant improvements to the traditional cascade PI controller.

To explain the detailed strategy, the mathematical packs for the driven calculations are explored to calculate error values for the PI regulators and prepare them as a cost function in terms of the PI gains. The presented cost function is based on both the PMSM and FOC strategy models, making it independent of their transfer functions. The optimization procedure of an improved adaptive velocity particle swarm is used offline to select the best PI parameters that correspond to the minimum errors.

There are two editions of the traditional FOC scheme. First, the improved anti-windup technique is used in the PI regulator to reduce the overrun caused by the integral controller. Secondly, feedforward of the compensated duty cycles is employed to calculate the required magnetizing current compared to the flux-weakening regulators. Thus, the enhanced FOC becomes independent of PMSM machine parameters. Thirdly, the low-pass filter is employed to reject high-frequency oscillation of the d-axis reference current in the flux-weakening regulator required to limit the flux level. Finally, the proposed improved FOC strategy is validated through theoretical performance analysis, comprehensive simulations, and experimental results. Additionally, the dynamic performance of the control algorithm is compared to that of the improved FOC algorithm presented in [2] to demonstrate the strength of the new scheme. However, the strategy presented in [2] depends on line modulation to calculate the d-axis reference current.

It can also support a faster transition between the constant-torque region and the weak flow region and vice versa, independently of the motor parameters. Table 3.1 illustrates the main contrasts between the conventional flux-weakening control (CFWC) strategy

DOI: 10.1201/9781003320128-3

**TABLE 3.1**
**Contrasts between PFWC and CFWC**

|  | PFWC | CFWC |
|---|---|---|
| The low-pass filter of the flux regulator | true | false |
| AWPI technique of the FOC regulators | true | false |
| Optimal PI gains of the FOC regulators | true | false |

and the new flux-weakening control (PFWC). The main advantages of the improved FOC scheme proposed in this chapter relative to the traditional FOC are as follows:

The results demonstrate that the novel strategy can enhance the speed control bandwidth and the speed stability while operating in both the constant-torque and flux-weakening regions. Furthermore, the novel strategy can diminish the settling time, steady-state error, and power loss. It can also facilitate a faster transition between the constant-torque and the weak-flow regions and vice versa, independently of the motor parameters. Table 3.1 depicts the main differences between the conventional flux-weakening control (CFWC) strategy and the proposed flux-weakening control (PFWC) scheme. The principal benefits of the proposed FOC scheme outlined in this chapter compared to the conventional FOC are as follows:

(1) Proposing a cost function based on the error values with respect to the PI parameters of all regulators in one FOC scheme to choose the optimal parameters in a single decision that corresponds to the minimum error. Hence, it can be used instead of the trial-and-error method, which involves modifying the parameters until the motor speed behavior attains an acceptable override in the transient period and a fluctuation in the steady-state period.
(2) The parameters of the proposed FOC scheme depend on the PMSM parameters. However, the feedforward can independently control the PMSM, regardless of its parameters, unlike the CFWC.
(3) The proposed FOC scheme is applicable to the salient PMSM.
(4) The proposed FOC scheme depends on the flux-weakening control loop, which is supported by the low-pass filter to reject the high-frequency signal. Thus, the filter prevents the sudden decrease of the average current of d-axis. The subsequent sections describe the proposed adaptive flux-weakening control scheme, the existing flux-weakening control strategy of [2], and a discussion of the simulation and experimental results. .

## 3.2 IMPROVED ADAPTIVE FLUX-WEAKENING CONTROL STRATEGY

The objectives of this section are as follows:

(1) The schematic diagram of the new FOC strategy is presented in this section.
(2) The flux-weakening control loop is clarified based on the compensated duty cycles.

(3) Where the low-pass filter is placed in the new FOC strategy and the reason to use it are discussed.

(4) The output saturation of the PI regulators is shown separately.

(5) The technique of anti-windup in the PI regulator is illustrated, along with the reason to use it.

(6) The mathematical model of the space vector pulse-width modulation technique is provided.

(7) The matrix of the error values for the inherent PI regulators of the proposed FOC strategy is shown.

(8) The new fitting function for modelling based on the matrix of error values is explained.

(9) The implementation of the offline optimization procedure is explained, based on the adaptive velocity of the particle swarm algorithm with the aid of a flow chart.

(10) The improved flux-weakening control strategy of [2] based on the traditional FOC is explained to clarify its advantages and the feedforward based on the maximum line modulation.

### 3.2.1 ADAPTIVE CONTROL STRATEGY DESIGN

Figure 3.1 shows a block diagram of the improved FOC strategy. It can be noted from this figure that the non-salient type of PMSM is the controlled motor. Meanwhile, the traditional space vector modulation technique adapts the compensated voltages to control the PMSM through a two-level voltage source inverter and six pulses. However, the new control strategy is implemented in a digital control card, which depends on the speed sensor to count the mechanical position to calculate the rotor speed. It also depends on the current sensors to measure the stator currents. Lastly, the DC-link voltage of the voltage source inverter is measured as well, based on the voltage sensor. Thus, the input vector of the control strategy contains the measured speed of the PMSM rotor and the stator currents.

It is necessary to study the average model of the two-level voltage source inverter because it is a section in the drive system, which can be defined as [3–7]

$$I_{dc} = C_F \frac{dV_{dc}}{dt} + \frac{1}{3} d_{dqs}^T i_{dqs} \tag{3.1}$$

where $C_F$ is a capacitor of the DC-link circuit. The electrical characteristic model can be defined as

$$\frac{di_{dqs}}{dt} = \frac{1}{3L_s} d_{dqs} V_{dc} - \frac{\omega_e}{3L_s} \lambda_{dqs} + \begin{bmatrix} \dfrac{-R_s}{L_s} & \omega_e \\ -\omega_e & \dfrac{-R_s}{L_s} \end{bmatrix} i_{dqs} \tag{3.2}$$

where $i_{dqs}$ is the stator current vector of the $d$-$q$ axes, $d_{dqs}$ is the duty cycles of the $d$-$q$ axes, and $\lambda_{dqs}$ is the stator fluxes of the $d$-$q$ axes, which can be written as

$$i_{dqs} = \begin{bmatrix} i_{ds} \\ i_{qs} \end{bmatrix} \tag{3.3}$$

$$d_{dqs} = \begin{bmatrix} d_{ds} \\ d_{qs} \end{bmatrix} \tag{3.4}$$

$$\lambda_{dqs} = \begin{bmatrix} \lambda_{ds} \\ \lambda_{qs} \end{bmatrix} \tag{3.5}$$

It can be seen from (3.7) that the electromotive force is increased as the velocity of the PMSM shaft increases. As a result, the compensated stator voltage cannot control the electromagnetic torque and stator currents. This is the issue behind the flux-weakening region; the electromagnetic torque becomes weak and highly oscillated. Therefore, the $d$-axis stator current should be injected in the negative trajectory to set the flux level in order to increase the bandwidth of the speed control.

To introduce a solution to this flux-weakening issue, block diagrams are shown in Figure 3.1, Figure 3.2, and Figure 3.3, which introduce the improved field-oriented control strategy, considering the operation in the constant-torque region and flux-weakening region as well. First, Figure 3.1 shows the general scheme of the improved FOC. It can be noted from this figure that the inner controller has a saturation block to indirectly limit the compensated voltages. The reference duty cycles of the $d$-$q$ axes are calculated by dividing the compensated voltages by the DC-link voltage value. Thus, the reference duty cycles are employed as a feedforward in the flux-weakening control loop. Therefore, the saturation block of the inner controller has upper and lower values of one.

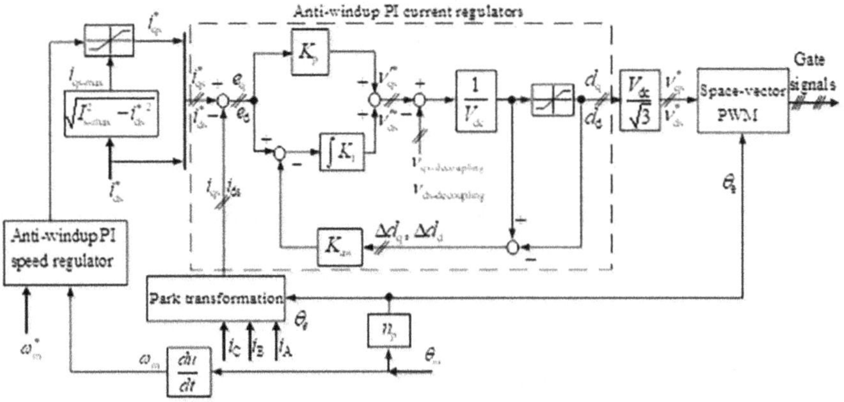

FIGURE 3.1   Speed and inner regulators of the advanced control strategy.

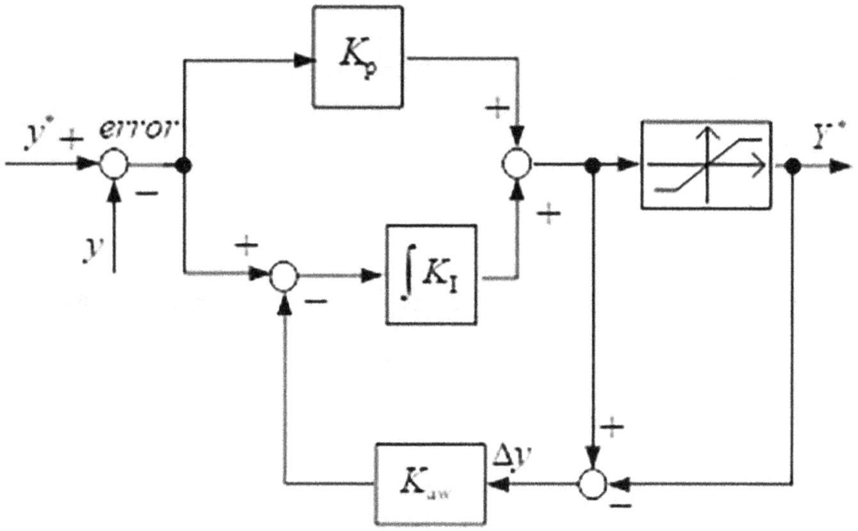

**FIGURE 3.2** Proposed anti-windup algorithm.

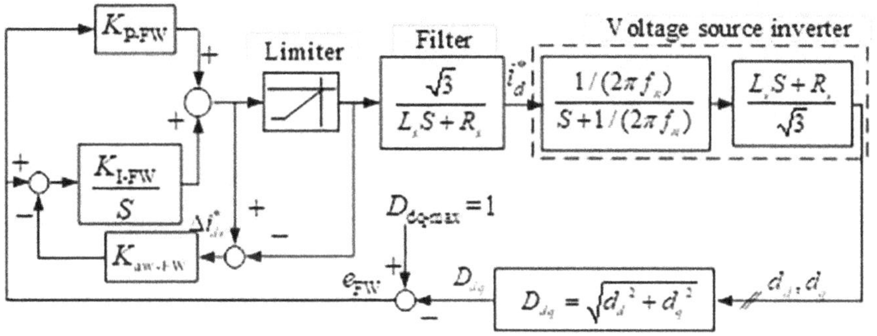

**FIGURE 3.3** Proposed adaptive control loop for the flux-weakening region.

Meanwhile, it can be seen from (3.6), and (3.7) that the decoupling of the non-linear terms is inappropriate to control the stator current based on the field-oriented control strategy. Thus, the decoupling nonlinear terms have to be removed from the compensated stator voltages of the $d$-$q$ axes as shown in Figure 3.1, and the variables of $v_{ds\_decoupling}$ and $v_{qs\_decoupling}$ can be defined by [8]

$$v_{ds\text{-decoupling}} = -L_s \omega_e i_{qs} \tag{3.6}$$

$$v_{qs\text{-decoupling}} = L_s \omega_e i_{ds} + \omega_e \lambda_m \tag{3.7}$$

Finally, the outer controller of the velocity regulator has a saturation block to consider the maximum current of the proposed drive system ($I_{s\text{-max}}$) that the PMSM and inverter can withstand. As a result, the upper and lower values of the last saturation can be determined based on the values of the reference stator current of the $d$-axis in Figure 3.3. Thus, the maximum magnitude of the $q$-axis stator current vector ($i_{qs\text{-max}}$) can be determined based on the maximum current of the drive system and the reference $d$-axis stator current ($i^*_{ds}$), as

$$i_{qs\text{-max}} = \sqrt{I^2_{s\text{-max}} - i^{*\,2}_{ds}} \tag{3.8}$$

It can be seen from the previous equation that the $q$-axis stator current decreases with the increase of the stator current of the $d$-axis. Therefore, the electromagnetic torque will decrease. However, the demagnetizing stator current of the $d$-axis should be as dispirited as possible to increase the induced torque.

Secondly, Figure 3.2 shows the anti-windup technique to overcome the overshoot caused by the integral controller. For the behavior of the integral controller, it accumulates the input error and then strengthens it at each sample time. When the measured signal gets close to the reference signal, the value of the integral controller will be added to the output signal, causing the value to be exceeded. To explain the anti-windup technique in Figure 3.2, the feedback shown is subtracted from the value of the input error ($e$) to be inserted in the integral controller. However, the feedback shown ($\Delta y$) is the summation of the input and the negative output ($-Y^*$) of the saturation block, increased by a factor of $K_{aw}$. Meanwhile, a value of the anti-windup gain ($K_{aw}$) can be defined as the ratio of the integral gain to the proportional gain. It can be noted that* indicates the reference value of the signal, and $\Delta$ indicates that the quantity is the result of the summation of the input and the negative output of the saturation block. Finally, the anti-windup technique can reduce the overrun impact of the saturation issue.

Based on Figures 3.1 and 3.2, the error values of the inner controller of the PI current regulators of the $d$-$q$ axes can be defined in terms of the proportional and integral parameters ($K_{I\text{-}d}$, $K_{I\text{-}q}$, $K_{I\text{-}d}$, and $K_{I\text{-}q}$) of the PI current regulators of the $d$-$q$ axes, respectively, as

$$e_q = \frac{v^*_{qs}}{K_{P\text{-}q}} - \frac{K_{I\text{-}q}}{K_{P\text{-}q}} E_q + \int K^2_{aw\text{-}q} \Delta d^*_{qs} - \frac{L_s n_p}{K_{P\text{-}q}} i_{ds} \omega_m - \frac{\lambda_m n_p}{K_{P\text{-}q}} \omega_m - i^*_{qs} + i_{qs} \tag{3.9}$$

$$e_d = \frac{v^*_{ds}}{K_{P\text{-}d}} - \frac{K_{I\text{-}d}}{K_{P\text{-}d}} E_d + \int K^2_{aw\text{-}d} \Delta d^*_{ds} + \frac{L_s n_p}{K_{P\text{-}d}} i_{qs} \omega_m - i^*_{ds} + i_{ds} \tag{3.10}$$

where $K_{aw\text{-}q}$ and $K_{aw\text{-}d}$ are the anti-windup variables of the current regulators of the $q$-$d$ axes, $E_q$ and $E_d$ are the values of the error integrations of the current regulators of the $q$-$d$ axes, and $e_q$ and $e_d$ are the error values of the current regulators of the $q$-$d$ axes, respectively.

For the outer speed regulator, the error value can be calculated as

$$e_\omega = \frac{i^*_{qs}}{K_{P\text{-}\omega}} - \frac{K_{I\text{-}\omega}}{K_{P\text{-}\omega}} E_\omega + \int K^2_{aw\text{-}\omega} \Delta i^*_{qs} - \omega^*_m + \omega_m \tag{3.11}$$

where $K_{P-\omega}$ and $K_{I-\omega}$ are the constant parameters of the proportional and integral controllers of the PI velocity regulator, $K_{aw-\omega}$ is the anti-windup variable of the velocity regulator, $e_\omega$ is the error variable of the velocity regulator, and $E_\omega$ is the value of the error integration of the velocity regulator.

To calculate the demagnetizing current, Figure 3.3 shows how the required $d$-axis stator current can be calculated. The significant principle is feedforwarding the calculated reference duty cycles to compare its maximum value with its compensated vector ($D_{dq}$). Therefore, as shown in Figures 3.1 and 3.3, the reference vector of the duty cycles can be calculated as [9]

$$d_q = (v_{qs}'^* - v_{qs\text{-decoupling}})/V_{dc}, -1 \le d_q \le 1 \tag{3.12}$$

$$d_d = (v_{ds}'^* - v_{ds\text{-decoupling}})/V_{dc}, -1 \le d_d \le 1 \tag{3.13}$$

$$D_{dq} = \sqrt{d_d^2 + d_q^2}, D_{dq} \in \left[0, \sqrt{2}\right] \tag{3.14}$$

where $v_{qs}'^*$ and $v_{ds}'^*$ are the reference stator voltages of the $q$-$d$ axes for the stator current PI controller as its output ports. This comparison provides the $d$-axis reference current to automatically set the flux level even if the DC-link voltage unexpectedly releases down.

The reason for the use of the flux-weakening control loop is to determine the sudden rise in the flux weakening. The advantage of this outer control loop for the flux-weakening region is that it is enabled automatically and obtains a high dynamic performance under parameter variation of the PMSM machine and load torque trouble. As a result, the driver obtains a stable motor speed and pushes up the maximum limit.

For the operating principle of the flux-weakening control loop, if the compensated vector of the duty cycle is equal to or greater than its defined maximum value ($D_{dq\text{-max}} = 1$), the control loop is automatically enabled. As a result, the PI regulator of the flux-weakening loop considers the error value between them to calculate the required stator current of the $d$-axis to avoid the predicted saturation for the PI regulators of the $d$-$q$ axes stator current (inner controller). Otherwise, the required demagnetizing current of the $d$-axis trajectory can be set to zero if the compensated vector of the duty cycle is less than the defined maximum value.

The matrix of the error values should be completed with the error value of the PI regulator for the flux-weakening region, which can be calculated as

$$e_{FW} = \frac{i_{ds}^*}{K_{P\text{-FW}}} - \frac{K_{I\text{-FW}}}{K_{P\text{-FW}}} E_{FW} + \int K_{aw\text{-FW}}^2 \Delta i_{ds}^* - D_{dq\text{-max}} + D_{dq} \tag{3.15}$$

where $K_{aw\text{-FW}}$ is the anti-windup variable of the PI regulator of the flux-weakening region, $e_{FW}$ is the error value of the PI regulator for the flux control loop, $E_{FW}$ is the value of the error integration of the PI regulator of the flux-weakening region, and $K_{P\text{-FW}}$ and $K_{I\text{-FW}}$ are the value of the proportional and integral parameters of the flux PI regulator, respectively.

Lastly, the summation operation to calculate the error signal of the reference duty-cycle vector generates high oscillation attached to the error signal because the condition of enabling this flux loop occurs frequently when the PMSM is operating in the flux-weakening region, and thus, the required $d$-axis current is suddenly down to zero. Therefore, the proposed improved control scheme can reject the high oscillation by obtaining the required $d$-axis stator current through the low-pass filter as shown in Figure 3.3 [10]. The filter increases the mean value of the $d$-axis stator current in the negative trajectory. As a result, this loop increases the root mean square value of the DC-link supply voltage.

### 3.2.2 Cost Function for Adjusting the PI Parameters

The proposed control drive must depend on the best PI parameters to obtain stable speed and current control in the flux-weakening region. Therefore, this chapter begins with a novel tuning technique to adjust the optimum PI parameters for the cascade PI regulators. This new tuning technique does not depend on the transfer functions of the PMSM or the control strategy. It depends on the mathematical model of the controlled PMSM and control strategy as illustrated earlier. The PI gains that minimize the equilibrium point in terms of the error values of (3.9), (3.10), (3.11), and (3.15) are the optimal parameters. The equilibrium point terms can be obtained in terms of the Taylor series expansion, which can be defined as

$$\vec{f}(\vec{x},\vec{u}) \cong \vec{f}(\vec{x}_0,\vec{u}_0) + \frac{\partial \vec{f}(\vec{x}_0,\vec{u}_0)}{\partial \vec{x}}(\vec{x}-\vec{x}_0) + \frac{\partial \vec{f}(\vec{x}_0,\vec{u}_0)}{\partial \vec{u}}(\vec{u}-\vec{u}_0) +$$
$$\frac{1}{2!}\left[\begin{array}{l}\frac{\partial^2 \vec{f}(\vec{x}_0,\vec{u}_0)}{\partial \vec{x}^2}(\vec{x}-\vec{x}_0)^2 + \frac{\partial^2 \vec{f}(\vec{x}_0,\vec{u}_0)}{\partial \vec{x}\partial \vec{u}}(\vec{x}-\vec{x}_0)(\vec{u}-\vec{u}_0) \\ + \frac{\partial^2 \vec{f}(\vec{x}_0,\vec{u}_0)}{\partial \vec{u}^2}(\vec{u}-\vec{u}_0)^2\end{array}\right] \quad (3.16)$$
$$+ \dots$$

where $0$ indicates that the equilibrium points to the number of the variables defined from the flux-weakening region.

To define the proposed cost function, the mathematical model of the controlled PMSM in the previous chapter for (2.6), (2.7), and (2.12), and the mathematical model of the control strategy explained earlier, can be expressed in terms of the Taylor series expansion to obtain the cost function as presented in (3.17) and (3.18).

The cost function can be abbreviated as

$$\text{Minimize}\left\{\sum_{j=1}^{4} e_j\right\} \quad (3.17)$$

The cost function can be evaluated online on a digital microprocessor, or an offline optimization procedure is possible on simulation software based on any optimization technique to obtain the optimal PI gains corresponding to the drive. The following section explains how the adaptive velocity particle swarm optimization technique can be employed to evaluate the cost function [11].

$$
\begin{bmatrix} \dot{i}_{qs0} \\ \dot{i}_{ds0} \\ \dot{\omega}_{m0} \\ e_{q0} \\ e_{d0} \\ e_{\omega0} \\ e_{FW0} \end{bmatrix} = \begin{bmatrix} -R_s/L_s & -0.5n_p\omega_{m0} \\ 0.5n_p\omega_{m0} & -R_s/L_s \\ (3n_p\lambda_m)/(2J) & 0 \\ 1 & (L_s n_p\omega_{m0})/(2K_{P\text{-}q}) \\ (L_s n_p\omega_{m0})/(2K_{P\text{-}d}) & 1 \\ 0 & 0 \\ 0 & 0 \end{bmatrix}
$$

$$
\begin{bmatrix} -n_p(0.5i_{ds0}+(\lambda_m/L_s)) & 0 & 0 & 0 & 0 \\ 0.5n_p i_{qs0} & 0 & 0 & 0 & 0 \\ -B/J & 0 & 0 & 0 & 0 \\ (L_s n_p i_{ds0})/(2K_{P\text{-}q})+(n_p\lambda_m/K_{P\text{-}q}) & -K_{aw\text{-}q} & 0 & 0 & 0 \\ (L_s n_p i_{qs0})/(2K_{P\text{-}d}) & 0 & -K_{aw\text{-}d} & 0 & 0 \\ 1 & 0 & 0 & -K_{aw\text{-}\omega} & 0 \\ 0 & 0 & 0 & 0 & -K_{aw\text{-}FW} \end{bmatrix} \begin{bmatrix} i_{qs0} \\ i_{ds0} \\ \omega_{m0} \\ E_{q0} \\ E_{d0} \\ E_{\omega0} \\ E_{FW0} \end{bmatrix} + \quad (3.18)
$$

$$
\begin{bmatrix} v_{qs0}/L_s \\ v_{ds0}/L_s \\ -T_L/J \\ (v_{qs0}^*/K_{P\text{-}q})-i_{qs0}^*+\int K_{aw\text{-}q}^2\Delta d_{q0} \\ (v_{ds0}^*/K_{P\text{-}d})-i_{ds0}^*+\int K_{aw\text{-}d}^2\Delta d_{d0} \\ (i_{qs0}^*/K_{P\text{-}\omega})-\omega_{m0}^*+\int K_{aw\text{-}\omega}^2\Delta i_{qs0}^* \\ (i_{ds0}^*/K_{P\text{-}FW})-D_{dq\text{-}max0}+\int K_{aw\text{-}FW}^2\Delta i_{ds0}^* \end{bmatrix}
$$

### 3.2.3 COST FUNCTION EVALUATION BASED ON ADAPTIVE VELOCITY PARTICLE SWARM OPTIMIZATION ALGORITHM

Many stochastic optimization techniques have been improved and developed in recent years to solve a convex fitness function due to their effectiveness and simplicity. The main idea of the particle swarm optimization algorithm depends on the global model, in which the velocity of each particle is affected by the best neighbor in the evaluation value and the best particle of the swarm in the evaluation value in terms of the current iteration.

The cost function considered in this chapter has eight PI parameters of the PI regulators gains, which need optimization to obtain the minimum evaluation of the cost function. The AVPSO begins to initialize the position vector signal with them as

$$
\hat{p}_n(1\times8) = \begin{bmatrix} K_{I\text{-}q}, K_{P\text{-}q}, K_{I\text{-}d}, K_{P\text{-}d}, K_{I\text{-}s}, K_{P\text{-}s}, K_{I\text{-}FW}, K_{P\text{-}FW} \end{bmatrix}, n = \begin{bmatrix} 1:N \end{bmatrix} \quad (3.19)
$$

where $\hat{p}_n$ is the position vector of the particle swarm, $n^{th}$ indicates one particle of the swarm, and $N$ is the swarm length. It can be seen from (3.19) that each $n^{th}$ particle has many individual locations equal to eight of the drive system. The net swarm of the particle $n^{th}$ equals $N$, and the vector of their velocities should be initialized and set to zero as

$$\hat{v}_n(1\times8) = [0,0,0,0,0,0,0,0], n = [1:N] \qquad (3.20)$$

where $\hat{v}_n$ is the velocity vector of the particle swarm. Lastly, the vector of the optimum position of the particle swarm should be initialized and set to zero as

$$\hat{b}_n(1\times8) = [0,0,0,0,0,0,0,0], n = [1:N] \qquad (3.21)$$

where $\hat{b}_n$ is the historical local optimum location vector of the particle swarm. Afterwards, the AVPSO technique evaluates each particle of the swarm, and the cost function needs the eight PI parameters from each particle of $n^{th}$ to set the corresponding vector of the error summation as

$$\vec{g}_n = \left[ \text{evaluation of} \hat{p}_n(1\times8) \right], n = [1:N] \qquad (3.22)$$

where $\vec{g}_n$ is the evaluation vector of the particle swarm. Thus, the minimum evaluation of the cost function can be determined from the vector of $\vec{g}_n$ to set as the global solution symbolized by $g_{best}$, corresponding to the $n^{th}$ particle of $\hat{P}_n$, called the optimum global particle location symbolized by $p_{global}$. This ends the initialization steps before starting the search procedure of the global optimum location into the defined number of the generation.

To search for the optimal solution, the particle swarm locations should be updated to generate different ratings. The upgraded locations of the particle swarm by the AVPSO algorithm can be defined as

$$\hat{P}_n = \hat{V}_n + \hat{p}_n \qquad (3.23)$$

where $\hat{P}_n$ is the updated location vector of the $n^{th}$ particle of the swarm. It can be seen from (3.23) that the updated location vector depends on the updated velocity of the $n^{th}$ particle symbolized by $\hat{V}_n$. The AVPSO algorithm defines the adaptive velocity as

$$\hat{V}_n = \omega_A \hat{v}_n + c_1 r_1 \left( \hat{b}_n - \hat{p}_n \right) + c_2 r_2 \left( p_{global} - \hat{p}_n \right) \qquad (3.24)$$

$$\omega_A = (g/G)(\sum_1^{L_p} \left| p_{global} - \hat{b}_n \right| / L_p) \qquad (3.25)$$

where $c_1$ and $c_2$ are the acceleration coefficients, and $r_1$ and $r_2$ are positive random numbers. The value of the acceleration factors and random numbers can be defined as

$$c_1, c_2 = 2 \tag{3.26}$$

$$0 \le r_1, r_2 \le 1 \tag{3.27}$$

Meanwhile, the adaptive velocity of (3.24) is affected by the following:

- The local optimum location of the $n^{\text{th}}$ particle swarm $\hat{b}_n$
- The last decided global optimum particle location of $p_{\text{global}}$
- The current location of the $n^{\text{th}}$ particle swarm $\hat{p}_n$
- The variable weight factor of $\omega_A$

It can be seen from (3.25) that the variable weight factor depends on the following:

- The total number of generations is symbolized by $G$.
- The current iteration is symbolized by $g$, which stands for the try to find the best particle location.
- The last decided global optimum particle location ($p_{\text{global}}$) refers to the best solution of $g_{\text{best}}$.
- The local optimum location of the $n^{\text{th}}$ particle swarm.
- The length of the $n^{\text{th}}$ particle swarm is symbolized by $L_P$.

The optimization procedure for the PSO algorithm will stop if the optimal location of the PI parameters has not changed for up to 12 iterations. Otherwise, it will stop after 100 iterations.

There are unknown variables of the equilibrium points for $E_{q0}$, $E_{d0}$, $E_{\omega0}$, $E_{FW0}$, $i^*_{qs0}$, and $i^*_{ds0}$. The variables in the cost function of (3.17) are symbolized by $E_{q0}$, $E_{d0}$, $E_{\omega0}$, $E_{FW0}$, $i^*_{qs0}$, and $i^*_{ds0}$. These unknown variables face the optimization procedure as a constraint while evaluating the cost function. A program called the AVPSO technique will be recalled separately each time the cost function is evaluated. It appears that the outer AVPSO technique evaluates the cost function to find the optimum PI parameters. However, there is an internal AVPSO technique that will determine the unknown variables to get the value of the cost function in terms of the specified PI parameters.

The location vector of the $k^{\text{th}}$ particle swarm of the internal AVPSO technique can be defined as

$$\tilde{p}_k\,(1 \times 6) = \left[ E_{q0}, E_{d0}, E_{\omega0}, E_{FW0}, i^*_{qs0}, i^*_{ds0} \right], k = [1 : N_1] \tag{3.28}$$

where $\tilde{p}_k$ is the position vector of the particle swarm, $k^{\text{th}}$ indicates one particle of the swarm, and $N_1$ is the swarm length, which equals 200. It can be seen from (3.28) that each $k^{\text{th}}$ particle has many individual locations equal to six of the variables that need to be determined. The net swarm of the particle $k^{\text{th}}$ equals $N_1$. Also, the vector of their velocities should be initialized and set to zero as

$$\tilde{v}_k \left(1 \times 6\right) = \left[0,0,0,0,0,0\right], k = \left[1 : N_1\right] \tag{3.29}$$

where $\tilde{v}_k$ is the velocity vector of the particle swarm. Lastly, the vector of the optimum position of the particle swarm should be initialized and set to zero as

$$\tilde{b}_k \left(1 \times 6\right) = \left[0,0,0,0,0,0\right], k = \left[1 : N_1\right] \tag{3.30}$$

where $\tilde{b}_k$ is the historical-local optimum location vector of the particle swarm. Afterwards, the internal AVPSO technique evaluates each particle of the swarm; the evaluation vector of the particle swarm can be defined as

$$\vec{\tilde{g}}_{n_1} = \left[\text{evaluation of } \tilde{p}_k \left(1 \times 6\right)\right], n_1 = \left[1 : N_1\right] \tag{3.31}$$

Therefore, the global solution can be set as the minimum value of the $\vec{\tilde{g}}_{n_1}$ vector. The initialization steps are complete; the search for the global optimum location in the defined number of the generation can start. The updated equation of the particles' swarm and their velocities of (3.23) to (3.27) are the same as those used in the internal AVPSO technique. The upgraded locations of the particle swarm by the internal AVPSO algorithm can be defined as

$$\tilde{P}_k = \tilde{V}_k + \tilde{p}_k \tag{3.32}$$

where $\tilde{P}_k$ is the updated location vector of the $k^{\text{th}}$ particle of the internal AVPSO swarm. It can be seen from (3.32) that the updated location vector depends on the updated velocity of the $k^{\text{th}}$ particle symbolized by $\tilde{V}_k$. As a result, the internal AVPSO algorithm defines the adaptive velocity of $\tilde{V}_k$ as

$$\tilde{V}_k = \omega_{A_1} \tilde{v}_k + c_1 r_1 \left(\tilde{b}_k - \tilde{p}_k\right) + c_2 r_2 \left(P_{\text{global-1}} - \tilde{P}_k\right) \tag{3.33}$$

$$\omega_{A_1} = (g_1 / G_1)\left(\sum_{1}^{L_{p_1}} \left|P_{\text{global-1}} - \tilde{b}_k\right| / L_{p_1}\right) \tag{3.34}$$

where $p_{\text{global-1}}$ is the optimum global particle location of the internal AVPSO.

Finally, a flow chart is introduced to illustrate how the main AVPSO technique evaluates the cost function to select the best eight PI parameters for the control scheme as shown in Figure 3.4. Figure 3.5 introduces a flow chart to illustrate how the inner AVPSO technique is recalled for determining the values of $E_{q0}$, $E_{d0}$, $E_{\omega0}$, $E_{\text{FW0}}$, $i^*_{qs0}$, and $i^*_{ds0}$.

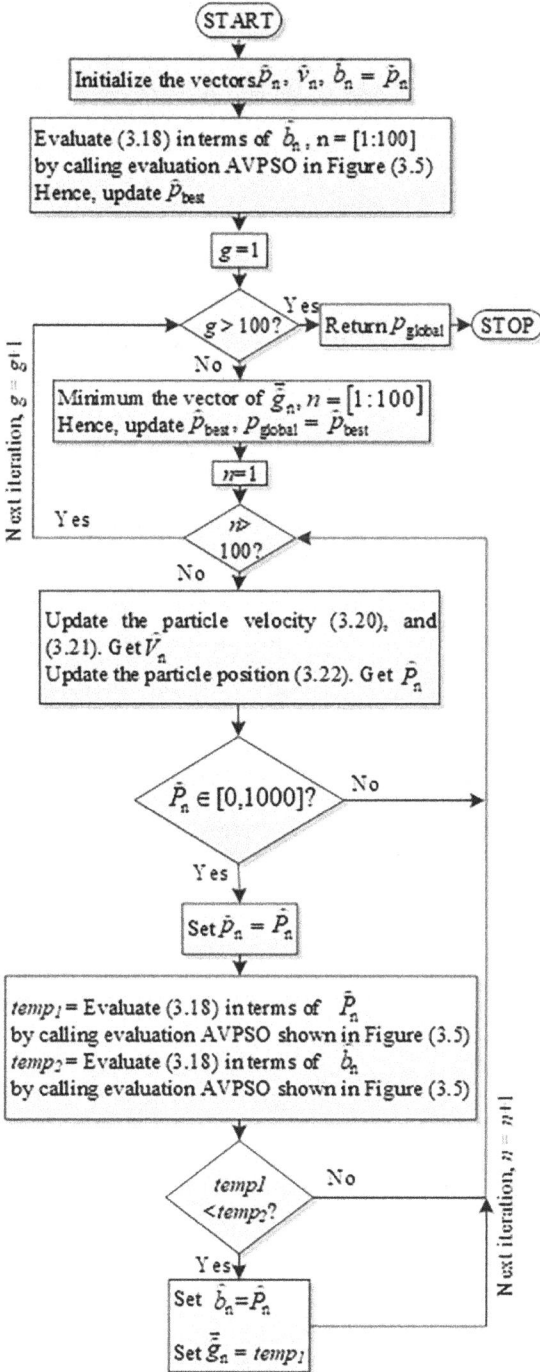

**FIGURE 3.4**   Implementation steps of the AVPSO algorithm to select the best PI regulator gains.

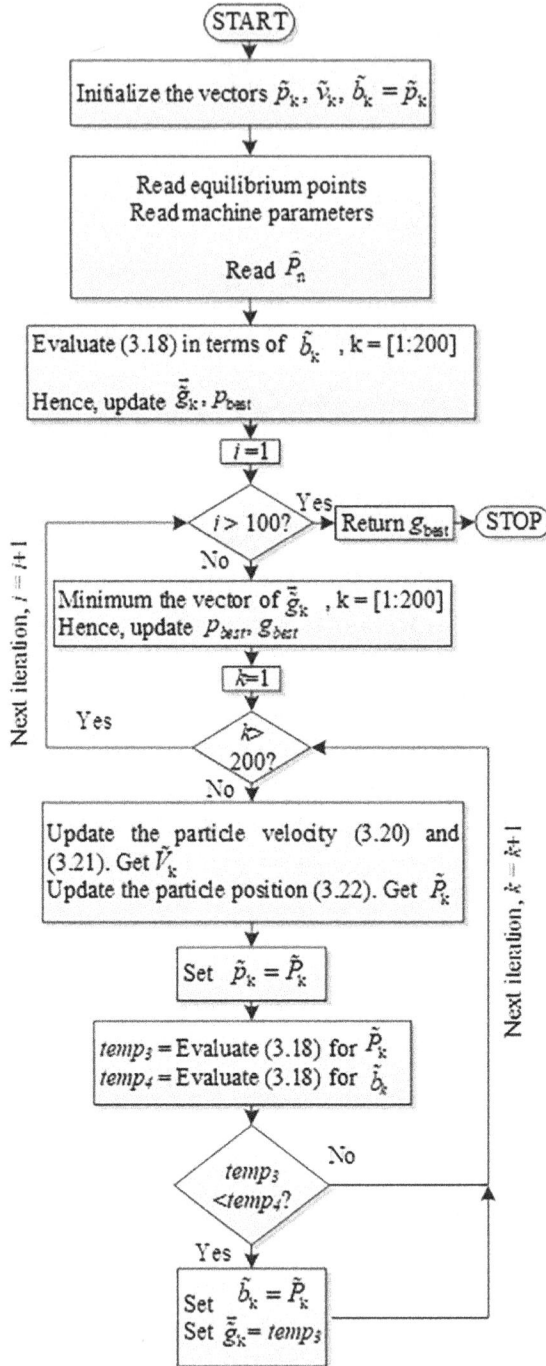

**FIGURE 3.5**   Implementation steps of the internal AVPSO algorithm to determine the unknown variables.

### 3.2.4 IMPROVED ADAPTIVE FLUX-WEAKENING CONTROL STRATEGY BASED ON MAXIMUM LINE MODULATION

The feedforward is from the FOC strategy output to the input to compare it with the maximum value to adjust the field level. This feedforward can depend on many states of the compensated voltages, compensated duty cycles, and modulation index of the voltage source inverter. The outer regulator of the adaptive field-decreasing control loop was proposed in [2] in a traditional cascade presentation with the $d$-axis stator current PI regulator. Meanwhile, the researchers in [2] depend on maximizing the line-modulation ratio of the inverter to set the flux level of the controlled PMSM while it is operating in the flux-weakening region, as shown in Figure 3.6.

In [2], the line modulation indexes are employed instead of the traditional space vector pulse-width modulation. Thus, the challenge of this control strategy is to obtain the line modulations of three phases symbolized by $\delta_a$, $\delta_b$, and $\delta_c$. If the maximum line modulation is greater than or equal to one, the flux-weakening control loop will be started to provide the required $d$-axis stator current in the negative trajectory.

By converting the compensated stator voltages of the $d$-$q$ axes of $v^*_{ds}$, and $v^*_{qs}$ using the inverse Park transformation, it can obtain the reference line voltages $u^*_{ac}$ and $u^*_{bc}$, which can be employed to obtain the values of the modulation indexes, $m_{ac}$ and $m_{bc}$ as

$$m_{ac} = u^*_{ac} / V_{dc} \qquad (3.35)$$

$$m_{bc} = u^*_{bc} / V_{dc} \qquad (3.36)$$

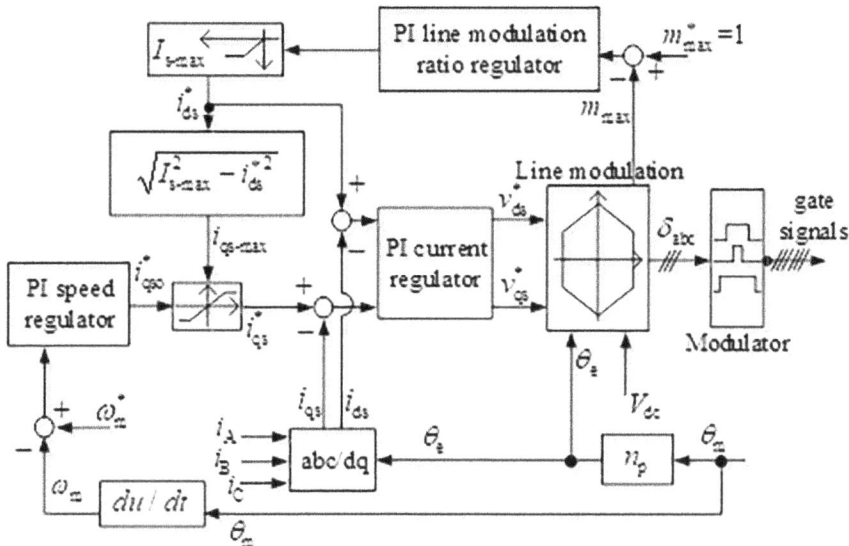

**FIGURE 3.6**   Improved flux weakening control method based on maximum line-modulation [2].

Thus, the line modulations of $\delta_a$, $\delta_b$, and $\delta_c$, based on (3.35) and (3.36), are

$$\delta_a = m_{ac} + \delta_c \tag{3.37}$$

$$\delta_b = m_{bc} + \delta_c \tag{3.38}$$

Therefore, the challenge of the line modulations becomes how to calculate $\delta_c$. The range of the line modulation can be defined as follows:

$$0 \le \delta_a, \ \delta_b, \ \delta_c \le 1 \tag{3.39}$$

Hence, the following case must be met:

$$\delta_{max} - \delta_{min} \le 1 \tag{3.39}$$

where

$$\delta_{max} = \max\{\delta_a, \ \delta_b, \ \delta_c\} \tag{3.40}$$

$$\delta_{min} = \min\{\delta_a, \ \delta_b, \ \delta_c\} \tag{3.41}$$

As a result, the maximum compensated stator voltage can be defined based on (3.40) as

$$V_{max} = m_{max} V_{dc} \tag{3.42}$$

where $m_{max}$ is the maximum line-modulation index, which can be defined as [12]

$$m_{max} = \delta_{max} - \delta_{min} \tag{3.43}$$

Thus, the maximum line-modulation voltage can be determined based on (3.35) and (3.36) as

$$u_{xc} = \max\{u_{ac}, \ u_{bc}, \ u_{cc}\} \tag{3.44}$$

$$u_{yc} = \min\{u_{ac}, \ u_{bc}, \ u_{cc}\} \tag{3.45}$$

$$m_{max,c} = \max\left\{m_{ac}, \ m_{bc}, \ \frac{u_{xc}}{V_{dc}}\right\} \tag{3.46}$$

$$m_{min,c} = \min\left\{m_{ac}, \ m_{bc}, \ \frac{u_{yc}}{V_{dc}}\right\} \tag{3.47}$$

Finally, the challenge of determining $\delta_c$ can be completed as

$$\delta_{c\_min} = -m_{min,c} \qquad (3.48)$$

$$\delta_{c\_max} = 1 - m_{max,c} \qquad (3.49)$$

$$\delta_c = \frac{\delta_{c\_min} + \delta_{c\_max}}{2} \qquad (3.50)$$

where $\delta_{c\_max}$ and $\delta_{c\_min}$ are the maximum and minimum values of the duty cycle of phase $C$, respectively. Finally, the gate signals will be generated by comparing the calculated duty cycles of $\delta_a$, $\delta_b$, and $\delta_c$ with triangles waves.

## 3.3 SIMULATION CASES ANALYSIS

This section introduces the stability identification of the proposed flux-weakening control strategy in terms of the bod diagrams. Simulation scenarios of the dynamic performance of the advanced control strategy are compared to the classical control strategy to prove the necessity of the new drive system. The simulation study is based on the MATLAB/Simulink software [13]. The parameters of the controlled surface-mounted PMSM and the inherent power inverter are listed in Table 2.3 in Chapter 2.

### 3.3.1 STABILITY IDENTIFICATION OF THE CONTROL SCHEME

A Simulink model is designed to install the AVPSO algorithm that evaluates the cost function (3.17) and (3.18). Figure 3.7 introduces the generation of the AVPSO technique and the corresponding evaluation of the cost function. The number of generations and the population size are 100 and 50, respectively. It can be seen from this figure that the 39th generation obtains the minimum value of the cost function, which equals $1.10134 \times 1e^{-13}$. The AVPSO technique stops at the 51st generation based on the stop criteria mentioned earlier. The optimal parameters of the proportional and integral gains of the PI regulators for the proposed FOC strategy are presented in Table 3.2.

Figures 3.8 and 3.9 introduce the stability analysis of the drive system based on the bode diagram presenting the frequency analysis.

The open-loop response of the speed regulator is shown in Figure 3.8 as the bode plot standard. The crossover frequency of 0.0387 kHz refers to the minimum stability limit at the 112° phase margin. The zero crossover frequency refers to the maximum gain of 484 dB. The open-loop response of the current regulator is shown in Figure 3.9 as the bode plot standard as well. The crossover frequency of 0.647 kHz refers to the minimum stability limit at the 95.7° phase margin and -324 dB gain margin. The zero crossover frequency refers to the maximum gain of 436 dB.

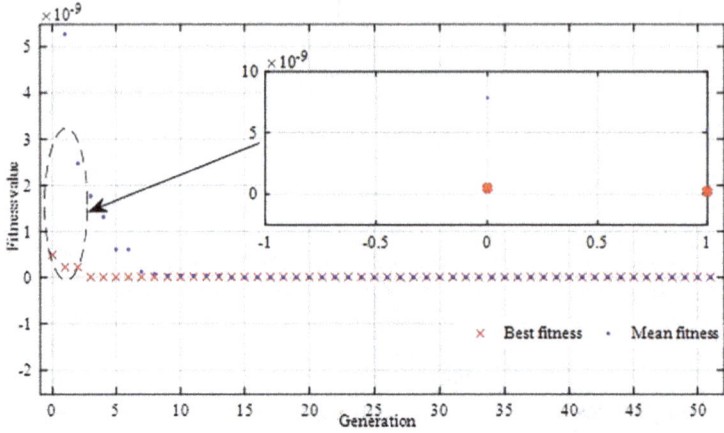

**FIGURE 3.7**   Fitness value of the cost function based on the AVPSO algorithm.

**TABLE 3.2**
**Optimal PI Gains of the Control Strategy**

| | | |
|---|---|---|
| Speed regulator | $K_{P\text{-}\omega}$ | 0.1 |
| | $K_{I\text{-}\omega}$ | 13.6 |
| $q$-axis current regulator | $K_{P\text{-}q}$ | 22.5 |
| | $K_{I\text{-}q}$ | 319.3 |
| $d$-axis current regulator | $K_{P\text{-}d}$ | 19.6 |
| | $K_{I\text{-}d}$ | 528.6 |
| Flux-weakening regulator | $K_{P\text{-}FW}$ | 14.9 |
| | $K_{I\text{-}FW}$ | 614.7 |

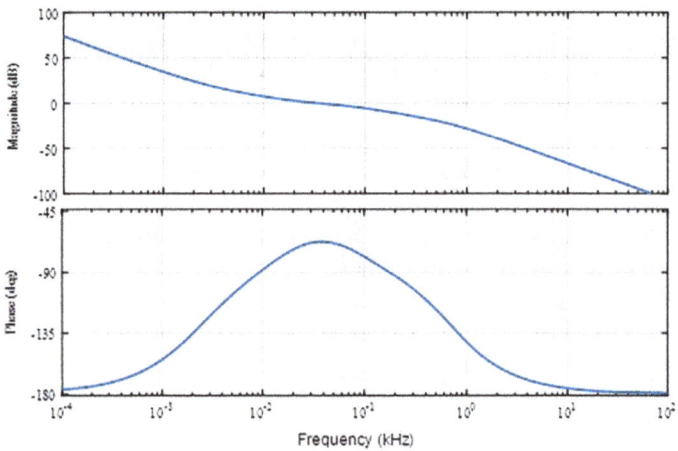

**FIGURE 3.8**   Open-loop response of the speed regulator.

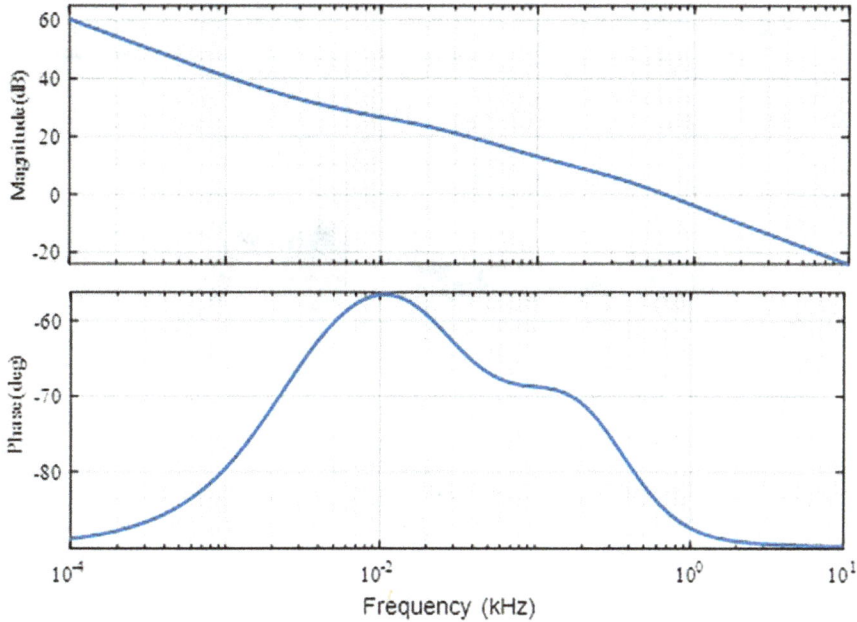

**FIGURE 3.9**   Open-loop response of the current regulator.

## 3.3.2 Dynamic Performance Analysis in the Presence of Torque Disorder

The following scenario studies the dynamic performance of a surface-mounted PMSM that is controlled around the applied load torque in a step-up model while the motor is operating in the flux-weakening region. In this case, the DC-link voltage is set at 400 V. The switching frequency of the traditional pulse space vector pulse-width modulation is a constant mode of 5 kHz. The control strategy drives the controlled PMSM to a reference speed of 2300 rpm in the flux-weakening region and employs an initial load torque of about 8.5 Nm, which is changed to 4.25 Nm at the time of 2 s. The dynamic performance is shown in Figures 3.10 to 3.15: the advanced control strategy (PFWC) compared to the conventional control strategy (CFWC). Besides the advanced control strategy, the standard proportional-integral controller structure (SPIS) technique is also employed to compare the dynamic performance relative to the anti-wind technique, which shows the effectiveness of that advanced technique, as shown in Figures 3.10 to 3.15.

Figure 3.10 shows three-speed signals of the classical control and advanced control strategies. The advanced control strategy based on the AWS technique limits the overshoot to 3.39%. However, the classical control strategy provides a speed with a higher overshoot of about 6.4%.

The advanced control strategy based on the SPIS technique obtains the same overshoot; depending on the AWS technique, it delays entering the steady-state period. The SPIS-based control strategy achieves the steady-state period at 1.262 s while the AWS-based control strategy achieves the steady-state period at 0.5642 s. The

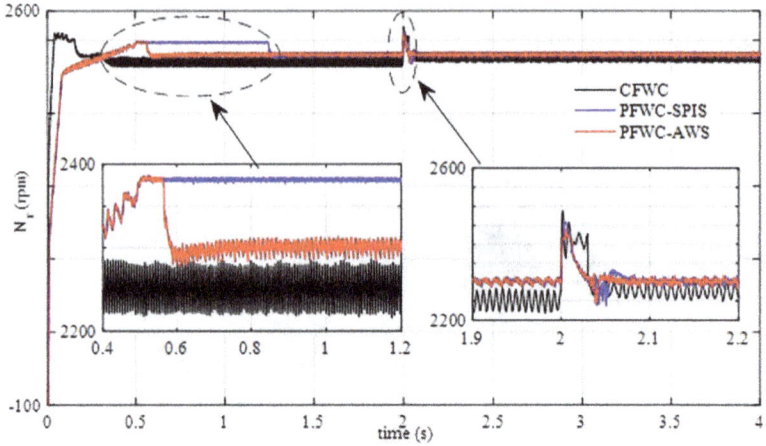

**FIGURE 3.10** Measured speed of the PMSM under the control of both strategies in case of load torque disturbance in the flux-weakening region with a reference speed of 2300 rpm.

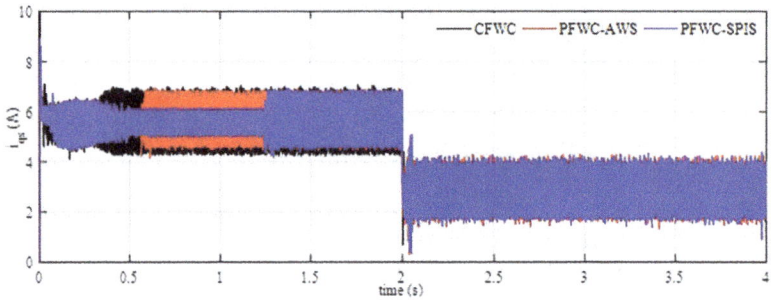

**FIGURE 3.11** Measured $q$-axis stator current under the control of both strategies in case of load torque disturbance in the flux-weakening region with a reference speed of 2300 rpm.

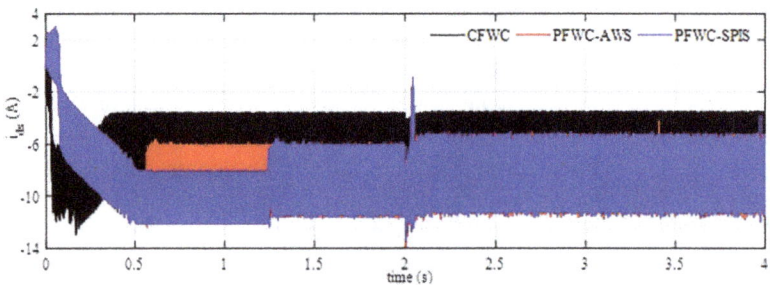

**FIGURE 3.12** Measured $d$-axis current under the control of both strategies in case of load torque disturbance in the flux-weakening region with a reference speed of 2300 rpm.

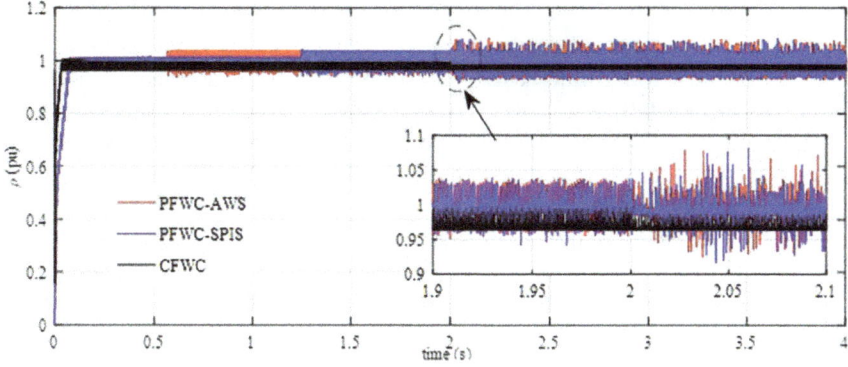

**FIGURE 3.13**   The $\rho$ vector under the control of both strategies in case of load torque disturbance in the flux-weakening region with a reference speed of 2300 rpm.

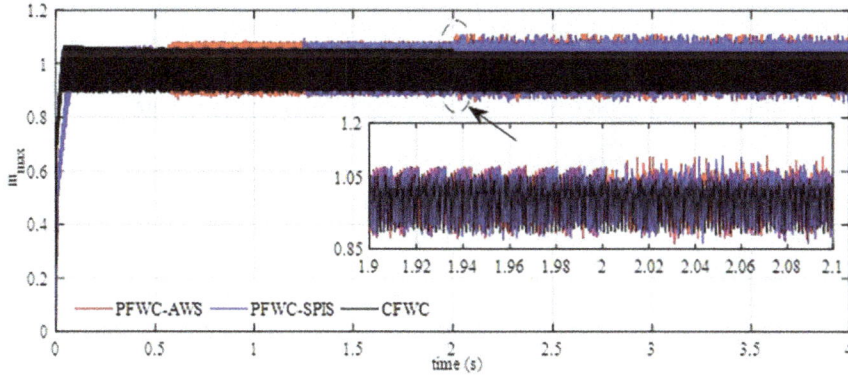

**FIGURE 3.14**   The $m_{max}$ ratio under the control of both strategies in case of load torque disturbance in the flux-weakening region with a reference speed of 2300 rpm.

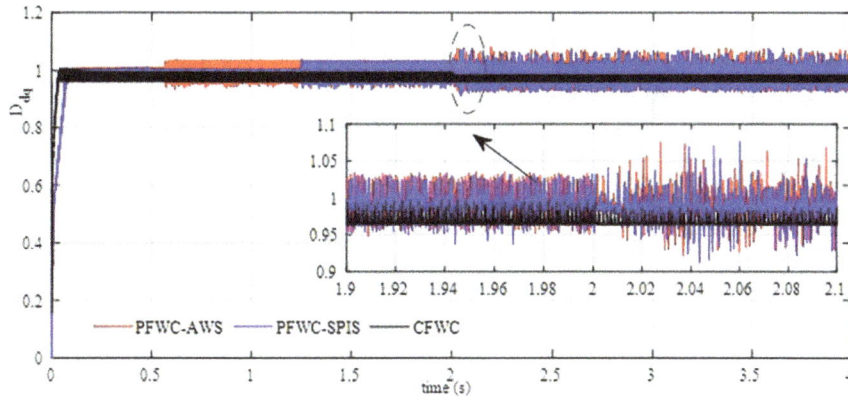

**FIGURE 3.15**   $D_{dq}$ magnitude under the control of both strategies in case of load torque disturbance in the flux-weakening region with a reference speed of 2300 rpm.

classical control strategy achieves the steady-state period at 0.4 s. This is the reason for its higher overshoot—the shorter transient time corresponds to a higher overshoot. Thus, the settling time at the starting period of the advanced and classical control strategies is 0.07 s and 0.03 s, respectively.

Also, it can be noticed from this figure that a higher oscillation peak is attached to the speed signal of the traditional control strategy than that of the advanced control strategy. Thus, the oscillation peak values for them are 31 rpm and 7.5 rpm, respectively. The values of the steady-state error of the advanced control strategy and the classical control strategy are 1.5 rpm and 48 rpm, respectively. Finally, for the measured speed, the advanced control strategy based on the anti-windup technique can reduce the steady-state error, overshoot, and the oscillation peak value relative to the classical control strategy by about 96.88%, 46.94%, and 75.81%, respectively.

The load torque suddenly changes, causing the measured velocity to be exceeded in the flux-weakening region, as shown in Figure 3.10. Thus, the overshoot of the classical control strategy and advanced control strategies based on SPIS and anti-wind technique are 8.22%, 7%, and 6.26%, respectively. The oscillation peak in the measured velocity is 25.5 rpm for the traditional control strategy and 16.5 rpm for the advanced control strategy. As a result, the value of the steady-state error is 22.5 rpm of the traditional control strategy. However, the advanced control strategy can eliminate the steady-state error after the transient period of the load torque variation. Finally, it can be concluded that the advanced flux-weakening control strategy based on the anti-windup technique can reduce the oscillation peak value and the overshoot value of the measured speed by 35.3% and 23.8%, respectively, relative to the classical control strategy.

Figure 3.11 shows the measured $q$-axis stator current of the PMSM under the control of both strategies beside the measured $q$-axis current under the control of the advanced control strategy based on the SPIS. As a result, it can be seen that the mean values during the pre-period of the torque disturbance are 5.66 A, 5.6 A, and 5.66 A for the advanced control strategy of the AWS technique, traditional control strategy, and advanced control strategy of the SPIS technique, respectively. However, the mean values during the post-period of the torque disturbance decrease to 3 A, 2.8 A, and 2.82 A, respectively.

Moreover, the value of the current peak at the pre-period of the torque disturbance is 1.2 A for the advanced control strategy based on the AWS technique and 1.35 A for the traditional control strategy. Meanwhile, the value of the current peak at the post-period of the torque disturbance is 1.3 A for the advanced control strategy based on the AWS technique and 1.44 A for the advanced control strategy based on the SPIS technique. Thus, it can be concluded that the advanced control strategy based on the AWS technique can reduce the peak of the $q$-axis stator current by 11% relative to the traditional control strategy. Moreover, it can reduce the peak of the $q$-axis stator current by 7% relative to the advanced control strategy based on the SPIS technique.

Figure 3.12 shows the measured $d$-axis stator current of the PMSM under the control of both strategies beside the measured $d$-axis current under the control of the advanced control strategy based on the SPIS. As a result, it can be seen that the mean values in the negative trajectory during the pre-period of the torque disturbance are -8.8 A, -6.54 A, and -8.8 A for the advanced control strategy of the AWS technique, traditional control strategy, and advanced control strategy of the SPIS technique, respectively. However, the mean values during the post-period of the torque disturbance slightly decrease on a negative trajectory to be -8.35 A, -6.53 A, and -8.33 A, respectively. Thus, it can be concluded that the advanced control strategy can increase the mean $d$-axis current by 34.66% relative to the traditional control strategy. Meanwhile, this increase is the reason behind the higher quality of the speed wave, as explained in Figure 3.10. The reason for the higher mean value of the $d$-axis current is that the advanced control strategy is based on the low-pass filter AWS technique for the flux-weakening outer control loop.

Figure 3.13 shows the per-unit voltage vector of the PMSM under the control of both strategies beside the per unit for the voltage vector under the control of the advanced control strategy based on the SPIS. As a result, it can be seen that the mean values during the pre-period of the torque disturbance are 0.99, 0.98, and 0.99 for the advanced control strategy of the AWS technique, traditional control strategy, and advanced control strategy of the SPIS technique, respectively. However, the mean values during the post-period of the torque disturbance slightly increase to 1.0, 0.97, and 1.0, respectively. It can be concluded that the advanced control strategy can increase the DC-link utilization by 0.9% and 2.56% for the pre-period and post-period of the torque disturbance relative to the traditional control strategy, respectively.

Figure 3.14 shows the maximum line modulation of the PMSM under the control of both strategies beside the maximum line-modulation under the control of the advanced control strategy based on the SPIS. As a result, it can be seen that the mean values during the pre-period of the torque disturbance are 0.99, 0.98, and 0.99 for the advanced control strategy of the AWS technique, traditional control strategy, and advanced control strategy of the SPIS technique, respectively. However, the mean values during the post-period of the torque disturbance slightly increase to 1, 0.98, and 0.99, respectively. It can be concluded that the advanced control strategy can increase the DC-link utilization by 0.1% and 2.1% for the pre-period and post-period, respectively, of the torque disturbance relative to the traditional control strategy. As a result, the mean value of the speed is higher than that of the conventional control strategy.

Generally, the duty cycles vector is a criterion to evaluate the utilization of the DC-link voltage under the control of the defined control strategy and can be calculated as

$$D_{dq} = \left( \sqrt{v_{ds}^{*\,2} + v_{qs}^{*\,2}} \right) / V_{dc} \tag{3.51}$$

Figure 3.15 shows the duty-cycle vector of the PMSM under the control of both strategies beside the duty-cycle vector under the control of the advanced strategy based on the SPIS. As a result, it can be seen that the mean values during the pre-period of the torque disturbance are 0.99, 0.98, and 0.98 for the advanced control strategy of the AWS technique, traditional control strategy, and advanced control strategy of the SPIS technique, respectively.

However, the mean values during the post-period of the torque disturbance slightly increase to 1.0, 0.97, and 1.0, respectively. It can be concluded that the advanced control strategy can increase the DC-link utilization by 0.6% and 2.87% for the pre-period and post-period, respectively, of the torque disturbance relative to the traditional control strategy. Thus, it can be concluded from Figures 3.13, 3.14, and 3.15 that the mean value of the *d*-axis current of the advanced control strategy is higher than that of the traditional control strategy. As a result, the output power and efficiency of the advanced control strategy are higher than the traditional control strategy.

### 3.3.3 Dynamic Performance Analysis in the Presence of the PMSM Parameters Variation

This section introduces the dynamic performance comparison between the advanced control strategy and the traditional control strategy in the case of variation of the controlled machine parameters of PMSM. The previous scenario of the reference speed of 2300 rpm and the step-down of load torque variation of 8.5 Nm to 4.25 Nm are employed in this section. The dynamic performance comparison is shown in Figures 3.16 to 3.20 to clarify to what range the advanced control strategy depends on the machine parameters of the resistance and reactance of the stator winding [14]. The stator resistance would be changed at 122 °C for the environment temperature from the rated value of 0.8 Ω at 20 °C to 1.12 Ω, which can be modeled as [15]

$$R_b = \left(R_a\left(t_b + k_1\right)\right)/\left(t_a + k_1\right) \tag{3.52}$$

where $R_a$ is the stator resistance measured at the winding temperature of $t_a = 20$ °C, $R_b$ is the corrected resistance at the winding temperature of $t_b$, and $k_1$ is the 100% conductivity copper coefficient determined by 234.5. The temperature is measured in degrees Celsius, and the electrical resistance is measured in the unit of the ohm.

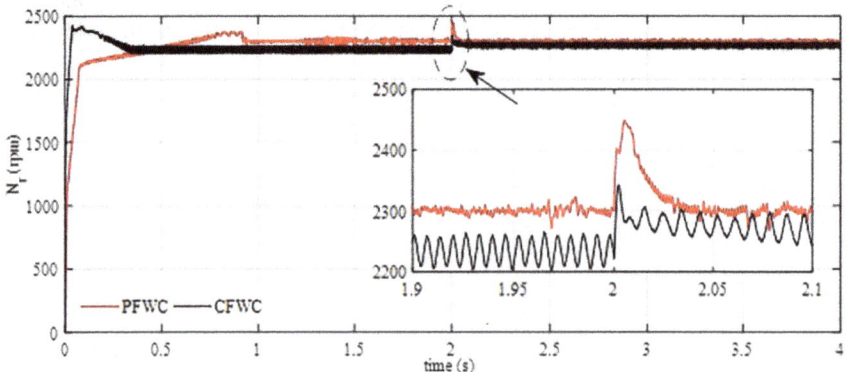

**FIGURE 3.16** Measured speed under the control of both strategies in case of the parameters variation of the controlled PMSM and load torque disturbance in the flux-weakening region with a reference speed of 2300 rpm.

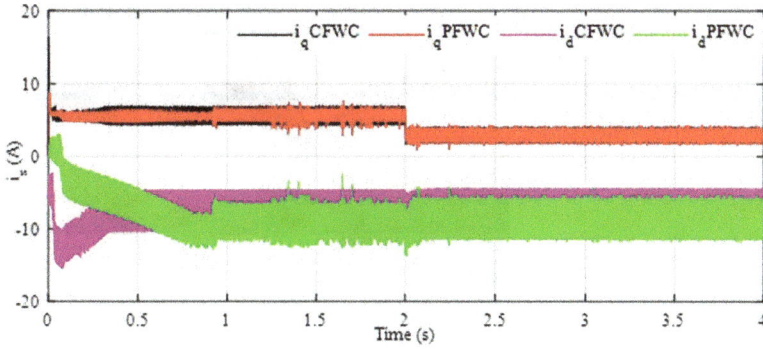

**FIGURE 3.17**  Measured $d$-$q$ axes stator currents under the control of both strategies in case of the parameters variation of the controlled PMSM and load torque disturbance with a reference speed of 2300 rpm.

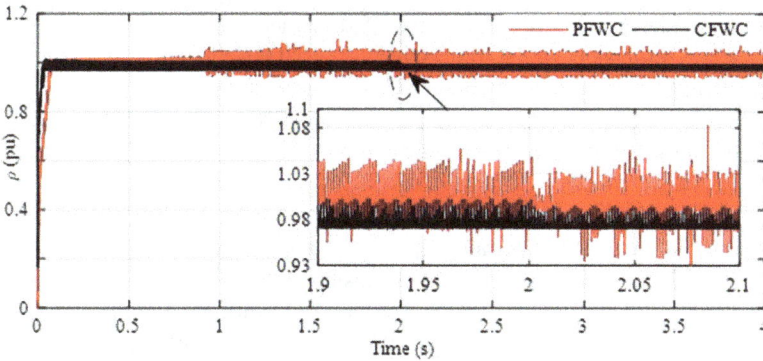

**FIGURE 3.18**  The $\rho$ vector under the control of both strategies in case of the parameters variation of the controlled PMSM and load torque disturbance in the flux-weakening region with a reference speed of 2300 rpm.

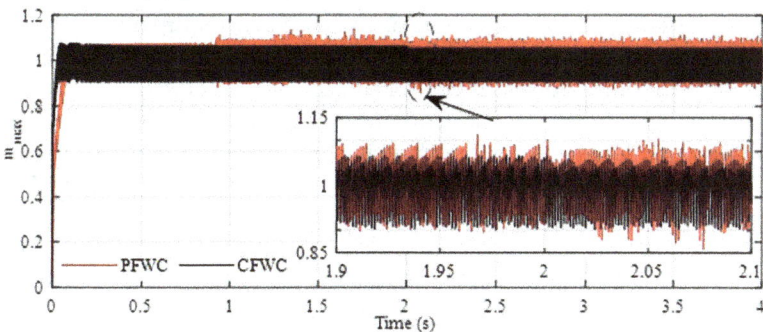

**FIGURE 3.19**  The $m_{\max}$ ratio under the control of both strategies in case of the parameters variation of the controlled PMSM and load torque disturbance in the flux-weakening region with a reference speed of 2300 rpm.

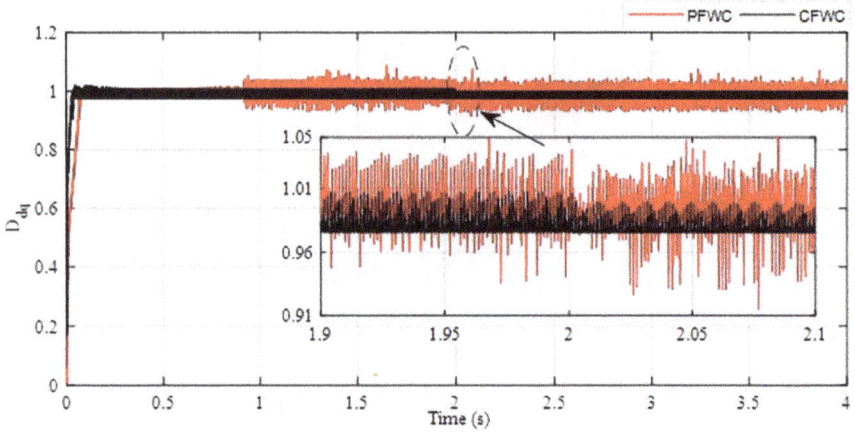

**FIGURE 3.20** $D_{dq}$ vector under the control of both strategies in case of the parameters variation of the controlled PMSM and load torque disturbance in the flux-weakening region with a reference speed of 2300 rpm.

However, the winding inductance would drop to 40% of its rated value during the occurrence of magnetic saturation [16].

Figure 3.16 shows two-speed signals of the advanced control strategy and the traditional control strategy. The advanced control strategy limits the overshoot to 3.17%. The classical control strategy provides a speed with a higher overshoot of about 5.3%. Thus, the overshoot of the transient period can be reduced under the advanced control strategy by 40.16% relative to the traditional control strategy. However, the settling time of the advanced control strategy is slightly increased by 2.15% relative to the traditional control strategy. For detailed values, the values of the settling time are 0.08 s, and 0.03 s, respectively.

Meanwhile, it can be noticed from this figure that a higher oscillation peak is attached to the speed signal of the traditional control strategy than that of the advanced control strategy. Thus, the oscillation peak values are 25 rpm and 30 rpm for the pre-period from the load torque disturbance and 15.5 rpm and 22.5 rpm for the post-period from the load torque disturbance, respectively. The values of the steady-state error of the advanced control strategy and the classical control strategy are 4 rpm and 69 rpm for the pre-period from the load torque disturbance and 3.5 rpm and 31.5 rpm for the post-period from the load torque disturbance, respectively. Thus, the advanced control strategy can reduce the values of the oscillation peak and steady-state error relative to the traditional control strategy by about 17%, 94% for the pre-period from the load torque disturbance, and 31% and 90% for the post-period from the load torque disturbance, respectively. Lastly, the high quality of the stable speed for the advanced control strategy is the clue that it can run the controlled PMSM with the independence of its parameters.

Figure 3.17 shows the measured waves of the $d$-$q$ axes of the stator currents for the advanced control strategy and the traditional control strategy. It can be seen from this figure that the mean values of the $q$-axis stator current of the advanced control strategy and the traditional control strategy are 5.68 A and 5.62 A for the pre-period from the load torque disturbance, respectively. Additionally, they are 2.86 A and 2.88 A for the post-period from the load torque disturbance, respectively. Thus, the advanced control strategy can increase the mean value of the $q$-axis current relative to the conventional control strategy by about 1.1% for the pre-period from the load torque disturbance.

However, it is approximately the same during the post-period from the load torque disturbance. This increase in the mean value yields more output power, increasing the efficiency and decreasing the copper losses of the controlled PMSM.

For comparison to the current ripple peak, the peak values of the $q$-axis current ripple for the pre-period from the load torque disturbance are 0.99 A of the advanced control strategy and 1.41 A of the conventional control strategy. For the post-period from the load torque disturbance, they are approximately the same: about 1.1 A. Thus, the advanced control strategy can reduce the peak value of the $q$-axis current ripple relative to the conventional control strategy by about 29.7% for the pre-period from the load torque disturbance. This reduction in the peak value yields less torque ripple and increases the life of the controlled PMSM. Analysis of the $d$-axis current can be seen in Figure 3.17; the mean values of the $d$-axis stator current in the negative trajectory of the advanced control strategy and the traditional control strategy are -9 A, and -7.51 A for the pre-period from the load torque disturbance, respectively. They are also 8.42 A, and 7.46 A for the post-period from the load torque disturbance, respectively. Thus, the advanced control strategy can increase the mean value of the $d$-axis current relative to the conventional control strategy by about 19.83% for the pre-period from the load torque disturbance and about 12.92% for the post-period from the load torque disturbance. This increase in the mean value yields more output power, increasing efficiency and decreasing the copper losses of the controlled PMSM.

For comparison to the current ripple peak, the peak values of the $d$-axis current ripple for the pre-period from the load torque disturbance are 2.7 A of the advanced control strategy and 2.9 A of the conventional control strategy. For the post-period from the load torque disturbance, they are approximately the same at about 3.0 A. Thus, the advanced control strategy can reduce the peak value of the $d$-axis current ripple relative to the conventional control strategy by about 7.52% for the pre-period from the load torque disturbance. Also, this reduction in the peak value yields less torque ripple and increases the life of the controlled PMSM.

Figure 3.18 shows the per-unit voltage vector of the PMSM under the control of the advanced and traditional strategies. The mean values during the pre-period of the torque disturbance are 0.996, and 0.987 for the advanced control strategy and the traditional control strategy, respectively. However, the mean values during the post-period of the torque disturbance are about 1.0, and 0.982, respectively. It can be concluded that the advanced control strategy can increase the DC-link utilization by 0.89% for the pre-period from the torque disturbance relative to the traditional control strategy.

Figure 3.19 shows the maximum line modulation of the PMSM under the control of both the advanced and conventional strategies. It can be seen that the mean values during the pre-period of the torque disturbance are 0.9968 and 0.9832 for the advanced control strategy and traditional control strategy, respectively. However, the mean values during the post-period of the torque disturbance increase slightly to 0.9914, and 0.9794, respectively.

It can be concluded that the advanced control strategy can increase the DC-link utilization by 1.38% and 1.23% for the pre-period and post-period of the torque disturbance relative to the traditional control strategy, respectively. As a result, the mean value of the speed is higher than that of the conventional control strategy.

Figure 3.20 shows the duty-cycle vector of the PMSM under the control of the advanced and traditional strategies. The mean values during the pre-period of the torque disturbance are 1.01 and 0.98 for the advanced control strategy and the conventional control strategy, respectively. However, the mean values during the post-period of the torque disturbance increase slightly to 1.0, and 0.99, respectively.

It can be concluded that the advanced control strategy can increase the DC-link utilization by 0.89% and 1.82% for the pre-period and post-period of the torque disturbance relative to the traditional control strategy, respectively. This is why the mean value of the $d$-axis current of the advanced control strategy is higher than that of the traditional control strategy. As a result, the output power and efficiency of the advanced control strategy are higher than that of the traditional control strategy.

## 3.4 EXPERIMENTAL CASES ANALYSIS

The effectiveness of the dynamic performance of the proposed flux-weakening control strategy is experimentally validated compared to the dynamic performance of the classical control drive. A photograph of the experimental laboratory used is shown in Figure 2.24. The parameters of the controlled PMSM and power inverter are listed in Table 2.4 in Chapter 2. It should be mentioned that the DC-link voltage is saturated at 150 V. Moreover, the sample time of the control strategy is 100 μs. Lastly, the hardware control unit provides a further ripple attached to the measured signal.

### 3.4.1 Validating the Dynamic Performance of the Anti-Windup Technique

The scenario of this section is to drive the controlled PMSM at the constant-torque region based on the two versions of the advanced control strategy, as shown in Figures 3.21 to 3.25.

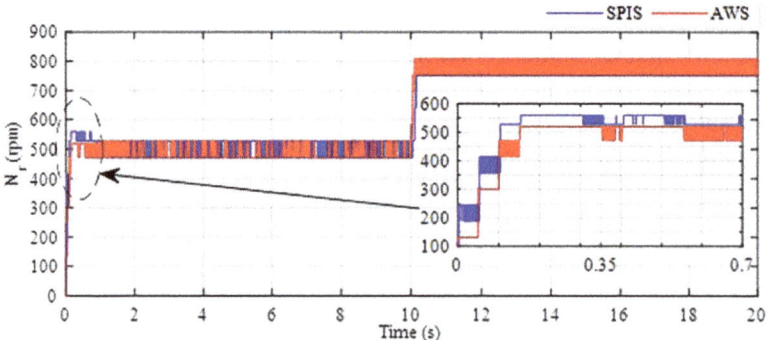

**FIGURE 3.21**    Measured speed under the control of the advanced strategy in case of the reference speed variation in the constant-torque region with a load torque of about 7.5 Nm.

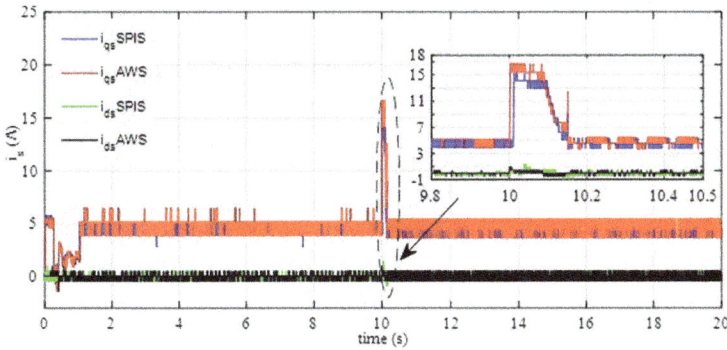

**FIGURE 3.22**   Measured $d$-$q$ axes stator currents under the control of the advanced strategy in case of the reference speed variation in the constant-torque region with a load torque of about 7.5 Nm.

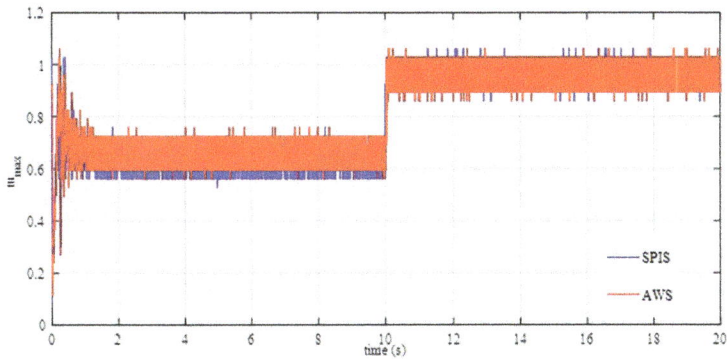

**FIGURE 3.23**   The $m_{max}$ ratio under the control of the advanced control strategy in case of the reference speed variation in the constant-torque region with a load torque of about 7.5 Nm.

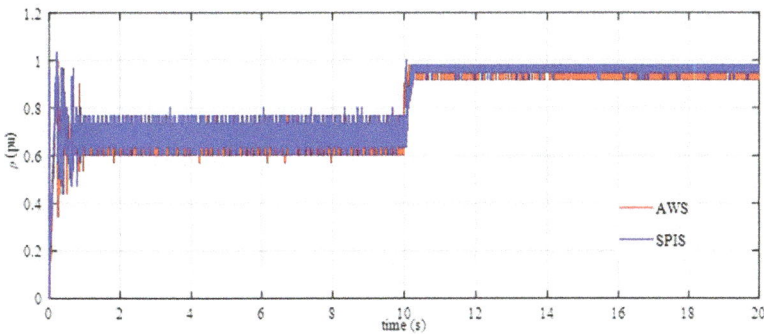

**FIGURE 3.24**   The $\rho$ vector under the control of the advanced control strategy in case of the reference speed variation in the constant-torque region with a load torque of about 7.5 Nm.

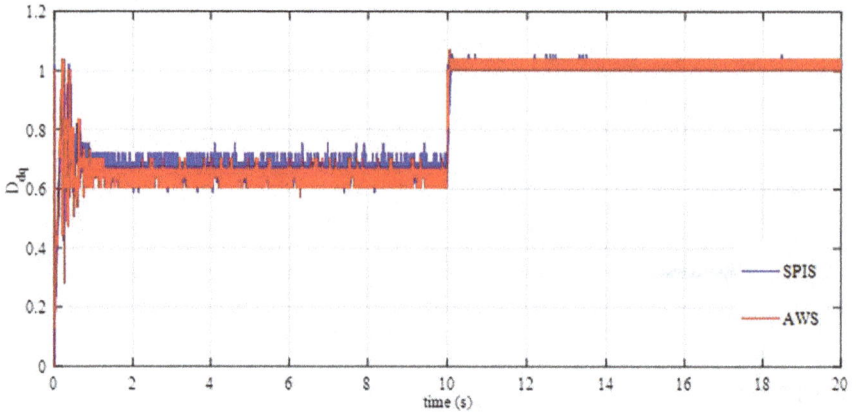

**FIGURE 3.25** $D_{dq}$ magnitude under the control of the advanced control strategy in case of the reference speed variation in the constant-torque region with a load torque of about 7.5 Nm.

The first one depends on the anti-windup technique. The second one is the standard PI technique. The advanced control strategy drives at the initial reference speed of 500 rpm and 7.5 Nm of the initial load torque. Moreover, the initial reference speed is increased in the step-up model to achieve the maximum limit at the constant-torque region.

Figure 3.21 shows two signals of the measured speed the standard PI technique and the anti-windup technique for the advanced control strategy. It can be seen from this figure that the mean values of the AWS technique and the SPIS technique are the same at about 499.9 rpm. The values of the overshoot rate are 4% and 12%, respectively, and the values of the settling time are 0.1037 s, and 0.1496 s, respectively. Therefore, the AWS technique can reduce the overshoot at the transient period relative to the SPIS technique by 66.67%. However, the settling time slightly increases by 44.26%. The proposed drive system suddenly increased the reference speed to a maximum limit of the constant-torque region. As a result, the mean values of the speed maximum limit are 780.7 rpm for the AWS technique, and 754.3 rpm for the SPIS technique. Thus, the AWS technique can raise the maximum limit of the speed by 3.5% relative to the SPIS technique of the advanced control strategy.

Figure 3.22 shows the measured waves of the $d$-$q$ axes of the stator currents of both the AWS and SPIS techniques for the advanced control strategy. It can be seen from this figure that the mean values of the $q$-axis stator current of the AWS and SPIS techniques are approximately the same at about 4.5 A for the pre-period from the reference speed variation.

They do not change after the post-period from the reference speed variation; they are approximately the same during the post-period from the reference speed variation as the previous value.

To analyze the $d$-axis current, it can be seen from Figure 3.22 that the mean values of the $d$-axis stator current of the AWS technique and the SPIS technique are the same at zero A for the pre-period and the post-period from the reference speed variation because the PMSM is controlled at the region of constant torque.

Figure 3.23 shows the maximum line modulation of the PMSM under the control of both techniques of the advanced strategy, the AWS and the SPIS techniques. It can be seen that the mean values during the pre-period of the torque disturbance are 0.67 and 0.63 for the AWS technique and SPIS technique, respectively. However, the mean values during the post-period of the torque disturbance increase to 0.96 and 0.95, respectively. It can be concluded that the advanced control strategy based on the AWS technique can increase the DC-link utilization by 7.7% for the pre-period of the torque disturbance relative to the SPIS technique. As a result, the mean value of the speed is higher than that of the SPIS technique.

Figure 3.24 shows the per-unit voltage vector of the PMSM under the control of the advanced and traditional strategies. The mean values during the pre-period of the torque disturbance are 0.67 and 0.69 for the advanced control strategy and the traditional control strategy, respectively. However, the mean values during the post-period of the torque disturbance are about 0.94 and 0.97, respectively.

Figure 3.25 shows the duty-cycle vector of the PMSM under the control of the advanced and traditional strategies. The mean values during the pre-period of the reference speed variation are 0.64 and 0.67 for the advanced control strategy and the conventional control strategy, respectively. However, the mean values during the post-period of the reference speed variation increase to 1.02 for both.

### 3.4.2 Validating the Dynamic Performance of the Advanced Control Strategy in Four Control Case Quadrants

In this section, the dynamic performance of the PMSM under the advanced control strategy is compared to the traditional control strategy in the case of driving the reference speed to the maximum limit at the flux-weakening region in both reverse and forward, covering the four quadrant regions of the drive system. The initial load torque is 7.5 Nm. Figures 3.26 to 3.30 show that the speed direction is switched at 50 s to track the reference speed acting as a ramp model.

Figure 3.26 shows two signals of the measured speed for the advanced and traditional strategies. It can be seen from this figure that the mean values of the maximum limit of the advanced control strategy and the traditional control strategy are 1175 rpm and 1016 rpm for the reverse direction, respectively. They are 1175 rpm and 1016 rpm for the forward direction, respectively. Thus, the advanced control strategy can increase the maximum limit of the speed relative to the traditional control strategy by 15.65% for the reverse and forward directions. It can also be noted that the advanced control strategy achieves the steady-state period faster than the traditional control strategy. The values of the settling time are 2.187 s, and 19.59 s, respectively. Thus, the advanced strategy can reduce the settling time by about 88.84% relative to the traditional strategy.

Figure 3.27 shows the measured waves of the $d$-$q$ axes of the stator currents for the advanced control strategy and the traditional control strategy. It can be seen from this figure that the mean values of the $q$-axis stator current of the advanced and traditional strategies in the reverse direction are -4.762 A and -1.73 A for the pre-period from

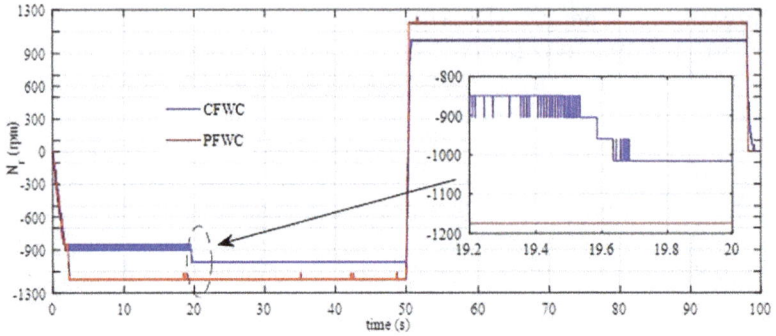

**FIGURE 3.26**  Measured speed under the control of both strategies in case of the forward-reverse direction in the flux-weakening region with a load torque of about 7.5 Nm.

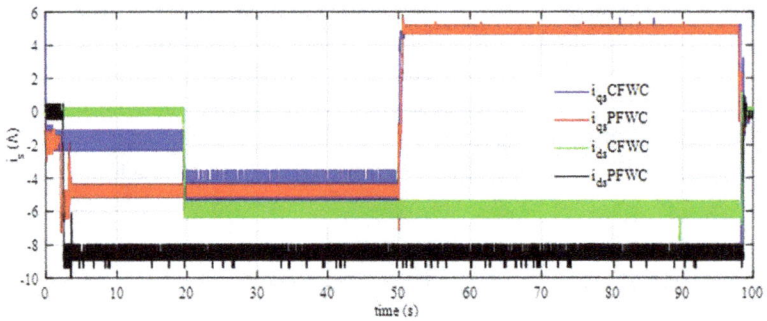

**FIGURE 3.27**  Measured $d$-$q$ axes stator currents under the control of both control strategies in case of the forward-reverse direction in the flux-weakening region with a load torque of about 7.5 Nm.

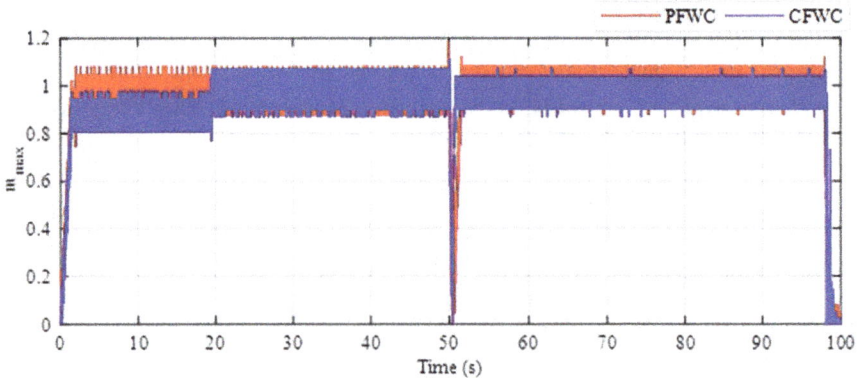

**FIGURE 3.28**  The $m_{max}$ ratio under the control of both strategies in case of the forward-reverse direction in the flux-weakening region with a load torque of about 7.5 Nm.

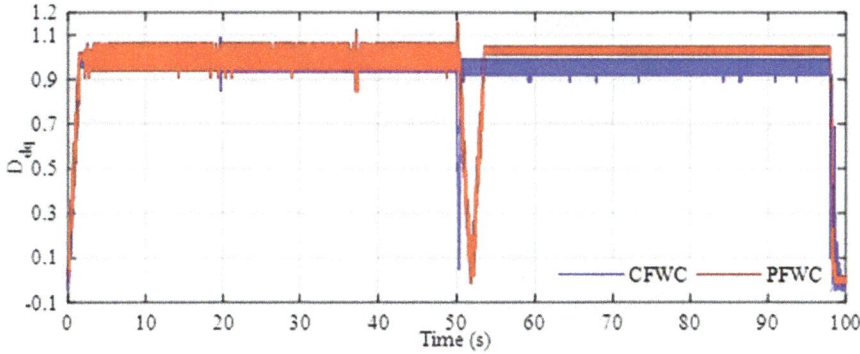

**FIGURE 3.29**   $D_{dq}$ magnitude under the control of both strategies in case of the forward-reverse direction in the flux-weakening region with a load torque of about 7.5 Nm.

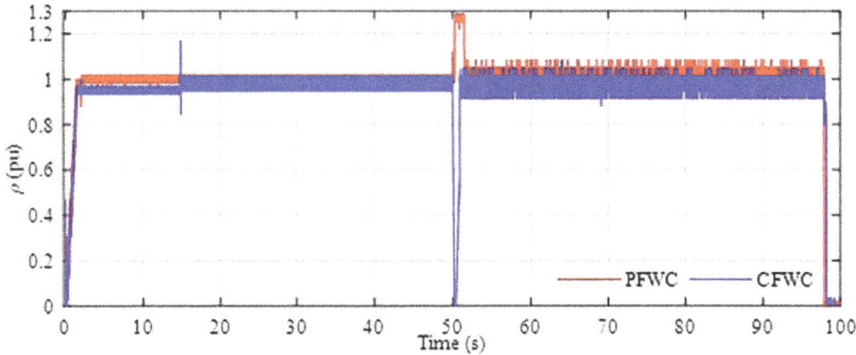

**FIGURE 3.30**   $\rho$ vector under the control of both strategies in case of the forward-reverse direction in the flux-weakening region with a load torque of about 7.5 Nm.

the reference speed variation, respectively. They are 4.9 A and 5 A for the post-period from the reference speed variation in the forward direction, respectively. Thus, the advanced control strategy can increase the mean value of the $q$-axis current relative to the conventional control strategy by about 176.46% and 1.98% for the pre-period and the post-period from the reference speed variation, respectively. This increase in the mean value yields more output power, increasing efficiency and decreasing the copper losses of the controlled PMSM.

For comparison to the current ripple peak, the peak values of the $q$-axis current ripple for the pre-period from the reference speed variation are 0.41 A for the advanced control strategy, and 0.61 A for the conventional control strategy. Besides, they are 0.23 A and 0.2 A for the post-period from the reference speed variation, respectively. Thus, the advanced control strategy can reduce the peak value of the q-axis current ripple relative to the conventional control strategy by about 32.4% and 17.13% for the pre-period and the post-period from the reference speed variation, respectively.

This reduction in the peak value yields less torque ripple and then increases the life of the controlled PMSM.

For analyzing the $d$-axis current, it can be seen from Figure 3.27 that the mean values of the $d$-axis stator current in the negative trajectory of the advanced control strategy and the traditional control strategy are -8.45 A, and -5.87 A for the pre-period from the reference speed variation, respectively. Besides, they are the same for the post-period as the pre-period from the reference speed variation. Thus, the advanced control strategy can increase the mean value of the $d$-axis current relative to the conventional control strategy by about 44%. This increase in the mean value yields more output power, increasing efficiency and decreasing the copper losses of the controlled PMSM.

For comparison to the current ripple peak, the peak values of the $d$-axis current ripple for the pre-period from the reference speed variation are 0.46 A for the advanced control strategy and 0.47 A for the conventional control strategy. They are approximately the same for the post-period as the pre-period from the reference speed variation. Thus, the advanced control strategy can reduce the peak value of the $d$-axis current ripple relative to the conventional control strategy by about 1.6%. Also, this reduction in the peak value yields less torque ripple and increases the life of the controlled PMSM.

Figure 3.28 shows the maximum line modulation of the PMSM under the control of both the advanced and conventional strategies. The mean values during the pre-period of the torque disturbance are 0.963 and 0.97 for the advanced control strategy and traditional control strategy, respectively. However, the mean value during the post-period of the torque disturbance increases to 0.997.

However, the conventional control strategy has a constant mean value. It can be concluded that the advanced control strategy can increase the DC-link utilization by 2.77% for the post-period of the torque disturbance relative to the traditional control strategy. The mean value of the speed is higher than that of the conventional control strategy.

Figure 3.29 shows the duty-cycle vector of the PMSM under the control of the advanced and control strategies. The mean values during the pre-period of the reference speed disturbance are 1.0 and 0.95 for the advanced control strategy and the conventional control strategy, respectively. However, the mean values during the post-period of the reference speed disturbance increase slightly to 1.03 and 0.96, respectively. It can be concluded that the advanced control strategy can increase the DC-link utilization by 5.27% and 7.8% for the pre-period and post-period of the reference speed disturbance relative to the traditional control strategy, respectively. This is why the mean value of the $d$-axis current of the advanced control strategy is higher than that of the traditional control strategy. The output power and efficiency of the advanced control strategy are higher than the traditional control strategy.

Figure 3.30 shows the per-unit voltage vector of the PMSM under the control of the advanced and traditional strategies. The mean values during the pre-period of the reference speed variation are 1 and 0.98 for the advanced control strategy and the traditional control strategy, respectively. The mean values do not change during the post-period of the reference speed variation.

It can be concluded that the advanced control strategy can increase the DC-link utilization by 2.0% relative to the traditional control strategy.

### 3.4.3 VALIDATING THE DYNAMIC PERFORMANCE OF THE ADVANCED CONTROL STRATEGY IN TORQUE DISTURBANCE CONTROL CASE

In this section, the DC-link voltages are increased to 400 V to break the speed limit of both regions of the constant torque and flux weakening of the previous case shown in Figures 3.26 to 3.30, as mentioned in Chapter 2. Thus, Figures 3.31 to 3.35 show the dynamic performance comparison between the advanced control strategy and the traditional control strategy.

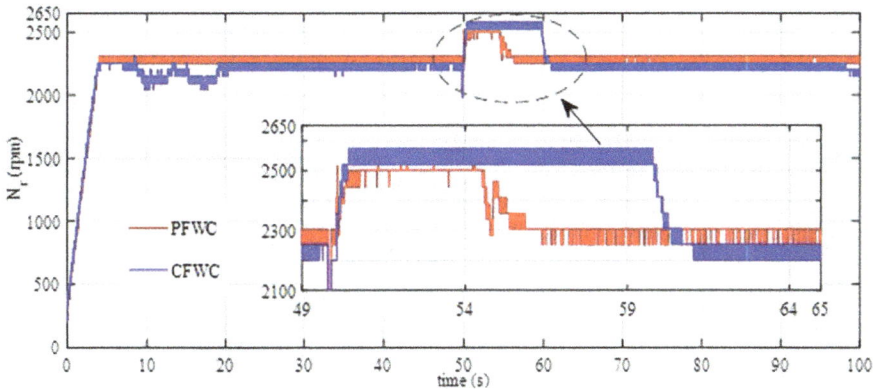

FIGURE 3.31    Measured speed under the control of both strategies in case of load torque disturbance from 3.5 to 0.5 Nm in the flux-weakening region with a reference speed of 2300 rpm.

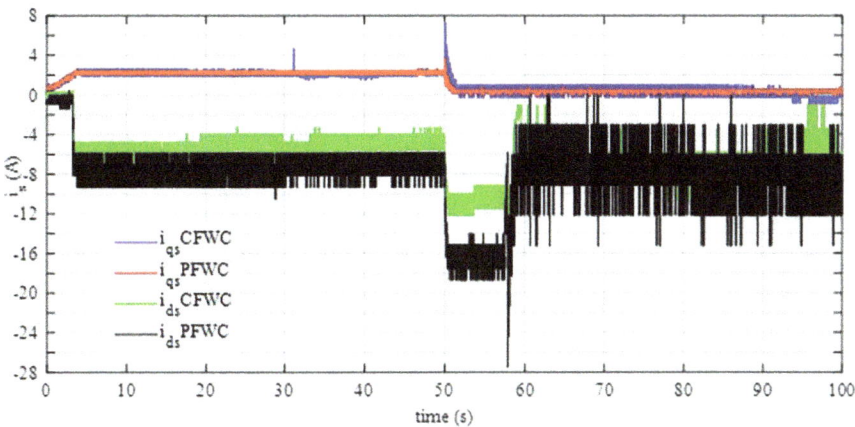

FIGURE 3.32    Measured $d$-$q$ axes stator currents under the control of both strategies in case of load torque disturbance from 3.5 to 0.5 Nm in the flux-weakening region with a reference speed of 2300 rpm.

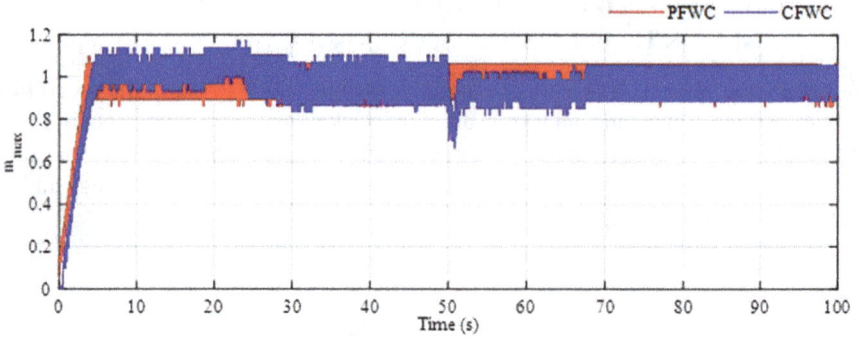

**FIGURE 3.33** The $m_{max}$ ratio under the control of both strategies in case of load torque disturbance from 3.5 to 0.5 Nm in the flux-weakening region with a reference speed of 2300 rpm.

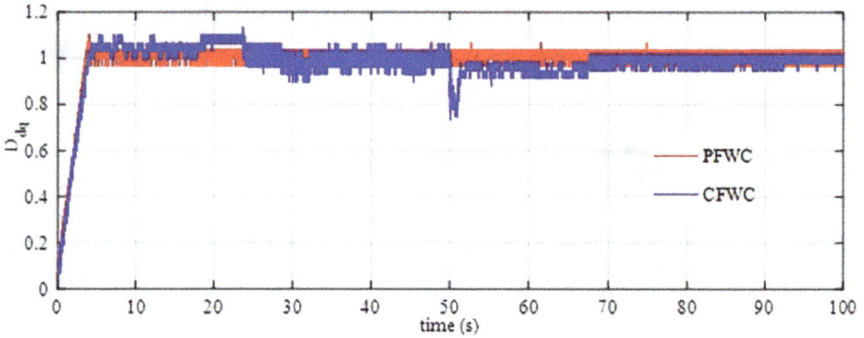

**FIGURE 3.34** The $D_{dq}$ magnitude under the control of both strategies in case of load torque disturbance from 3.5 to 0.5 Nm in the flux-weakening region with a reference speed of 2300 rpm.

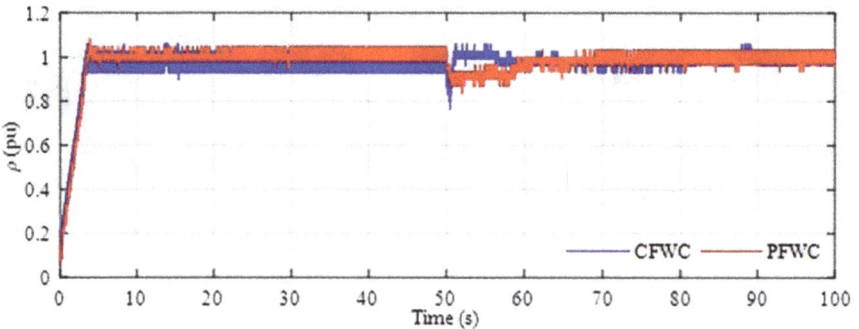

**FIGURE 3.35** The $\rho$ vector under the control of both strategies in case of load torque disturbance from 3.5 to 0.5 Nm in the flux-weakening region with a reference speed of 2300 rpm.

Additionally, the load torque is dropped suddenly from 3.5 to 0.5 Nm at 50 s while the control strategy tracks the reference speed of 2300 rpm.

Figure 3.31 shows two signals of the measured speed of the advanced and traditional strategies. It can be seen from this figure that the mean values of the speed of the advanced and traditional strategies are 2277 rpm and 2226.5 rpm for the pre-period and post-period of the load torque variation, respectively. As a result, the values of the steady-state error are slightly different by about 26 rpm and 26.5 rpm. There is also a speed drop in the speed trace of the traditional control strategy, indicating that the traditional control fails to track the reference speed and drops into the constant-torque region. This is because the required $d$-axis stator current is not efficient in this dropping period. The values of the overshoot during the load torque disturbance are 11.13% for the advanced control strategy and 12% for the traditional control strategy. Thus, the advanced control strategy can reduce the overshoot rate by about 6.9% relative to the traditional control strategy.

Figure 3.32 shows the measured waves of the $d$-$q$ axes of the stator currents for the advanced and traditional control strategies. It can be seen from this figure that the mean values of the $q$-axis stator current of the advanced and traditional control strategies are approximately the same at about 2.126 A for the pre-period from the load torque disturbance; they are 0.17 A and 0.32 A for the post-period from the load torque disturbance, respectively.

For comparison to the current ripple peak, the peak values of the $q$-axis current ripple for the pre-period from the load torque disturbance are 0.29 A for the advanced control strategy and 0.3 A for the conventional control strategy and 0.3 A and 0.58 A for the post-period from the load torque disturbance, respectively. Thus, the advanced control strategy can reduce the peak value of the $q$-axis current ripple relative to the conventional control strategy by about 4.32% and 47.9% for the pre-period and the post-period from the load torque disturbance, respectively. This reduction in the peak value yields less torque ripple and increases the life of the controlled PMSM.

To analyze the $d$-axis current, it can be seen from Figure 3.32 that the mean values of the $d$-axis stator current in the negative trajectory of the advanced and traditional strategies are -7.59 A and -5.9 A for the pre-period from the load torque disturbance, respectively and about -7.6 A and -7.1 A for the post-period from the load torque disturbance, respectively. Thus, the advanced control strategy can increase the mean value of the $d$-axis current relative to the conventional control strategy by about 28.6% and 7% for the pre-post and post-period from the load torque disturbance. Also, this increase in the mean value yields more output power, increasing efficiency and decreasing the copper losses of the controlled PMSM. For comparison to the current ripple peak, the peak values of the $d$-axis current ripple for the pre-period from the load torque disturbance are 1.76 A for the advanced control strategy and 1.12 A for the conventional control strategy.

Thus, the advanced control strategy can reduce the peak value of the $d$-axis current ripple relative to the conventional control strategy by about 57%. Also, this reduction in the peak value yields less torque ripple and increases the life of the controlled PMSM.

Figure 3.33 shows the maximum line modulation of the PMSM under the control of both the advanced and conventional strategies. It can be seen that the mean values during the pre-period of the torque disturbance are the same at about 0.98 for both strategies. However, the mean values during the post-period of the torque disturbance change to 0.99 and 0.97, respectively. It can be concluded that the advanced control strategy can increase the DC-link utilization by about 1% for the post-period of the torque disturbance relative to the traditional control strategy. As a result, the mean value of the speed is higher than that of the conventional control strategy.

Figure 3.34 shows the duty-cycle vector of the PMSM under the control of the advanced and the traditional strategies. The mean values during the pre-period of the torque disturbance are 1.0, and 0.98 for the advanced and conventional strategies, respectively. The mean values during the post-period of the torque disturbance do not change from those values of the pre-period. It can be concluded that the advanced control strategy can increase the DC-link utilization by 2.04%. This is why the mean value of the $d$-axis current of the advanced control strategy is higher than that of the traditional control strategy.

As a result, the outpower and efficiency of the advanced control strategy is higher than the traditional control strategy.

Figure 3.35 shows the per-unit voltage vector of the PMSM under the control of the advanced and traditional strategies. The mean values during the pre-period of the reference speed variation are 1.01 and 0.98 for the advanced and traditional strategies, respectively. The mean values do not change during the post-period of the reference speed variation. It can be concluded that the advanced control strategy can increase the DC-link utilization by 3.8% relative to the traditional control strategy.

## 3.5 SUMMARY

The primary objective of this chapter was to introduce a novel technique for adjusting the proportional and integral parameters of the many PI regulators that comprise the field-oriented control strategy. Additionally, the chapter introduced the anti-windup technique of the PI regulator structure. Finally, the flux-weakening control loop was improved based on the low-pass filter and the feedforward of the compensated duty cycles. The significant contributions of this work can be summarized as follows:

The proportional and integral parameters can be optimized corresponding to both the field-oriented control and the controlled PMSM models.

The PI gains of the regulators can be adjusted simultaneously without depending on the transfer function of any plants.

Additional parameters, such as the derivative controller parameter, can be included in the control strategy design for tuning purposes.

The proposed anti-windup model can reduce the signal overrun of the PI controller by controlling the integral controller's input.

The most significant contribution of this work is the development of a mathematical model for the cost function of the error matrix in terms of both the control strategy and the controlled PMSM models.

- The adaptive velocity of the proposed particle swarm optimization technique can significantly reduce the cost of the objective function.

- The anti-windup model can increase and maintain the mean value of the required d-axis current while the motor operates in the flux-weakening region.
- The proposed low-pass filter for the flux-weakening control loop leads to stable current because it can reduce the current oscillation by rejecting the high-frequency components of the reference duty cycle.
- The advanced control strategy can work independently of the machine parameters because the feedforward of the reference duty cycle can automatically set the flux level of the controlled PMSM.

Therefore, a theoretical study on the stability of the advanced drive system based on the optimal PI parameters is conducted, and relevant Bode diagrams with a high resolution are presented to control the motor speed and current. A performance comparison is analyzed relative to the conventional control strategy to validate the effectiveness of the dynamic performance of the advanced control strategy, which operates more stably in the flux-weakening region independent of the machine parameters. Meanwhile, the overshoot of the speed signal is lower than that of the conventional control strategy. The results indicate that the speed can shift smoothly between both regions of constant torque and constant power. Finally, the advanced control strategy can maximize the motor speed more than the conventional control strategy because it can utilize a higher mean value of the DC-link voltage to operate in the flux-weakening region.

## BIBLIOGRAPHY

[1]  K.-C. Kim. A novel magnetic flux weakening method of permanent magnet synchronous motor for electric vehicles. IEEE Transactions on Magnetics, October 2012, 48(11):4042–4045.

[2]  W. Wang, J. Zhang, and M. Cheng. Line-modulation-based flux-weakening control for permanent-magnet synchronous machines. IET Power Electronics, January 2018, 11(5):930–936.

[3]  S. Ahmed, Z. Shen, P. Mattavelli, D. Boroyevich, and K. J. Karimi. Small-signal model of voltage source inverter (VSI) and voltage source converter (VSC) considering the deadtime effect and space vector modulation types. IEEE Transactions on Power Electronics, June 2017, 32(6):4145–4156.

[4]  H. Bai, X. Wang, F. Blaabjerg, and P. C. Loh. Harmonic analysis and mitigation of low-frequency switching voltage source inverter with auxiliary VSI. IEEE Journal of Emerging and Selected Topics in Power Electronics, September 2018, 6(3):1355–1365.

[5]  B. Hoff and W. Sulkowski. Grid-connected VSI with LCL filter models and comparison. IEEE Transactions on Industry Applications, June 2014, 50(3):1974–1981.

[6]  K. Tabarraee, J. Iyer, H. Atighechi, and J. Jatskevich. Dynamic average-value modeling of 120° VSI-commutated brushless DC motors with trapezoidal back EMF. IEEE Transactions on Energy Conversion, June 2012, 27(2):296–307.

[7]  C. Cecati, A. O. D. Tommaso, F. Genduso, R. Miceli, and G. R. Galluzzo. Comprehensive modeling and experimental testing of fault detection and management of a non-redundant fault-tolerant VSI. IEEE Transactions on Industrial Electronics, June 2015, 62(6):3945–3954.

[8]  Yang, H., Lyu, S., Lin, H., Zhu, Z.Q., Peng, F., Zhuang, E., Fang, S. and Huang, Y. Stepwise magnetization control strategy for DC-magnetized memory machine. IEEE

Transactions on Industrial Electronics, June 2019, 66(6):4273–4285.

[9] Y. Jiang, W. Xu, C. Mu, and Y. Liu. Improved deadbeat predictive current control combined sliding mode strategy for PMSM drive system. IEEE Transactions on Vehicular Technology, January 2018, 67(1):251–263.

[10] H. Liu, Z. Zhu, E. Mohamed, Y. Fu, and X. Qi. Flux-weakening control of nonsalient pole PMSM having large winding inductance, accounting for resistive voltage drop and inverter nonlinearities. IEEE Transactions on Power Electronics, June 2011, 27(2): 942–952.

[11] N. Pragallapati, T. Sen, and V. Agarwal. Adaptive velocity PSO for global maximum power control of a PV array under nonuniform irradiation conditions. IEEE Journal of Photovoltaics, March 2017, 7(2):624–639.

[12] W. Wang, M. Cheng, B. Zhang, Y. Zhu, and S. Ding. A fault-tolerant permanent-magnet traction module for subway applications. IEEE Transactions on Power Electronics, April 2014, 29(4):1646–1658.

[13] C.-M. Ong. Dynamic simulation of electric machinery: using MATLAB/SIMULINK, Prentice Hall, November 1998.

[14] M. Tursini, E. Chiricozzi, and R. Petrella. Feedforward flux-weakening control of surface-mounted permanent-magnet synchronous motors accounting for resistive voltage drop. IEEE Transactions on Industrial Electronics, January 2010, 57(1):440–448.

[15] K. Tian, J. Wang, B. Wu, Z. Cheng, and N. R. Zargari. A virtual space vector modulation technique for the reduction of common-mode voltages in both magnitude and third-order component. IEEE Transactions on Power Electronics, March 2015, 31(1): 839–848.

[16] G. Feng, C. Lai, and N. C. Kar. A novel current injection-based online parameter estimation method for PMSMs considering magnetic saturation. IEEE Transactions on Magnetics, July 2016, 52(7):1–4.

# 4 Design and Optimization of Stator Current Regulators for Surface-Mounted Permanent Magnet Synchronous Motor Drives

## 4.1 INTRODUCTION

To illustrate the proposed optimal design, this chapter presents a novel cost function that includes three tasks: minimizing the controller's error, maximizing the control stability range, and considering the control stability limit. The error matrix for this advanced control strategy is deduced similarly to the previous chapter. However, the control stability in the stability range is achieved based on calculating the real eigenvalue of the state matrix of this control strategy and the PMSM model. The final part of this cost function is the control stability index to determine whether there are positive or zero values in the calculated eigenvalue. To illustrate the frequency-modulating algorithm, the frequency of the pulse-width modulation technique can be varied linearly with the predicted torque ripple and the required torque ripple. The proportional-integral gains of this advanced control strategy can be determined by minimizing the cost function based on the genetic optimization algorithm.

To observe the reduction in the peak for the induced torque ripple, the flux observer is employed based on the third-order generalized integral transfer function to estimate the fluxes and calculate the induced torque. The required negative stator current of the d-axis is calculated by the same outer control loop as in the previous chapter, based on the reference duty cycle's feedforward to avoid the current regulator's saturation limit.

The simulation and experimental results include aspects of the robustness issues of this driving system compared to the classical drive system. The results indicate that this driving algorithm can efficiently control the motor speed and stator current in the field-weakening range while reducing switching and copper losses compared to the classical control algorithm.

Table 4.1 provides a comparison between this advanced control algorithm and the classical algorithm, where the conventional control algorithm is the same algorithm discussed in Chapter 2 in the section on frequency modulation technique for torque ripple reduction. The proportional-integral (PI) parameters of this advanced control strategy are modified based on the fitness function, which can be considered an improvement to the fitness function as explained in Chapter 3. The difference lies in

DOI: 10.1201/9781003320128-4

**TABLE 4.1**

**Contrasts between the New and Conventional Driving Strategies**

| Items | New Algorithm | Classical Algorithm |
|---|---|---|
| Power inverter switching frequency is changed during motor operation (discussed in Chapter 2) | true | false |
| Parameters of PI regulators are optimized | true | false |
| Field-weakening operation is based on duty cycle's feedforward | true | false |

the q-axis current regulator, that is, the ability to control the region of the motor in which the field decreases, thus reducing the error between the measured value of the current and the reference q-axis value. The PI parameters of the traditional control strategy are modified using the trial-and-error algorithm, which involves changing the proportional gain until an acceptable transient period is obtained and changing the integral gain until an acceptable overshoot and steady-state error are obtained. The following sections present the improved torque ripple control strategy, which solves the implementation of the genetic algorithm (GA) optimization technique, and a discussion of the results.

The necessary points related to the control algorithm presented in this chapter can be summarized as follows:

- The control algorithm introduced in this chapter presents an improved cost function that can optimize the proportional-integral parameters of any traditional or modern field-oriented control (FOC) strategy offline or online. The contribution of this cost function is the driving system stability index, which is added with a decreasing procedure for the error values of the PI regulators to determine the optimal corresponding PI gains.
- The control algorithm does not increase the sample time of the selected digital control unit and has no further calculations except for the calculation related to the lookup table of the predicted peak of the stator current ripple.

## 4.2 COMBINED STRATEGY DESIGN FOR TORQUE RIPPLE REDUCTION OF SURFACE-MOUNTED PMSM DRIVE IN FLUX-WEAKENING OPERATION

Figure 4.1 shows the new drive system of a surface-mounted PMSM to reduce the torque peak ripple across the flux-weakening region. It is clear from this figure that the reference stator voltage is calculated using the same field-oriented control strategy used in Chapter 3, based on the outer loop of the duty-cycle vector.

The voltage source inverter is based on a variable upgraded frequency space-vector modulation technique, as shown in Figure 4.2. The six pulses are supplied to a two-level inverter. The inputs of the control strategy can be summarized as follows:

(1) The counted mechanical position of the PMSM rotor
(2) The three-phase stator currents of the controlled PMSM
(3) The DC voltage of the inherent inverter

This control strategy is mainly based on the previously known parameters of proportional-integral regulators. These parameters are set as a final decision of the offline optimization procedure that relies on a genetic algorithm optimization technique. The optimal values of the PI parameters are obtained based on the new objective function considered in this chapter.

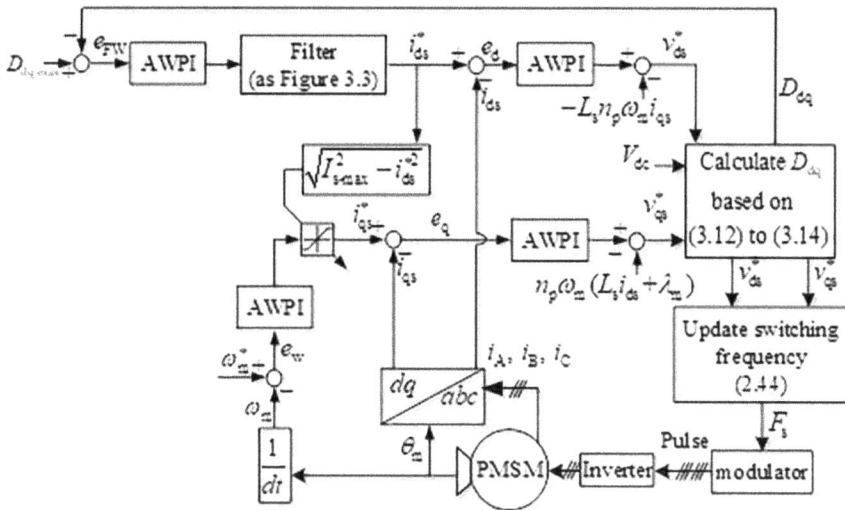

FIGURE 4.1   Combined algorithm for the ripple reduction of the stator current and torque.

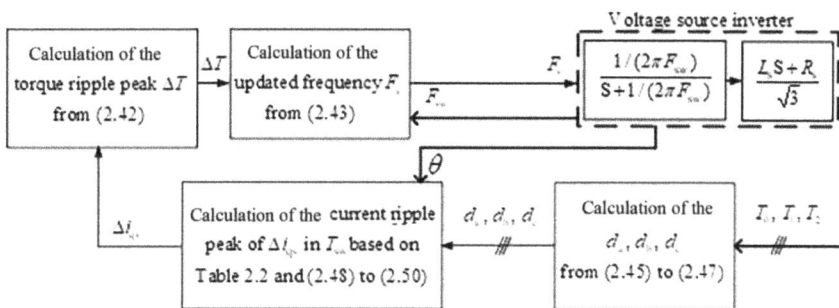

FIGURE 4.2   Calculation algorithm of the updated switching frequency.

The reference value of a $d$-axis stator current required to automatically set the reduction level of the rotor flux is calculated based on the outer flux loop of the duty-cycle vector. The limit reference value of the $q$-axis stator current can now be determined to calculate the compensated stator voltages of the $d$-$q$ axes that will be modulated using variable switching frequency.

### 4.2.1 Flux-Weakening Controller

FOC strategy cannot drive a PMSM over the base speed that is determined by the rated stator voltage and rated stator current because the back emf is greater than the terminal voltage. This problem can be solved by increasing the reference stator current of the $d$-axis in a negative direction, which reduces the rotor field linkage. It can be seen in Figure 4.1 that the PI regulator of the outer control loop for the required stator current of the $d$-axis depends on the feedback of the duty-cycle vector compared to its maximum value of a unit and can be defined as

$$\begin{vmatrix} d_{ds} \\ d_{qs} \end{vmatrix} = \frac{1}{V_{dc}} \begin{vmatrix} V_{ds}^* \\ V_{qs}^* \end{vmatrix} \tag{4.1}$$

where $d_{ds}$ and $d_{qs}$ $\varepsilon$ [-1 1]. This outer loop can be activated if the duty-cycle vector is greater than or equal to one. Otherwise, the required current of the $d$-axis is set to zero. Lastly, a low-pass filter is provided beyond the field PI regulator to cancel the high-frequency noise coming from the comparator. This has a positive impact on keeping the mean value of the $d$-axis current high continuously.

### 4.2.2 Modified Cost Function for PI Parameter Optimization Ensuring System Stability

The model of the control loops of the speed, flux, and $d$-$q$ axes currents for the proposed control strategy by neglecting the anti-windup technique can be represented as

$$e_\omega = \frac{i_{qs}^*}{K_{P\text{-}\omega}} - \frac{K_{I\text{-}\omega} E_\omega}{K_{P\text{-}\omega}} \tag{4.2}$$

$$e_{FW} = \frac{i_{ds}^*}{K_{P\text{-}FW}} - \frac{K_{I\text{-}FW} E_{FW}}{K_{P\text{-}FW}} \tag{4.3}$$

$$e_d = \frac{R_s i_{ds}}{K_{P\text{-}q}} + \frac{d}{dt} \frac{L_s i_{ds}}{K_{P\text{-}q}} - \frac{2n_p L_s i_{qs} \omega_r}{K_{P\text{-}q}} - \frac{K_{I\text{-}q} E_d}{K_{P\text{-}q}} \tag{4.4}$$

$$e_q = \frac{R_s i_{qs}}{K_{P\text{-}q}} + \frac{d}{dt} \frac{L_s i_{qs}}{K_{P\text{-}q}} + \frac{2n_p L_s i_{ds} \omega_r}{K_{P\text{-}q}} + \frac{2n_p \lambda_m \omega_r}{K_{P\text{-}q}} - \frac{K_{I\text{-}q} E_q}{K_{P\text{-}q}} \tag{4.5}$$

Recall that the state-space representation of the linear model of the d-q axes for the surface-mounted PMSM in (2.1) to (2.12) can be illustrated as shown when the non-linear cross-coupling of $\omega_r i_{ds}$ and $\omega_r i_{qs}$ is omitted:

$$\frac{1}{dt}\begin{vmatrix} i_{ds} \\ i_{qs} \\ \omega_r \end{vmatrix} = \begin{vmatrix} \dfrac{-R_s}{L_s} & .....0 & .....0 \\ 0 & \dfrac{-R_s}{L_s} & \dfrac{-\lambda_m n_p}{L_s} \\ 0 & \dfrac{1.5 n_p \lambda_m}{J} & \dfrac{-B_v}{J} \end{vmatrix}\begin{vmatrix} i_{ds} \\ i_{qs} \\ \omega_r \end{vmatrix} + \begin{vmatrix} \dfrac{1}{L_s} & 0 \\ 0 & \dfrac{1}{L_s} \\ 0 & 0 \end{vmatrix}\begin{vmatrix} v_{ds} \\ v_{qs} \end{vmatrix} + \begin{vmatrix} 0 \\ 0 \\ \dfrac{-1}{J} \end{vmatrix}|T_L| \qquad (4.6)$$

The linearized d-q axes model of the control strategy and the PMSM can be deduced based on a Taylor series expansion, which is defined as

$$f'(\delta x_1, \delta y_1) = f(X_{10}, Y_{10}) + \left(\frac{\partial f}{\partial x_1}\right)\bigg|_{X_{10}} \delta x_1 + \left(\frac{\partial f}{\partial y_1}\right)\bigg|_{Y_{10}} \delta y_1 \qquad (4.7)$$

where $x_1$ and $y_1$ are general state variables in a plant model. The symbols $0$ and $\delta$ indicate an equilibrium point value and a small signal value of state variables. Thus, the linearized model of the d-q axes for both the proposed control strategy and the controlled PMSM can be deduced as

$$S\,\delta i_{ds} = -\frac{R_s}{L_s}\delta i_{ds} + n_p \omega_{r0}\delta i_{qs} + n_p i_{qs0}\delta\omega_r + \frac{1}{L_s}\delta v_{ds} \qquad (4.8)$$

$$S\,\delta i_{qs} = n_p \omega_{r0}\delta i_{ds} - \frac{R_s}{L_s}\delta i_{qs} + \frac{-\lambda_m n_p}{L_s}\delta\omega_r + n_p i_{ds0}\delta\omega_r + \frac{1}{L_s}\delta v_{qs} \qquad (4.9)$$

$$S\,\delta\omega_r = \frac{1.5 n_p \lambda_m}{J}\delta i_{qs} - \frac{B_v}{J}\delta\omega_r - \frac{1}{J}\delta T_L \qquad (4.10)$$

$$S\,\delta E_\omega = \frac{1}{K_{P\text{-}\omega}}\delta i_{qs}^* - \frac{K_{I\text{-}\omega}}{K_{P\text{-}\omega}}\delta E_\omega \qquad (4.11)$$

$$S\,\delta E_{fw} = \frac{1}{K_{P\text{-}fw}}\delta i_{ds}^* - \frac{K_{I\text{-}fw}}{K_{P\text{-}fw}}\delta E_{fw} \qquad (4.12)$$

$$S\,\delta E_d = R_s\delta i_{ds} + S\frac{L_s}{K_{P\text{-}q}}\delta i_{ds} - \frac{2 n_p L_s \omega_{r0}}{K_{P\text{-}q}}\delta i_{qs} - \frac{2 n_p L_s i_{qs0}}{K_{P\text{-}q}}\delta\omega_r \qquad (4.13)$$

$$S \, \delta E_q = \frac{2n_p L_s \omega_{r0}}{K_{P\text{-}q}} \delta i_{ds} + \frac{R_s}{K_{P\text{-}q}} \delta i_{qs} + S \, \frac{L_s}{K_{P\text{-}q}} \delta i_{qs} + \frac{2n_p L_s i_{ds0}}{K_{P\text{-}q}} \delta \omega_r + \frac{2n_p \lambda_m}{K_{P\text{-}q}} \delta \omega_r \quad (4.14)$$

Thus, the cost function considered in this chapter can be formulated from the discussed linearized model. Meanwhile, the design steps of the cost function can be illustrated as follows:

(1) Generally, the stator current of the $q$-axis ripple causes a torque ripple for the surface-mounted PMSM, as mentioned in Chapter 2. For this reason, the PI parameters for the $d$-axis current controller are the same as those of the $q$-axis current controller in (4.13) and (4.14). The proposed cost function focuses on optimizing the PI parameters of the current controller of the $q$-axis rather than the other parameters by minimizing its error to zero. This is achieved by using the concept of the weight of each term of the cost function. The weight factor of the error for the $q$-axis stator current controller is greater than that of the other terms. The first release of the cost function is defined as follows:

$$F_C = e_{q0} + 0.5 \, e_{\omega 0} + 0.5 \, e_{FW0} \quad (4.15)$$

Based on the linearized model of the $d$-$q$ axes of the drive system of (4.8) through (4.14), the terms of the equilibrium points of (4.15) can be defined as follows:

$$e_{q0} = \frac{1}{K_{P\text{-}q}} (R_s i_{qs0} + 2n_p (L_s i_{ds0} + \lambda_m) \omega_{r0} - K_{I\text{-}q} E_{q0}) \quad (4.16)$$

$$e_{\omega 0} = \frac{1}{K_{P\text{-}\omega}} (\delta i_{qs0}^* - K_{I\text{-}\omega} E_{\omega 0}) \quad (4.17)$$

$$e_{FW0} = \frac{1}{K_{P\text{-}FW}} (\delta i_{ds0}^* - K_{I\text{-}FW} E_{FW0}) \quad (4.18)$$

(2)  The further contribution of the cost function considered in this chapter relative to that of Chapter 3 is the stability of the new drive system. A further term is added to the cost function to ensure that the selected PI parameters obtain a high stability limit. This term is evaluated by selecting PI parameters that cause a high real eigenvalue in the negative trajectory. Thus, the second release of the cost function can be updated as

$$F_C = (e_{q0} + 0.5 \, e_{\omega 0} + 0.5 \, e_{FW0}) + |max[real(\lambda_j)]|^{-1} \quad (4.19)$$

where $\lambda_j$ is the eigenvalue that can be calculated as follows:

$$det\left( A - \begin{vmatrix} 1 & 0 & 0 & 0 & 0 & 0 & 0 \\ 0 & 1 & 0 & 0 & 0 & 0 & 0 \\ 0 & 0 & 1 & 0 & 0 & 0 & 0 \\ 0 & 0 & 0 & 1 & 0 & 0 & 0 \\ 0 & 0 & 0 & 0 & 1 & 0 & 0 \\ 0 & 0 & 0 & 0 & 0 & 1 & 0 \\ 0 & 0 & 0 & 0 & 0 & 0 & 1 \end{vmatrix} \right) = 0.0 \tag{4.20}$$

where $A$ is the state matrix of the drive system that can be obtained from the state-space representation as follows:

$$S\,x(S) = A(7 \times 7)\,x(S) + B(7 \times 2)\,u(S) + \rho(S) \tag{4.21}$$

where $u, \rho, x, B,$ and $A$ can be defined as

$$u(S) = \begin{bmatrix} \delta v_{ds} & \delta v_{qs} \end{bmatrix}^T \tag{4.22}$$

$$\rho(S) = \begin{bmatrix} 0 & 0 & \rho_{31} & \rho_{41} & \rho_{51} & \dfrac{d}{dt}\rho_{61} & \dfrac{d}{dt}\rho_{71} \end{bmatrix}^T \tag{4.23}$$

$$x(S) = \begin{bmatrix} \delta i_{ds} & \delta i_{qs} & \delta\omega_r & \delta E_\omega & \delta E_{fw} & \delta E_d & \delta E_q \end{bmatrix}^T \tag{4.24}$$

$$B(7 \times 2) = \begin{bmatrix} \dfrac{1}{L_s} & 0 & & & & & \\ L_s & 1 & 0 & 0 & 0 & 0 & 0 \\ 0 & \dfrac{1}{L_s} & 0 & 0 & 0 & 0 & 0 \end{bmatrix}^T \tag{4.25}$$

$$A(7 \times 7) = \begin{bmatrix} a_{11} & a_{12} & a_{13} & 0 & 0 & 0 & 0 \\ a_{21} & a_{22} & a_{23} & 0 & 0 & 0 & 0 \\ 0 & a_{32}/J & a_{33}/J & 0 & 0 & 0 & 0 \\ 0 & 0 & 0 & a_{44} & 0 & 0 & 0 \\ 0 & 0 & 0 & 0 & a_{55} & 0 & 0 \\ a_{61} & a_{62} & a_{63}/K_{P\text{-}q} & 0 & 0 & a_{66} & 0 \\ a_{71} & a_{72} & a_{73}/K_{P\text{-}q} & 0 & 0 & 0 & a_{77} \end{bmatrix} \tag{4.26}$$

where the coefficients of the matrix $A$ can be defined as

$$a_{11} = -\frac{R_s}{L_s} \tag{4.27}$$

$$a_{12} = n_p \omega_{r0} \tag{4.28}$$

$$a_{13} = n_p i_{qs0} \tag{4.29}$$

$$a_{21} = n_p \omega_{r0} \tag{4.30}$$

$$a_{22} = -\frac{R_s}{L_s} \tag{4.31}$$

$$a_{23} = n_p i_{ds0} - \frac{\lambda_m n_p}{L_s} \tag{4.32}$$

$$a_{32} = 1.5 n_p \lambda_m \tag{4.33}$$

$$a_{33} = -B_v \tag{4.34}$$

$$a_{44} = -\frac{K_{I\text{-}\omega}}{K_{P\text{-}\omega}} \tag{4.35}$$

$$a_{55} = -\frac{K_{I\text{-}FW}}{K_{P\text{-}FW}} \tag{4.36}$$

$$a_{61} = \frac{R_s}{K_{P\text{-}q}} \tag{4.37}$$

$$a_{62} = -\frac{2 n_p L_s \omega_{r0}}{K_{P\text{-}q}} \tag{4.38}$$

$$a_{63} = -2 n_p L_s i_{qs0} \tag{4.39}$$

$$a_{66} = -\frac{K_{I\text{-}q}}{K_{P\text{-}q}} \tag{4.40}$$

$$a_{71} = \frac{2n_p L_s \omega_{r0}}{K_{P\text{-}q}} \tag{4.41}$$

$$a_{72} = \frac{R_s}{K_{P\text{-}q}} \tag{4.42}$$

$$a_{73} = 2n_p L_s i_{ds0} + 2n_p \lambda_m \tag{4.43}$$

$$a_{77} = -\frac{K_{I\text{-}q}}{K_{P\text{-}q}} \tag{4.44}$$

$$\rho_{31} = -\frac{1}{J}\delta T_L \tag{4.45}$$

$$\rho_{41} = \frac{1}{K_{P\text{-}\omega}}\delta i_{qs}^* \tag{4.46}$$

$$\rho_{51} = \frac{1}{K_{P\text{-FW}}}\delta i_{ds}^* \tag{4.47}$$

$$\rho_{61} = \frac{L_s}{K_{P\text{-}q}}\delta i_{ds} \tag{4.48}$$

$$\rho_{71} = \frac{L_s}{K_{P\text{-}q}}\delta i_{qs} \tag{4.49}$$

The second further term for ensuring stability is the penalty function. This is designed to reject the selected PI parameters that cause the zero and positive eigenvalues. Thus, the stability of the drive system cannot be unstable or marginally stable. The penalty function ($P$) can be defined as follows:

If
$$P = \infty \text{ at } max[real(\lambda_j)] \ge 0 \tag{4.50}$$

otherwise,
$$P = 0 \text{ .} \tag{4.51}$$

The final release of the cost function can be written as

$$F_C = (e_{q0} + 0.5\, e_{\omega 0} + 0.5\, e_{FW0}) + \left| max[real(\lambda_j)] \right|^{-1} + P(\lambda_j) \tag{4.52}$$

### 4.2.3 Genetic Optimization Algorithm and Implementation of the Drive System

The variables of the control system are optimized as an optimization problem into a certain search region depending on the artificial intelligence algorithms. Optimization genetic algorithms from the artificial intelligence strategies perform the optimization procedure of the convex and non-convex cost functions. In this chapter, the cost function presented in (4.52) has six manipulated variables (PI parameters) for the proposed control strategy. A genetic optimization algorithm aims to upgrade systematically, and initial population is based on the evolutionary concepts of natural selection and genetics programming. Therefore, a genetic algorithm was preferred to a particle swarm optimization technique due to its efficiency in optimizing the manipulated variables into a wide search space [1, 2]. The steps of a genetic optimization algorithm are as follows:

(1) Create the initial solutions for the manipulated variable.
(2) Compute the objective function.
(3) Evaluate the search regions to get better solutions within them.
(4) Create new offspring.
(5) Update new offspring.

The genetic algorithm used in this chapter to confirm the proposed cost function is summarized as follows:

(1) The algorithm randomly selects 100 initial populations respecting the determined upper and lower limits. A single particle of the population ($\hat{p}_n$) has a length of six of the PI parameters and can be written as

$$\hat{p}_n (1 \times 6) = \left[ K_{I\text{-}\omega}, K_{P\text{-}\omega}, K_{I\text{-}q}, K_{P\text{-}q}, K_{I\text{-}fw}, K_{P\text{-}fw} \right] \tag{4.53}$$

However, the genetic algorithm can consider each particle as a chromosome of a list of ones and zeros that can be written as

$$Chr = \sum_{i=0}^{40} \text{rand}(0,1) \, 2^i \tag{4.54}$$

where *Chr* stands for chromosome. As a result, this chromosome would be translated to a number in the range between the determined upper and lower limits by the equation

$$MBR = V_{LOW} + Chr \, (V_{UP} / (2^{40} - 1)) \tag{4.55}$$

where MBR is the particle member of $\hat{p}_n$ and $V_{LOW}$ and $V_{UP}$ are the lower and upper limits of these members, respectively. However, the saturated value in the optimization problems can be defined by

$$V_{LOW} = [0\ 0\ 0\ 0\ 0\ 0] \tag{4.56}$$

$$V_{UP} = [50\ 1\ 500\ 100\ 1000\ 50] \tag{4.57}$$

(2) The genetic algorithm evaluates the objective function (4.52) to calculate the current population. This evaluation step contains a linear system of equations each time, which can be solved by the Runge-Kutta strategy. The optimization procedure considers the linear system of equations when the cost function confirms the chromosome of the current population. As a result, the state variables become $E_{\omega 0}$, $E_{fw0}$, and $E_{q0}$.

(3) The particle probabilities $(P_n)$ are calculated by

$$P_n = \frac{f_n}{\sum_{N=1}^{100} f_N} \tag{4.58}$$

where $f_n$ refers to the particle $\hat{p}_n$ evaluation. The genetic algorithm selects the two best particles concerning the computed evaluation value of each particle and their selection chance.

(4) A genetic algorithm selects the parents of the chromosomes one by one to generate new offspring. A random junction bound is chosen to apply the crossover procedure in which the new particles can be produced by dividing each chromosome into two slits, then merging the broken parts from the front of one chromosome to the back of the other one, and so on.

(5) The new offspring is inverted or flipped from 0 to 1 and vice versa at each location to increase the mutation probability. Thus, the best particle in the population can be obtained by the new distribution.

(6) The determined limits the of $V_{LOW}$ and $V_{UP}$ bounds select between the new offspring and then position them in the new population.

(7) The population obtained can be utilized in an additional execution of the genetic algorithm. The previous steps are repeated, and the constraints are the same.

(8) The genetic optimization algorithm repeats all the discussed steps, seeking the optimal manipulated variables until the stop condition is achieved. The stop condition occurs when the summation of error of the cost function is not changed for 12 iterations, or the total number of generations reaches 100. If the stop criterion is reached, the optimization procedure returns the optimal selected PI parameters fitted by the modified cost function. Figure 4.3 shows the steps of the genetic algorithm implementation.

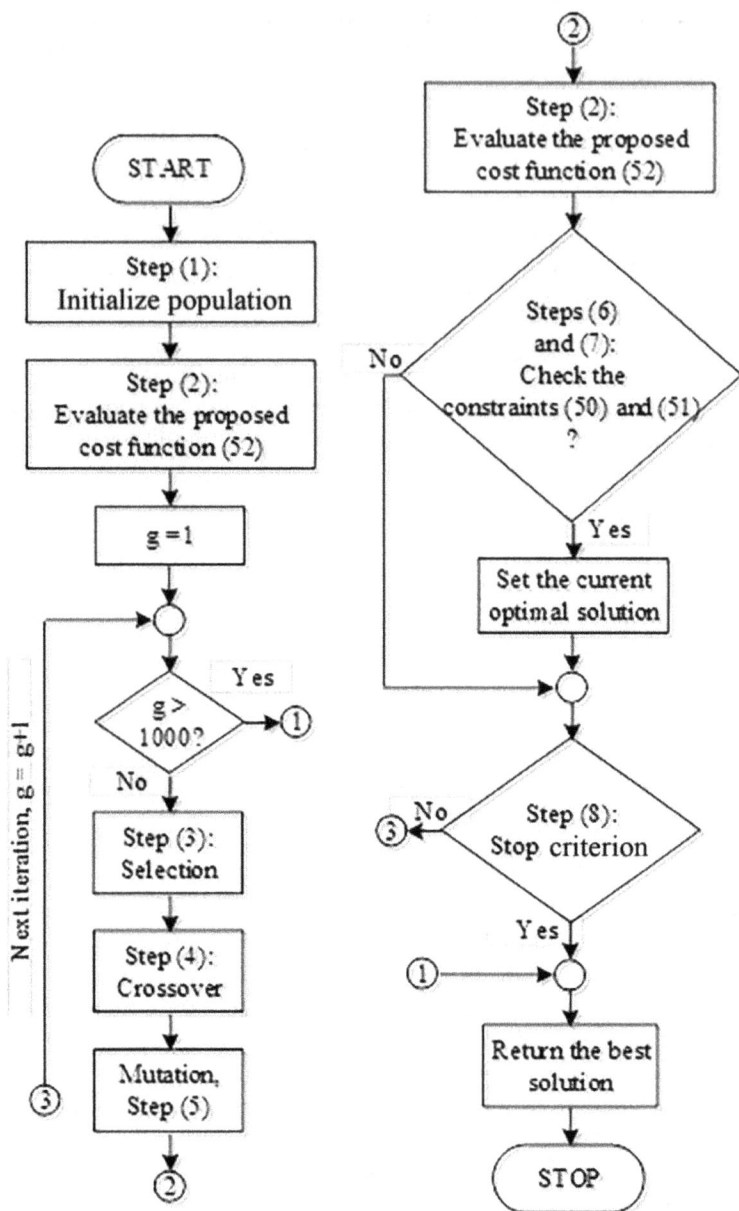

**FIGURE 4.3**　Implementing steps of the genetic algorithm.

### 4.2.4　Induced Torque Calculation Based on Third-Order Generalized Integral Flux Observer

The torque peak ripples of the surface-mounted PMSM in (2.41), (2.42), and (2.43) are not linear proportional with the ripple of the stator current of the $q$-axis. This is because of a certain harmonic of the motor structure, a harmonic due to the feed-back loop control, and an error of the orientation angle. The induced torque calculation is based on the flux estimation of either a first-order observer or a second-order observer. The DC offset and AC unwanted component waves affect a first-order observer due to mismatching machine parameters and the nonlinear behavior of the inherent control system drive. Thus, a third-order generalized integral flux observer is considered in this chapter to reduce the root mean square value of the unwanted components and DC offset.

The performance of the third-order generalized integrator (TOGI) observer was proven in [3] relative to the first- and second-order observers of generalized integrators in terms of the back-emf analysis of controlled PMSM by a Fourier series expansion. Meanwhile, the TOGI observer parameters are adjusted based on frequency analysis of the bode diagrams. This can be used to calculate the induced torque across the machine parameters mismatching and overall speed regions.

The estimation of the stator flux linkage of the PMSM in the $\alpha$-$\beta$ axes frame can be written as

$$\begin{vmatrix} \lambda_{\alpha s\text{-est}} \\ \lambda_{\beta s\text{-est}} \end{vmatrix} = \int \left( \begin{vmatrix} V^*_{\alpha s} \\ V^*_{\beta s} \end{vmatrix} - R_s \begin{vmatrix} i_{\alpha s} \\ i_{\beta s} \end{vmatrix} \right) dt \tag{4.59}$$

where $\alpha$ and $\beta$ stand for the stationary reference frame of $\alpha$-$\beta$ axes, and *est* indicates the estimated value. The electromagnetic torque can be determined based on the estimated flux linkages, as shown in (4.60):

$$T_{e\text{-TOGI}} = 1.5 n_p \left( \lambda_{\alpha s\text{-est}} i_{\beta s} - \lambda_{\beta s\text{-est}} i_{\alpha s} \right) \tag{4.60}$$

where $T_{e\text{-TOGI}}$ indicates the calculated induced torque. Thus, the discrete transfer function of the estimated stator fluxes of the $\alpha$-$\beta$ axes in terms of the TOGI observer can be written as [3]

$$\frac{\lambda_{\alpha s\text{-est}}}{V_{\alpha s}} = \frac{2W_0 \left( 1 - z^{-1} - z^{-2} + z^{-3} \right)}{1 + W_1 z^{-1} + W_2 z^{-2} + W_3 z^{-3}} \tag{4.61}$$

$$\frac{\lambda_{\beta s\text{-est}}}{V_{\beta s}} = \frac{\omega_c T_s W_0 \left( 1 - z^{-1} - z^{-2} + z^{-3} \right)}{1 + W_1 z^{-1} + W_2 z^{-2} + W_3 z^{-3}} \tag{4.62}$$

Lastly, the discrete implementation of TOGI is described in Figure 4.4, where $\omega_c$ is the center frequency, and $W_1$, $W_2$, $W_3$, and $W_0$ are integrator coefficients. References [3] and [4] provide additional information on TOGI transfer functions and how to define them.

126                     Permanent Magnet Synchronous Machines and Drives

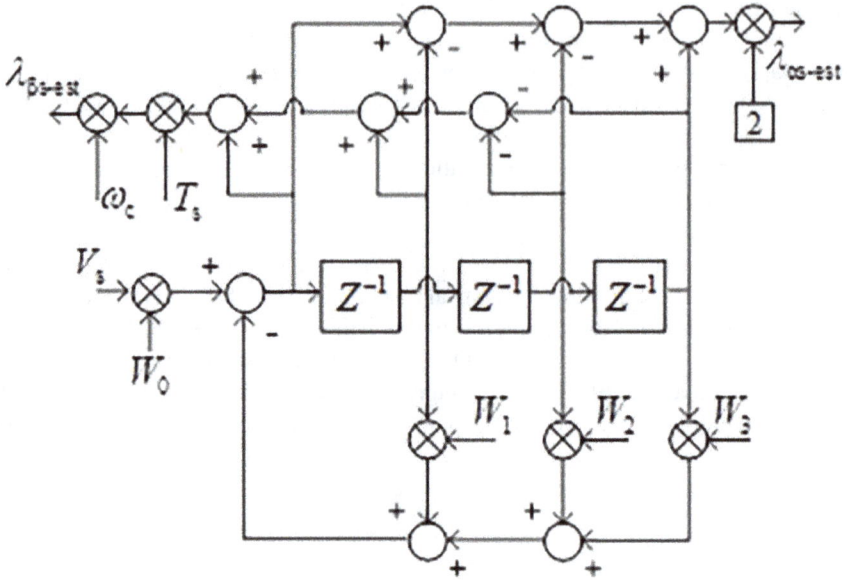

**FIGURE 4.4**  Discrete implementation of the TOGI flux observer [3].

## 4.3 SIMULATION RESULTS ANALYSIS

### 4.3.1 EVALUATION THE COST FUNCTION

As mentioned before, the optimum PI parameters of the control technique must be known in advance based on the offline optimization technique. This section shows the evaluation result of the cost function with the help of MATLAB simulation software. The equilibrium points are selected from the flux-weakening range. (The machine parameters and that of the inherent voltage source inverter are listed in Table 2.3 in Chapter 2.) Figure 4.5 presents the less-fitness evaluation and the corresponding iteration obtained in it. The minimum cost value is 0.0272661 and the total generation number is 51 iterations. The selected PI parameters are listed in Table 4.2.

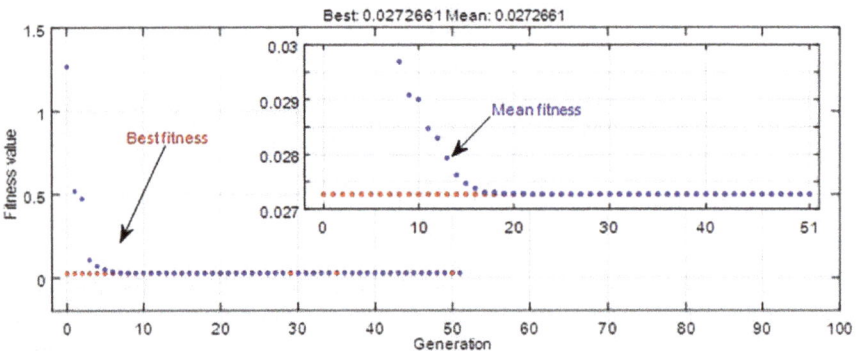

**FIGURE 4.5**  Cost function evaluation based on genetic optimization technique.

**TABLE 4.2**
**Optimal Selected PI Parameters Based on Genetic Algorithm Optimization Procedure**

| Outer speed controller | $K_{\text{I-}\omega}$ | 14.7113 |
|---|---|---|
| | $K_{\text{P-}\omega}$ | 0.1015 |
| $d$-$q$ axes current controllers | $K_{\text{I-q}}$ | 189.5323 |
| | $K_{\text{P-q}}$ | 61.7429 |
| Outer flux controller | $K_{\text{I-fw}}$ | 491.4330 |
| | $K_{\text{P-fw}}$ | 12.8351 |

### 4.3.2 VERIFY THE STABILITY OF THE DRIVE SYSTEM

This section presents the open-loop and closed-loop system responses of the stator current PI controllers of the $d$-$q$ axes for the drive system in terms of the bode diagrams to verify the system stability. Figures 4.6 and 4.7 show the bode diagrams of the inner PI controllers, where the output signal measures the $d$-axis stator current. The transfer function of the bode diagram shown in Figures 4.6 and 4.7 can be written as

$$\frac{\delta i_{ds}}{\delta i_{ds}^*} = (7129\,s^{11} + 1.6e8\,s^{10} + 1.2e12\,s^9 + 3.7e15\,s^8 + 3.5e18\,s^7 + 1.3e21\,s^6 + 2e23$$
$$s^5 + 1.2e25\,s^4 + 1.3e26\,s^3 + 5.7e26\,s^2 + 1.1e27\,s + 8.6e26)\,/\,den_1 \qquad (4.63)$$

$$\frac{\delta i_{ds}}{\delta i_{qs}^*} = (4.8e6\,s^{10} + 7e10\,s^9 + 2.8e14\,s^8 + 1.5e17\,s^7 + 2.5e19\,s^6 + 1.1e21\,s^5 + 8.9e21\,s^4 +$$
$$2.6e22\,s^3 + 2.6e22\,s^2 - 2.1e13\,s + 3e13)\,/\,den_1 \qquad (4.64)$$

where

$$den_1 = s^{12} + 2.9e4\,s^{11} + 3.3e8\,s^{10} + 1.8e12\,s^9 + 4.4e15\,s^8 + 3.8e18\,s^7 + 1.4e21\,s^6 + 2e23\,s^5$$
$$+ 1.2e25\,s^4 + 1.3e26\,s^3 + 5.8e26\,s^2 + 1.1e27\,s + 8.6e26 \qquad (4.65)$$

Meanwhile, Figures 4.8 and 4.9 show the bode diagrams of the inner PI controllers of the $d$-$q$ axes stator currents where the output signal measures the $q$-axis stator current. The transfer function of this bode diagram can be written as

$$\frac{\delta i_{qs}}{\delta i_{ds}^*} = -(4.8e6\,s^{10} + 7e10\,s^9 + 2.8e14\,s^8 + 1.4e17\,s^7 + 1.7e19\,s^6 + 1.5e20\,s^5 + 4.6e20\,s^4 + 5.2e20\,s^3$$
$$+ 8.5e19\,s^2 + 1.5e12\,s + 5.2e11)\,/\,den_1 \qquad (4.66)$$

$$\frac{\delta i_{qs}}{\delta i_{qs}^*} = (7129s^{11} + 1.6e8\,s^{10} + 1.2e12\,s^9 + 3.2e15\,s^8 + 1.5e18\,s^7 + 1.8e20\,s^6 + 2.1e21\,s^5 + 9.7e21\,s^4$$
$$+ 2e22\,s^3 + 1.8e22\,s^2 + 2.8e21\,s + 2.3e12)\,/\,den_1 \qquad (4.67)$$

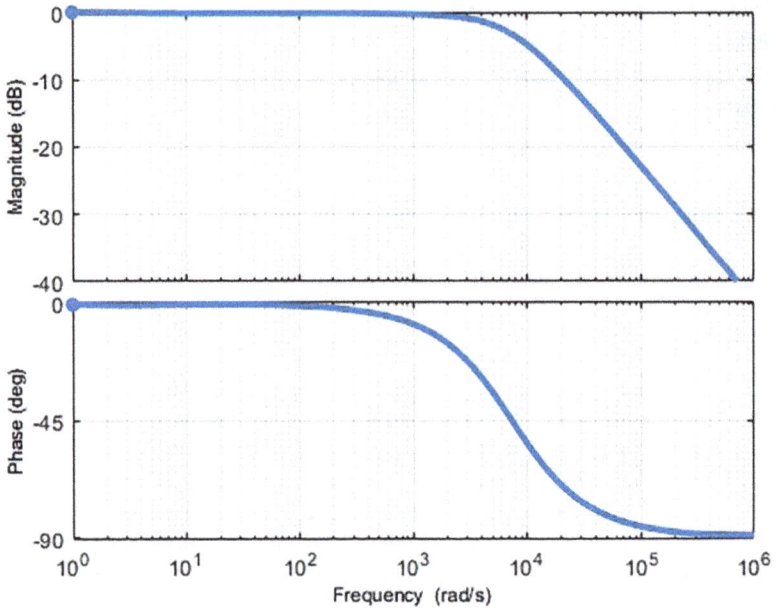

**FIGURE 4.6**    Bode diagram of $i_{ds}/i^*_{ds}$ for the closed-loop response of the current regulator of the $d$-$q$ axes.

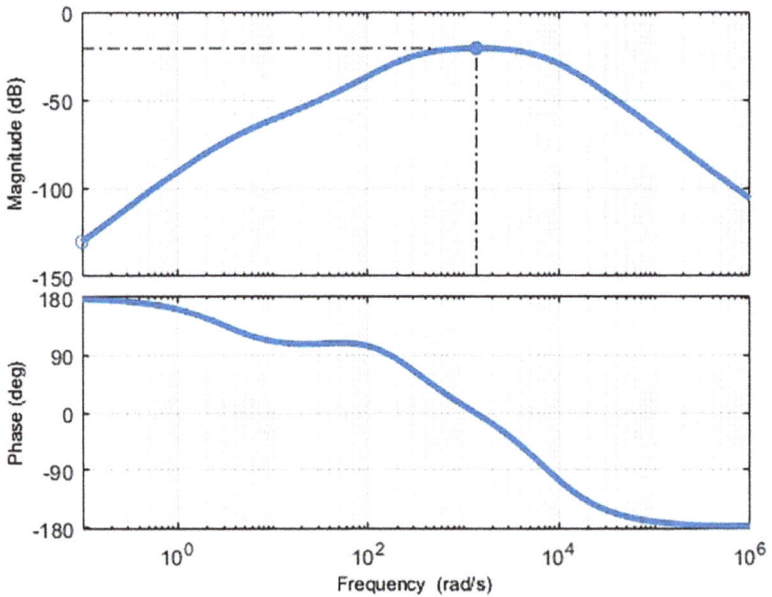

**FIGURE 4.7**    Bode diagram of $i_{ds}/i^*_{qs}$ for the closed-loop response of the current regulator of the $d$-$q$ axes.

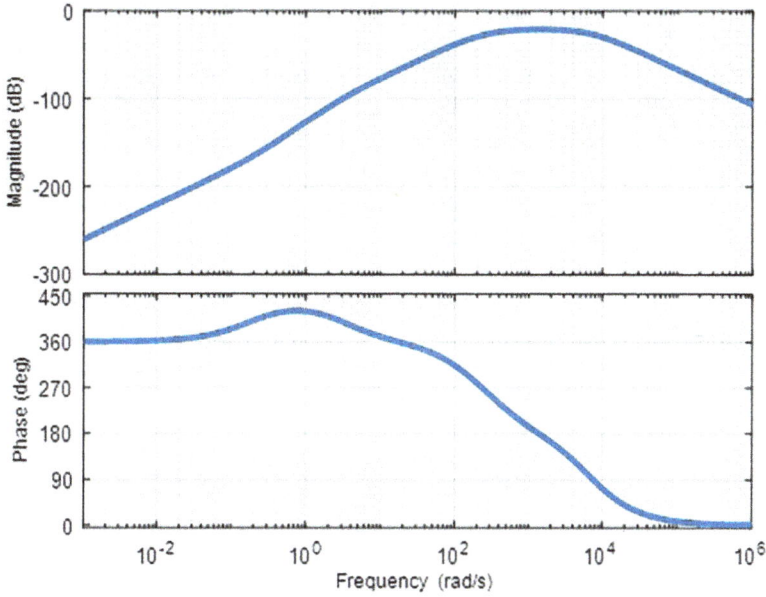

**FIGURE 4.8**   Bode diagram of $i_{qs}i^*_{ds}$ for the closed-loop response of the current regulator of the $d$-$q$ axes.

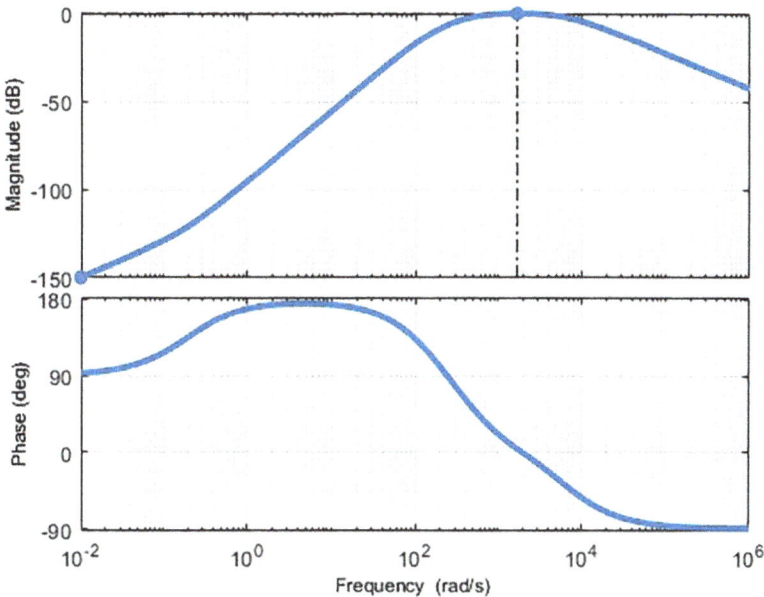

**FIGURE 4.9**   Bode diagram of $i_{qs}i^*_{qs}$ for the closed-loop response of the current regulator of the $d$-$q$ axes.

The state variables of the transfer functions consider the model of the motor and control loops. The transfer function has a high order in the Laplace domain, which can be obtained with the help of MATLAB control system algorithms. The input disturbance signal is the output of the PI current controllers of the inner controllers, and the output disturbance signal is the measured stator current of the $q$-axis. The control strategy can be executed in MATLAB simulation with reference points of 2100 rpm and 5 Nm [5]. Finally, the closed-loop transfer function can be achieved with computer help. For a state-space representation, the input vector of this transfer function is $[i^*_{ds}\ i^*_{qs}]$; $[\delta\omega_r\ \delta i_{ds}\ \delta i_{qs}\ \delta E_\omega\ \delta E_q\ \delta E_d]$ is the state vector.

The optimal PI parameters can achieve strong stability in the control system to drive the PMSM into the flux-weakening region, as can be seen in Figures 4.6, 4.7, 4.8, and 4.9. The gain margin and phase margin are high values in a positive trajectory that are (infinity, 179.9 deg.), (2.9E-13, infinity deg.), (11.2, infinity deg.), and (3.8E14, infinity deg.), respectively.

Figures 4.10 and 4.11 show the open-loop system response of the stator current PI controller of the $d$-axis for the drive system to verify the controller stability. The bode diagrams can be seen from these figures to describe robust stability thanks to the genetic optimization algorithm that sought the optimal PI parameters. The system response shown in these figures is based on the transfer function of the $d$-axis PI controller: that is, the $d$-axis measure current is the output, and the reference $d$-$q$ axes voltages are the input signals. The input signals present the variable states of the

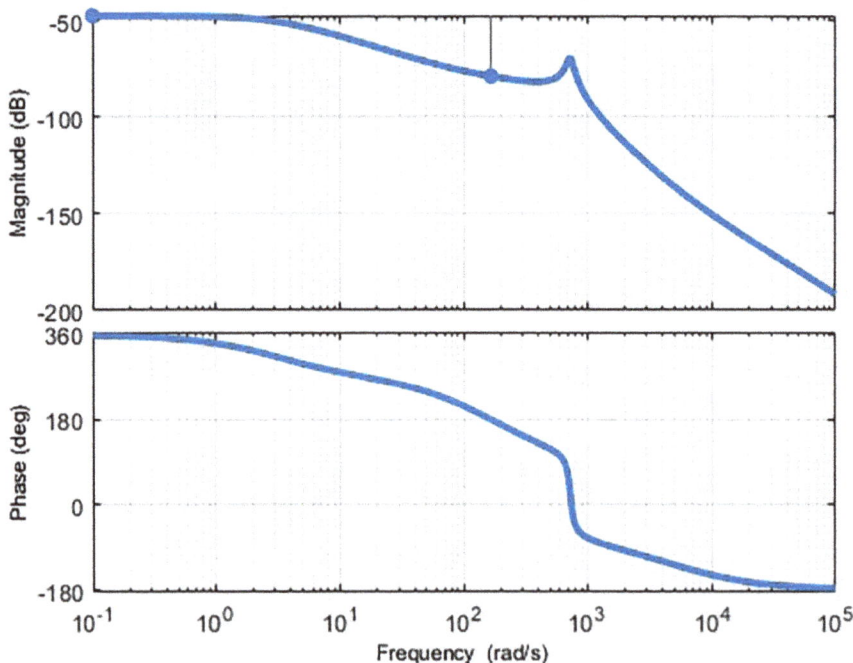

FIGURE 4.10 Bode diagram of $i_{ds}/v_{ds}$ for the open-loop response of the current regulator of the $d$-axis.

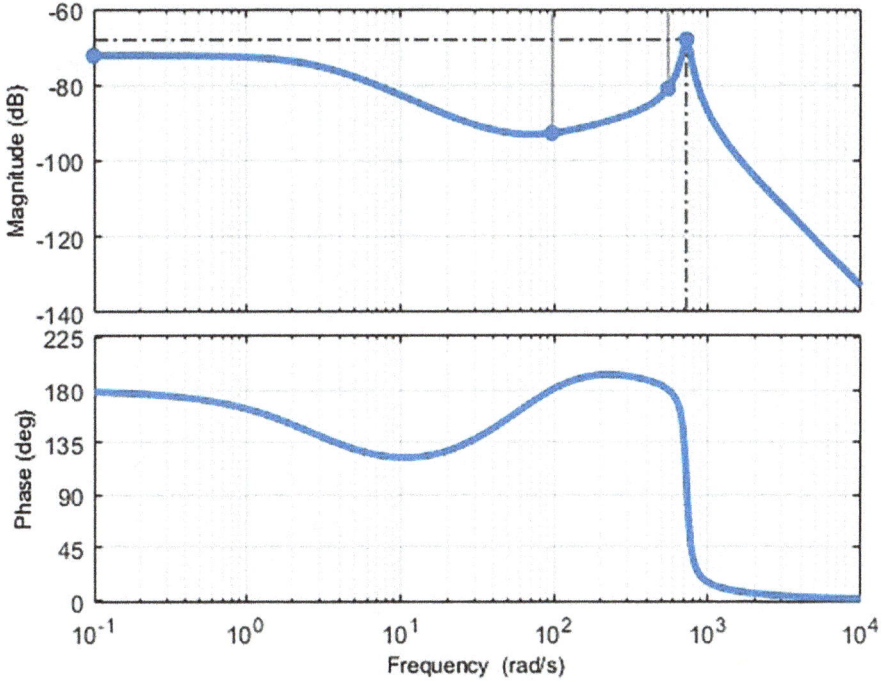

**FIGURE 4.11**    Bode diagram of $i_{ds}/v_{qs}$ for the open-loop response of the current regulator of the $d$-axis.

inverter model, and they can concern the modulation index. The transfer function of this bode diagram can be written as

$$\frac{\delta i_{ds}}{\delta v_{ds}} = (2.59\ s^2 - 1.44e4\ s + 1.71e6)\ /\ den_2 \tag{4.68}$$

$$\frac{\delta i_{ds}}{\delta v_{qs}} = (-21.88\ s^2 - 2980\ s - 9.835e4)\ /\ den_2 \tag{4.69}$$

where

$$den_2 = s^4 + 323.3\ s^3 + 5.506e5\ s^2 + 1.33e8\ s + 4.027e8 \tag{4.70}$$

Figure 4.12 and 4.13 shows the open-loop system response of the stator current PI controller of the $q$-axis for the drive system to verify the controller stability. The bode diagrams can be seen in these figures to describe the stability of the $q$-axis PI controller when the $q$-axis measure current is the output of the transfer function, and

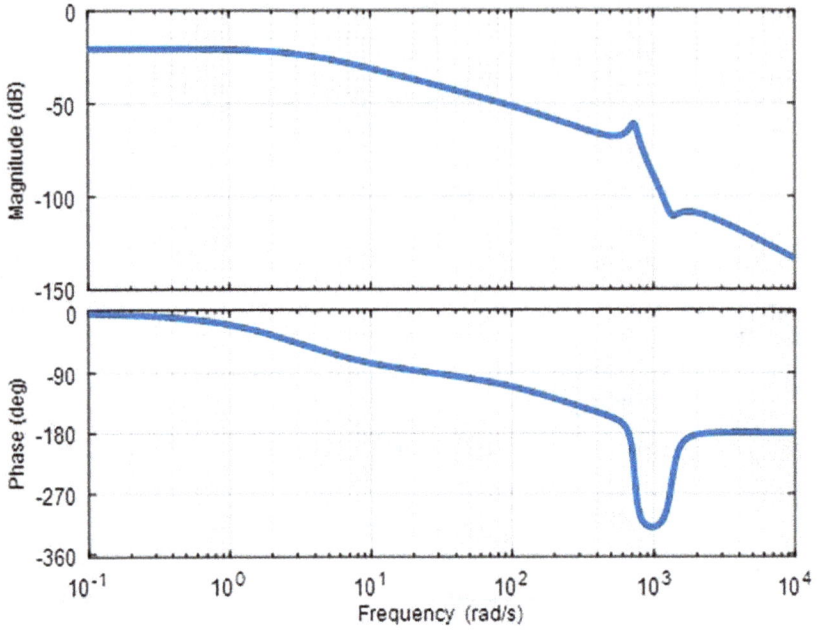

**FIGURE 4.12** Bode diagram of $i_{qs}/v_{ds}$ for the open-loop response of the current regulator of the $q$-axis.

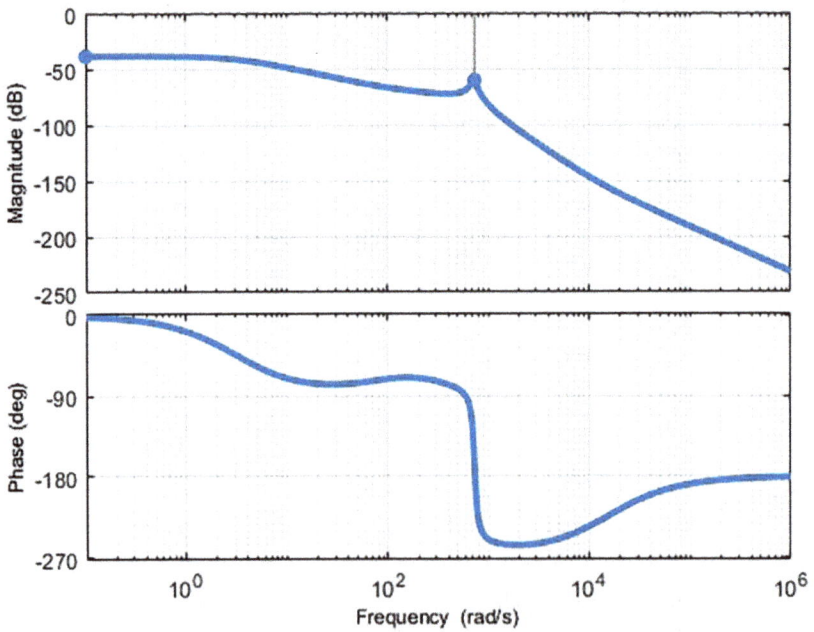

**FIGURE 4.13** Bode diagram of $i_{qs}/v_{qs}$ for the open-loop response of the current regulator of the $q$-axis.

the reference $d$-$q$ axes voltages are the input signals. The transfer function of this bode diagram can be written as

$$\frac{\delta i_{qs}}{\delta v_{ds}} = (21.88 \ s^2 + 5255 \ s + 3.83e7) \ / \ den_2 \qquad (4.71)$$

$$\frac{\delta i_{qs}}{\delta v_{qs}} = (2.59 \ s^2 - 4.013e4 \ s + 4.82e6) \ / \ den_2 \qquad (4.72)$$

Recall that the state-space representation can provide the transfer function represented in (4.68) to (4.72), as shown in (4.73)

$$TF = C \ (INV( \ diag \begin{bmatrix} 1 & 0 & 0 & 0 & 0 & 0 & 0 \\ 0 & 1 & 0 & 0 & 0 & 0 & 0 \\ 0 & 0 & 1 & 0 & 0 & 0 & 0 \\ 0 & 0 & 0 & 1 & 0 & 0 & 0 \\ 0 & 0 & 0 & 0 & 1 & 0 & 0 \\ 0 & 0 & 0 & 0 & 0 & 1 & 0 \\ 0 & 0 & 0 & 0 & 0 & 0 & 1 \end{bmatrix} S - A \ )) B \qquad (4.73)$$

where $TF$ stands for the transfer function, and the output matrix of $C$ can be defined as

$$C = \begin{bmatrix} 0 & 0 & 0 & 0 & 0 & 0 & 0 \\ 0 & 0 & 0 & 0 & 0 & 0 & 0 \\ 0 & 0 & 0 & 0 & 0 & 0 & 0 \\ 0 & 0 & 0 & 0 & 0 & 0 & 0 \\ 0 & 0 & 0 & 0 & 0 & 0 & 0 \\ 0 & 0 & 0 & 0 & 0 & 1 & 0 \\ 0 & 0 & 0 & 0 & 0 & 0 & 1 \end{bmatrix} \qquad (4.74)$$

In this transfer function, the set operating points are 225.15 rad/s, -8.4A, and 3A, which are equivalent to points in the flux-weakening range. The gain margins are 8626, 4095, 1698, and 1064 for the bode plots of Figures 4.10, 4.11, 4.12, and 4.13, respectively. Thus, the phase margins are growing to infinity. Lastly, it can depend on the optimal selected PI parameters to achieve the robust stability of the drive system.

### 4.3.3 DYNAMIC PERFORMANCE ANALYSIS IN THE PRESENCE OF TORQUE DISORDER AND VELOCITY VARIATION

The performance verification of the ripple reduction strategy presented in this section is compared to the ripple reduction technique mentioned in Chapter 2. This section also aims to validate the optimal obtained PI parameters during sudden changes in the reference speed and reference load torque. Figures 4.14 to 14.27 show the

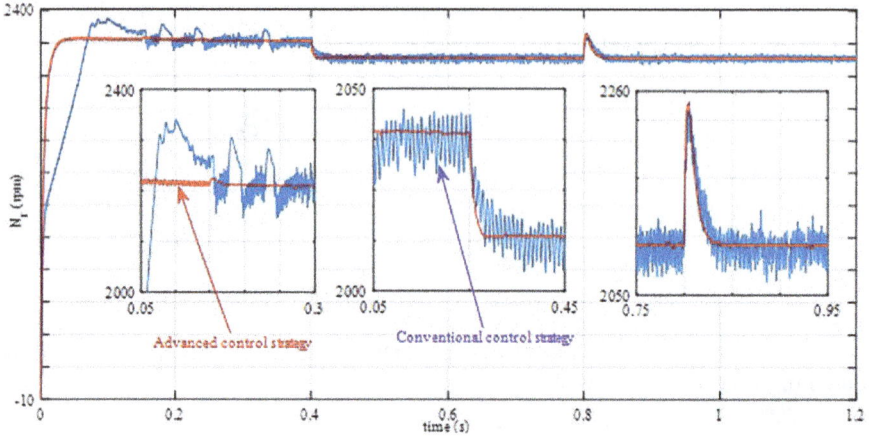

**FIGURE 4.14**　Rotational motor speed for the simulation case study of torque disorder and velocity variation.

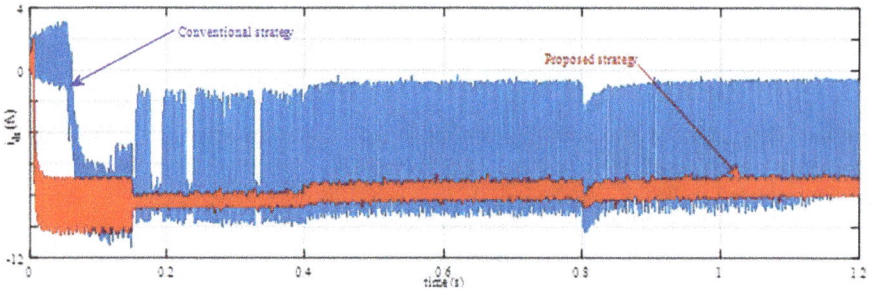

**FIGURE 4.15**　Stator currents of the $d$-axis for the simulation case study of torque disorder and velocity variation.

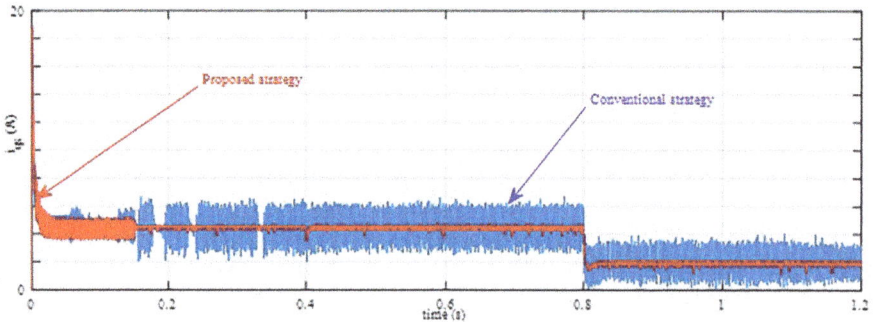

**FIGURE 4.16**　Stator currents of the $q$-axis for the simulation case study of torque disorder and velocity variation.

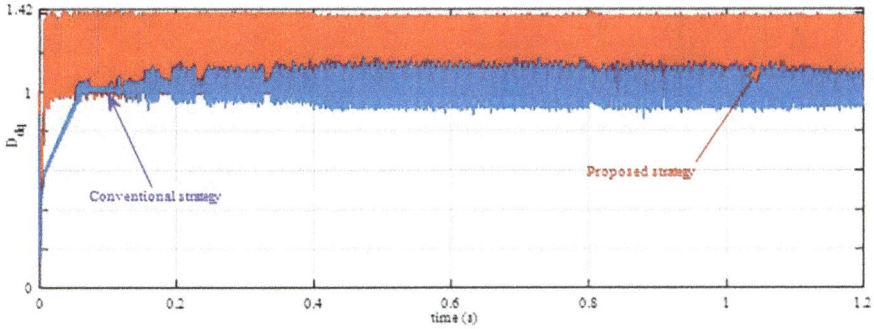

**FIGURE 4.17**  $D_{dq}$ vectors for the simulation case study of torque disorder and velocity variation.

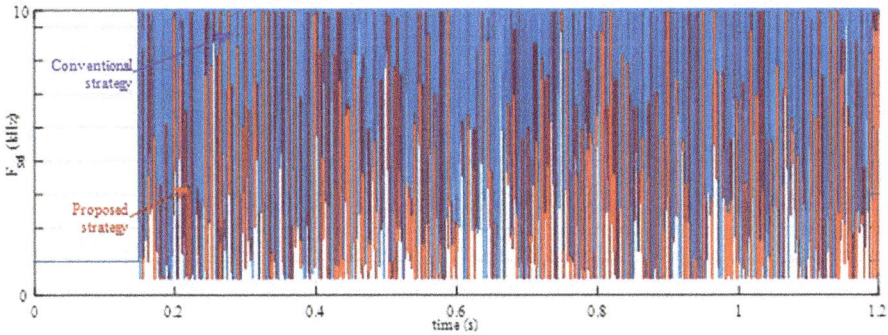

**FIGURE 4.18**  Variable switching frequency for the simulation case study of torque disorder and velocity variation.

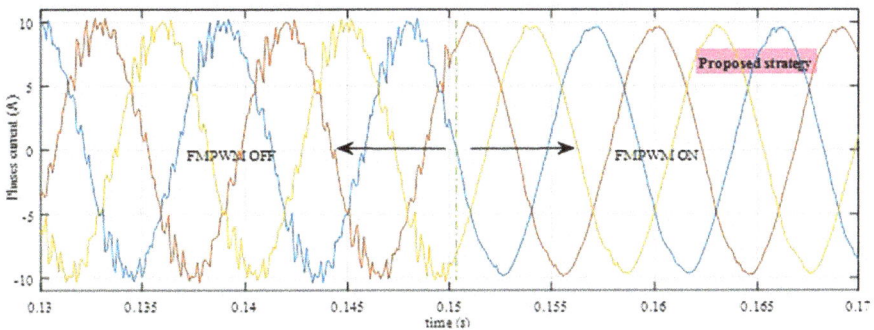

**FIGURE 4.19**  Stator phase currents of this control strategy for the simulation case study of torque disorder and velocity variation.

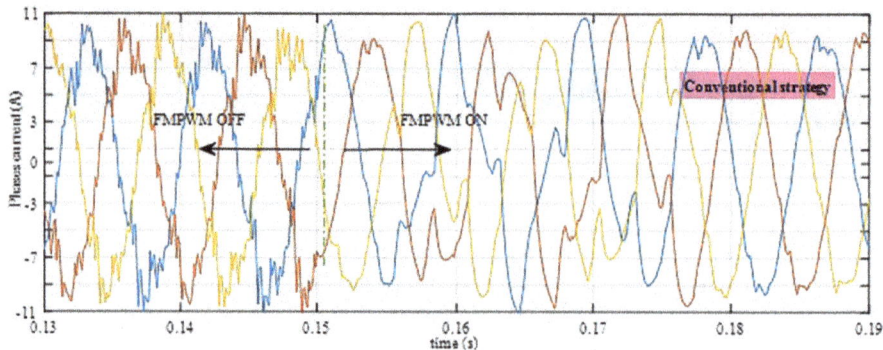

**FIGURE 4.20**  Stator phase currents of the conventional control strategy for the simulation case study of torque disorder and velocity variation.

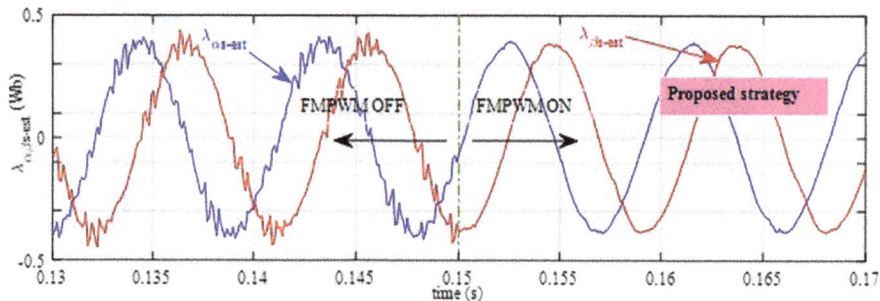

**FIGURE 4.21**  Estimated $\alpha$-$\beta$ axes fluxes based on the TOGI observer of this control strategy for the simulation case study of torque disorder and velocity variation.

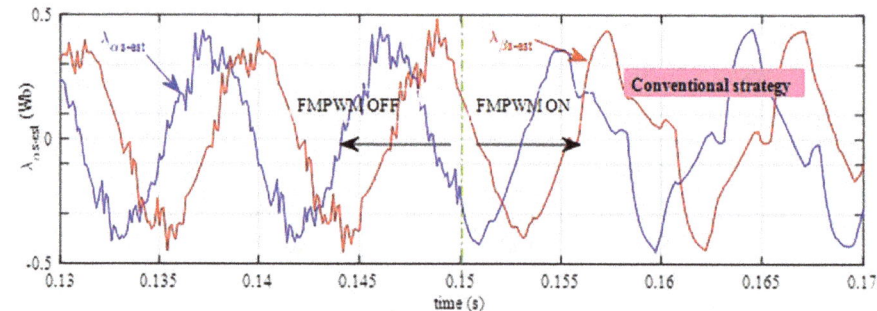

**FIGURE 4.22**  Estimated $\alpha$-$\beta$ axes fluxes based on the TOGI observer of the conventional control strategy for the simulation case study of torque disorder and velocity variation.

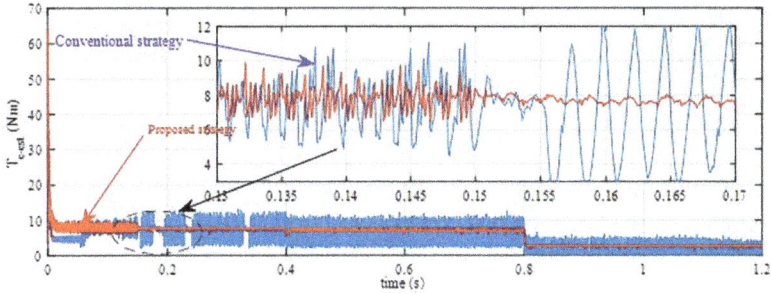

**FIGURE 4.23** Estimated electromagnetic torque based on the TOGI observer for the simulation case study of torque disorder and velocity variation.

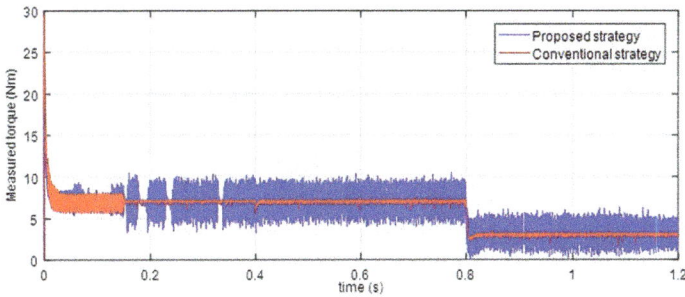

**FIGURE 4.24** Measured electromagnetic torque for the simulation case study of torque disorder and velocity variation.

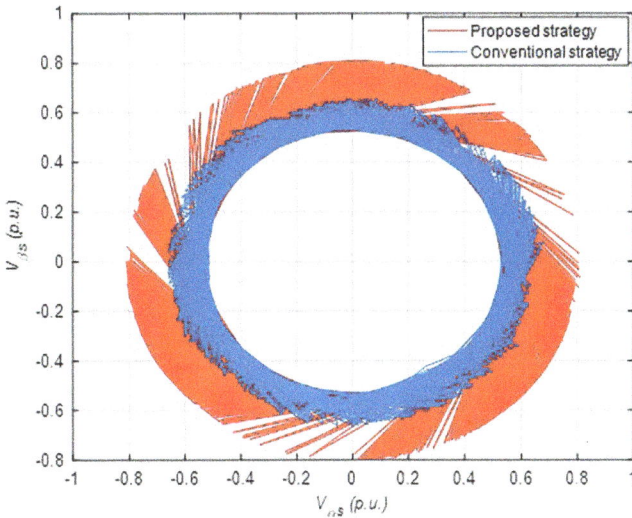

**FIGURE 4.25** Voltage trajectory rated value 400 V for the simulation case study of torque disorder and velocity variation.

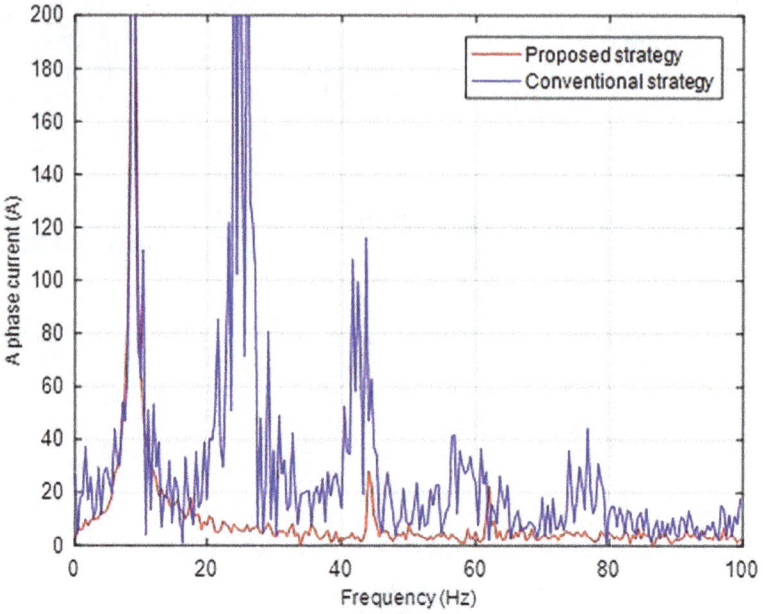

**FIGURE 4.26**  FFT analysis of the phase-A stator current for the simulation case study of torque disorder and velocity variation.

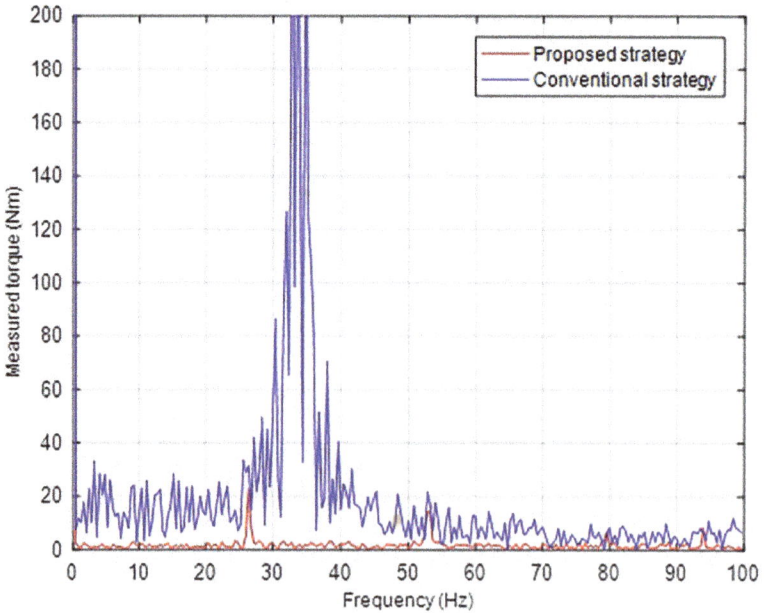

**FIGURE 4.27**  FFT analysis of the torque for the simulation case study of torque disorder and velocity variation.

simulation case scenario in which the frequency modulation is enabled in the steady-state periods of this and classical drive systems by about 0.15 s from the start time. The reference velocity of this case study is about 2200 rpm under 7 Nm of the premier load torque. The sudden step-down change of reference velocity reaches about 2100 rpm at about 0.4 s. Afterwards, the load torque disturbance steps down, reaching 3 Nm at about 0.8 s.

Figure 4.14 shows the measured motor speeds of this and classical control strategies. In this figure, the settling speed value of this control method advances on that of the classical control method by 0.3225 s. There is no overshoot of the motor speed with this method while 6.3% was recorded with the classical method. Many oscillations are observed under the classical motor speed and no ripple with the new method. For the period of the load torque disorder, a robust performance can be observed with the classical drive; there are no oscillations and an overshoot velocity lower by about 2.1%. The same strong stability can be observed during the speed disorder in terms of no overshooting velocity.

Figure 4.18 shows the variable switching frequencies of this and classical control strategies. This figure shows that the changing period of the switching frequency is from 2.5 to 10 kHz. The mean value can be reduced to 6.25 kHz. The performance of this control method appears on the waves of the stator currents of the $d$-$q$ axes, the measured torque, and the estimated electromagnetic torque as shown in the following figures. Figures 4.15 and 4.16 show the measured stator currents of the $d$-$q$ axes of this and the classical control strategies.

It can be observed that the ripple peak of the $d$-axis stator current under this control method is sufficiently mitigated from 20.7% to 6% while the ripple peak of the $q$-axis stator current under the control of the classical method is sufficiently mitigated from 18.6% to 7%. However, the ripple peaks for the same waves under the command of the classical control method are not reduced. Figure 4.15 shows that the mean current of the $d$-axis under the command of this control method is boosted from -8.7 A before the speed disorder period to -7.6 A after this disorder. However, the mean value is still constant at about -7.6 A. The mean value of the $d$-axis stator current under the control of the classical drive method is boosted from -8 to -4.9 A; the mean value is affected slightly during the reference torque disorder period.

Figure 4.16 shows that the mean values of the $q$-axis stator currents under the control of this and classical drive systems are still constant at about 4.4 A during the reference speed disorder. However, the mean values of the $q$-axis stator currents under the control of this and classical drive systems dropped from 4.4 A to about 2.0 A during the reference torque disorder.

Figures 4.21 and 4.22 present the estimated $\alpha$-$\beta$ axes fluxes based on the TOGI observer of this and classical drive methods, respectively. Figures 4.23 and 4.24 show the estimated electromagnetic torque and measured torque waves of this and classical control strategies that are estimated based on the estimated $\alpha$-$\beta$ axes fluxes and the stator current of the $q$-axis, respectively. The performance of the drive system can be observed from the torque ripple peak reduction that belongs to the stator current ripple reduction. The ripple peak of the electromagnetic torque can be minimized from about 39% to about 15%.

Figures 4.17 and 4.25 represent the duty-cycle vectors and per unit of the voltage trajectories with a rated value of 400 V of this and classical drive strategies, respectively. It can be observed that the voltage trajectory has been limited by about 0.58 $V_{dc}$ at the region of the flux, weakening by the modulation technique used in the two-level voltage source inverter as shown in Figure 4.25. It can also be noted that this drive system can profit from the mean value of the DC-link voltage relative to the conventional drive system. It can be seen from Figure 4.17 that the mean values of the duty-cycle vector are 1.19 and 1 for this and classical drive systems, respectively. This confirms what was mentioned about the utilization of this drive method of the DC-link voltage.

The operation of the controlled PMSM in the flux-weakening scope reduces the efficiency of the drive systems because the rated velocity for the motor did not achieve the rated torque. This drive system achieves the efficiency of the power inverter by 3.08 kW as the input real power and 2.539 kW as the output real power. However, the counterpart of the classical drive system has 3.26 kW as the input real power and 2.259 kW as the output real power. The loss of the real power of the power inverter is 42.853%, the switching losses for both this and classical drive systems. The real power that is consumed in the power inverter operation is considered the switching losses. The efficiency of this drive is higher than that of the classical drive as this drive delivers more real power. This drive system may operate based on the optimally selected PI parameters of the genetic optimization algorithm, which helps in delivering more real power. Both this and classical drive systems record 1.613 kW for the controlled PMSM as output real power. The efficiency of the PMSM under the control of this drive system is greater than that of the traditional drive system; they achieved 63.5% and 71.4%, respectively.

The real power of the voltage inverter delivered under the control of this and traditional drive systems is 2.35 kW and 1.992 kW, respectively; the real power delivered in the period of the post-reference torque disturbance is 2.108 kW and 1.788 kW, respectively. Therefore, this drive system achieves an efficiency of about 65.5% in the period of about 7 Nm and 31.3% in the period of about 3 Nm. However, the classical drive system achieves an efficiency of about 77.3% and 36.9%, respectively.

Figures 4.19 and 4.20 present the stator currents of the three phases for this and classical drive strategies. It can be seen from these figures that this drive system can reduce the stator current peak relative to the traditional drive method from about 10.73 A to 9.7 A. However, the classical drive system achieves greater efficiency of the controlled motor than that of this drive system due to the iron losses, which increase more than the copper losses in the scope of field decreasing [6].

Figure 4.26 shows the fast Fourier transform (FFT) analysis of the phase A stator current of this and classical control strategies. Figure 4.27 shows the FFT analysis of the torque of this and classical control strategies as well. It can be seen from these figures that the harmonic components can be significantly minimized under the control of this drive system compared to the traditional drive system if the frequency modulation technique is employed on the flux-weakening range with adjusted PI controllers by the optimal parameters of the genetic optimization algorithm.

## 4.4 EXPERIMENTAL RESULTS ANALYSIS

The advanced driving method as considered in this chapter and the traditional driving methods are implemented on the digital control unit of a TMS320F28335 float point digital signal processor (DSP) to control a surface-mounted PMSM separately. Studying its dynamic performance can confirm the effectiveness of the advanced control study.

A photograph of the experimental laboratory used is shown in Figure 2.24. The parameters of the controlled PMSM and power inverter are listed in Table 2.4 in Chapter 2. It should be mentioned that the DC-link voltage is saturated at 150 V, and the sample time of the control strategy is 100 μs. The hardware control unit provides a further ripple attached to the measured signal. In this experiment, it's difficult to measure the electromagnetic torque because the inherent torque sensor is employed to display the electromagnetic torque on a monitor. The following sections present the performance comparison between the advanced and classical driving strategies. The point of comparison is the robustness of the drive systems during sudden speed variation, reference torque disturbances, and measured speed oscillations.

### 4.4.1 DYNAMIC PERFORMANCE ANALYSIS IN THE EVENT OF A SUDDEN DROP IN VELOCITY

The dynamic performance comparison of the controlled motor is presented in Figures 4.28 through 4.61 under the command of strategies for both this and traditional controllers. These figures present the scenario of reference speed disorder from 2300 rpm to 2100 rpm while the controlled PMSM runs in the flux-weakening region under a reference load torque of 5.5 Nm. It can be seen from the figures that the velocity step-down variation occurs at 5.6 s, and the enabling time of the traditional frequency modulation technique is 2.8 s.

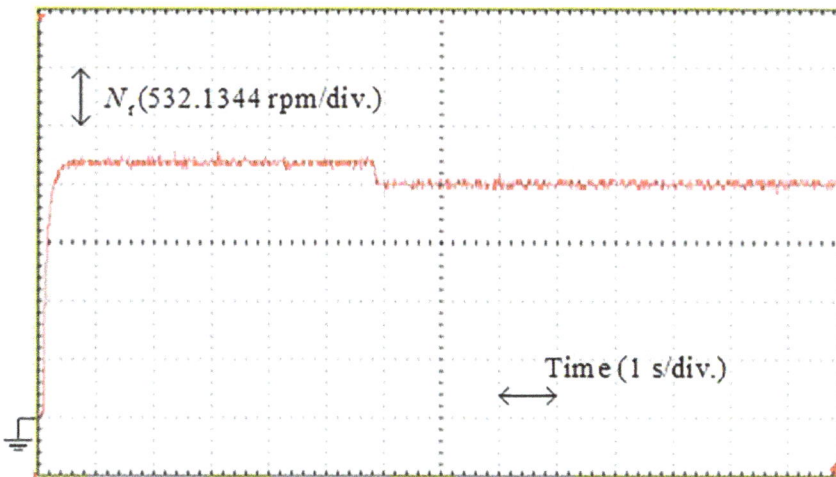

**FIGURE 4.28** Rotational motor speed under control of the advanced strategy for the case of a sudden drop in velocity.

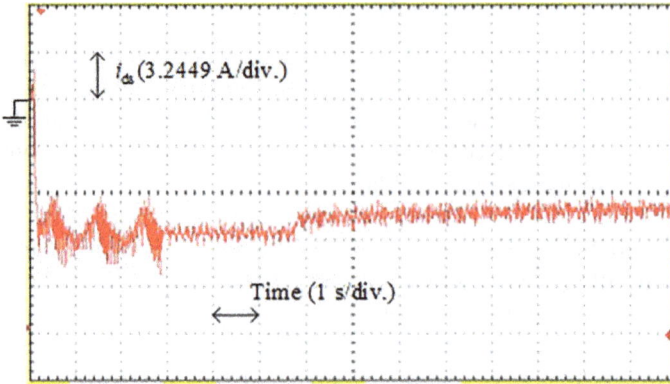

**FIGURE 4.29**  Stator current of the $d$-axis under control of the advanced strategy for the case of a sudden drop in velocity.

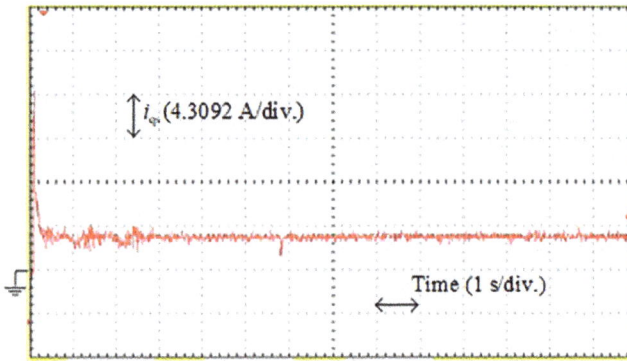

**FIGURE 4.30**  Stator current of the $q$-axis under control of the advanced strategy for the case of a sudden drop in velocity.

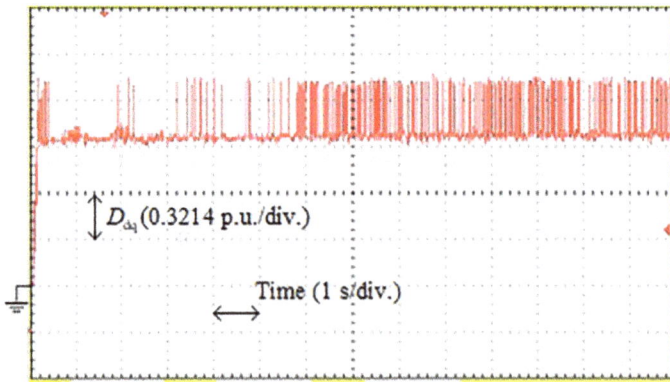

**FIGURE 4.31**  $D_{dq}$ vector under control of the advanced strategy for the case of a sudden drop in velocity.

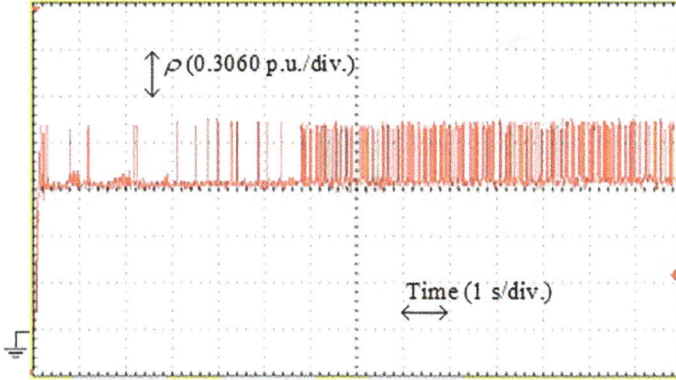

**FIGURE 4.32** Voltage vector magnitude under control of the advanced strategy for the case of a sudden drop in velocity.

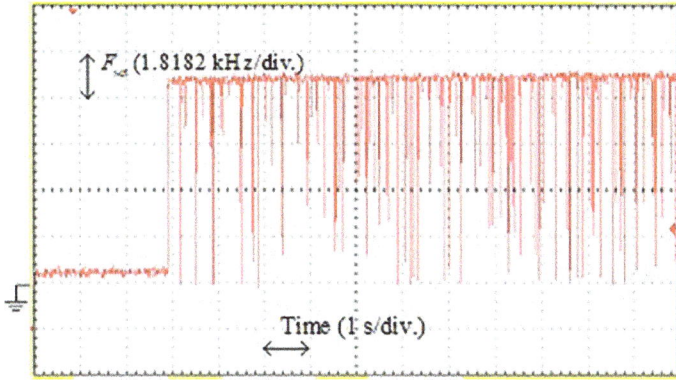

**FIGURE 4.33** Varying switching frequency under control of the advanced strategy for the case of a sudden drop in velocity.

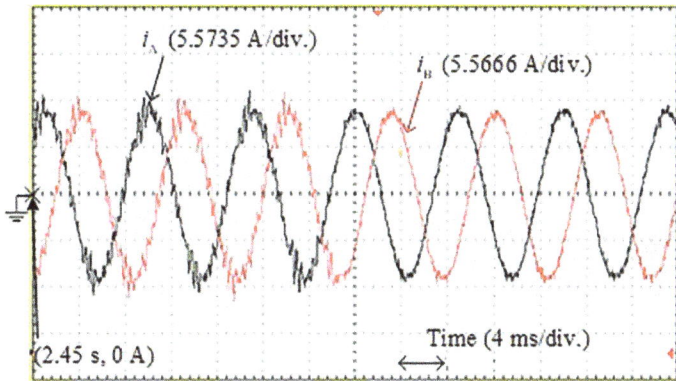

**FIGURE 4.34** Effect of the FMPWM algorithm in stator phase currents under control of the advanced strategy for the case of a sudden drop in velocity.

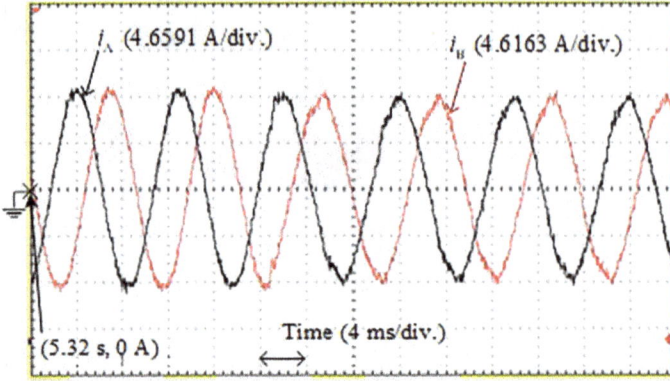

**FIGURE 4.35**  Effect of speed reduction in stator phase currents under control of the advanced strategy for the case of a sudden drop in velocity.

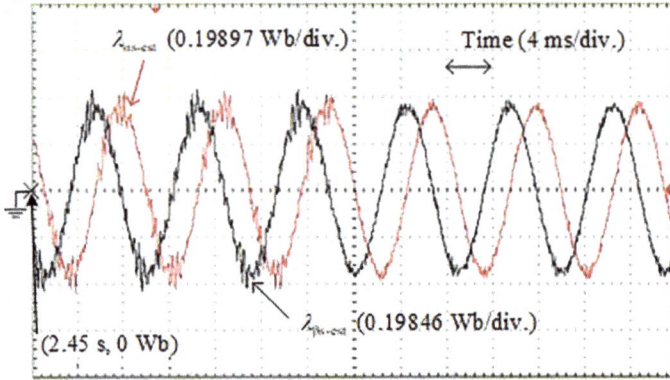

**FIGURE 4.36**  Estimated $\alpha$-$\beta$ axes fluxes based on the TOGI observer under control of the advanced strategy for the case of a sudden drop in velocity.

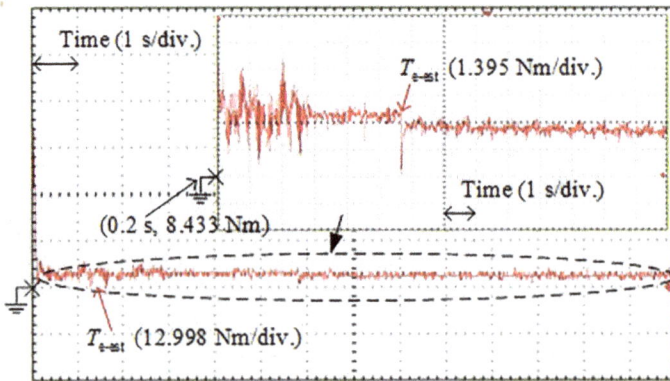

**FIGURE 4.37**  Estimated torque based on the TOGI observer under control of the advanced strategy for the case of a sudden drop in velocity.

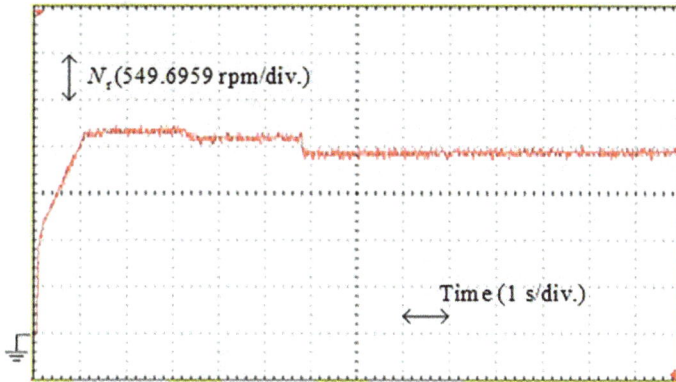

**FIGURE 4.38**  Rotational motor speed under control of the conventional strategy for the case of a sudden drop in velocity.

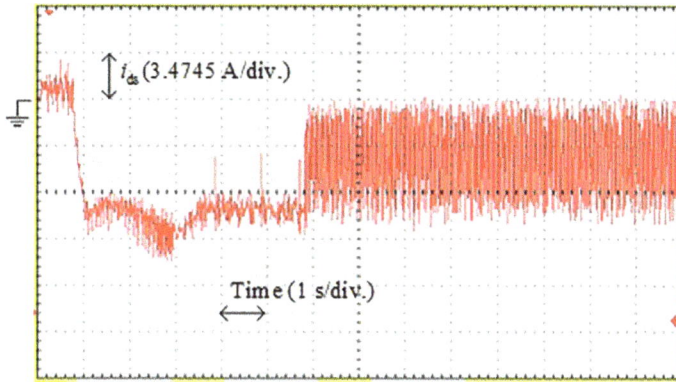

**FIGURE 4.39**  Stator current of the $d$-axis under control of the conventional strategy for the case of a sudden drop in velocity.

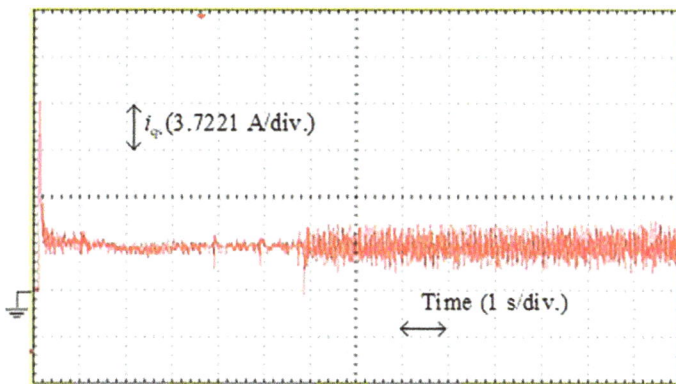

**FIGURE 4.40**  Stator current of the $q$-axis under control of the conventional strategy for the case of a sudden drop in velocity.

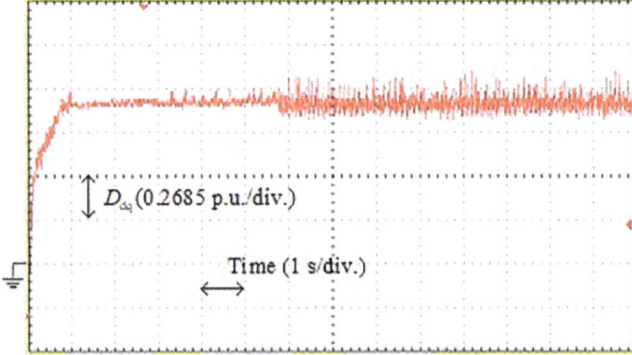

**FIGURE 4.41**  $D_{dq}$ vector under control of the conventional strategy for the case of a sudden drop in velocity.

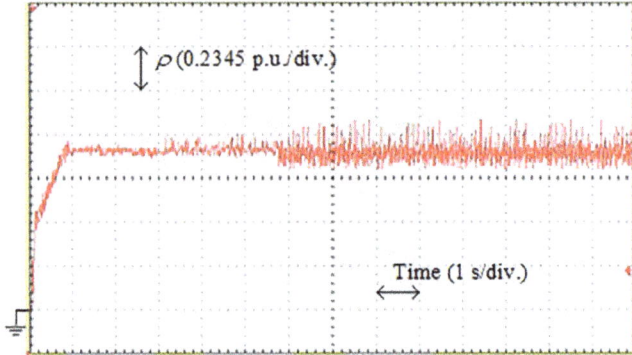

**FIGURE 4.42**  Voltage vector magnitude under control of the conventional strategy for the case of a sudden drop in velocity.

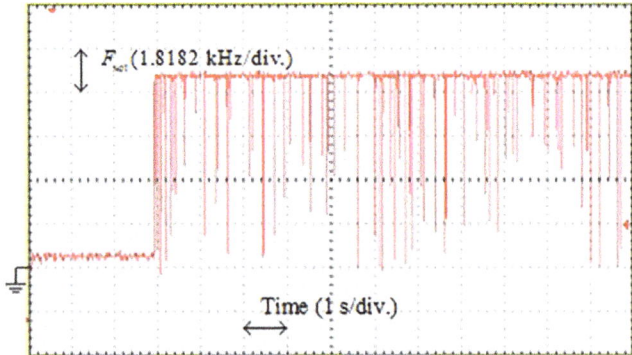

**FIGURE 4.43**  Varying switching frequency under control of the conventional strategy for the case of a sudden drop in velocity.

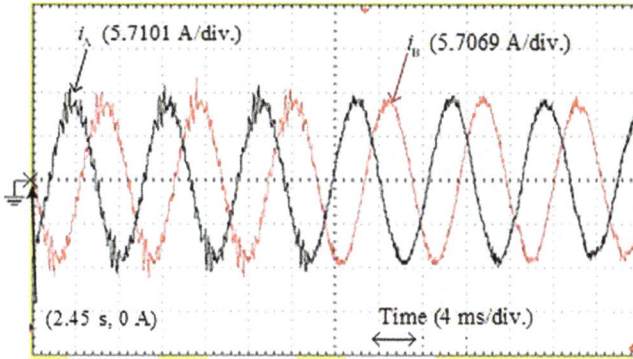

**FIGURE 4.44**  Effect of the FMPWM algorithm in stator phase currents under control of the conventional strategy for the case of a sudden drop in velocity.

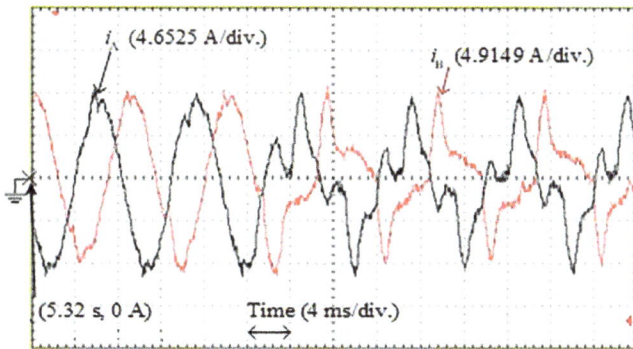

**FIGURE 4.45**  Effect of speed reduction in stator phase currents under control of the conventional strategy for the case of a sudden drop in velocity.

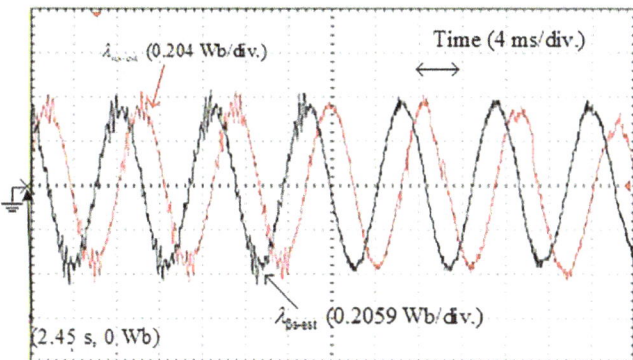

**FIGURE 4.46**  Estimated $\alpha$-$\beta$ axes fluxes based on the TOGI observer under control of the conventional strategy for the case of a sudden drop in velocity.

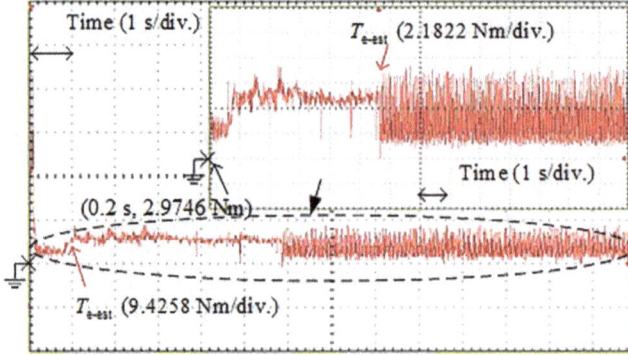

**FIGURE 4.47**  Estimated torque based on the TOGI observer under control of the conventional strategy for the case of a sudden drop in velocity.

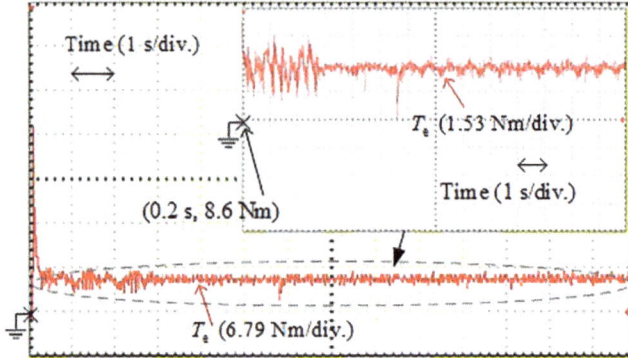

**FIGURE 4.48**  Measured torque under control of the advanced strategy for the case of a sudden drop in velocity.

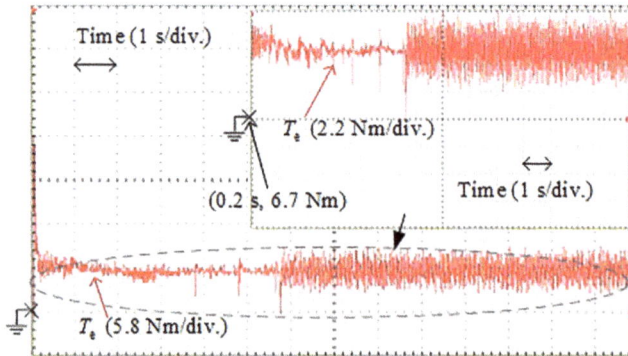

**FIGURE 4.49**  Measured torque under control of the conventional strategy for the case of a sudden drop in velocity.

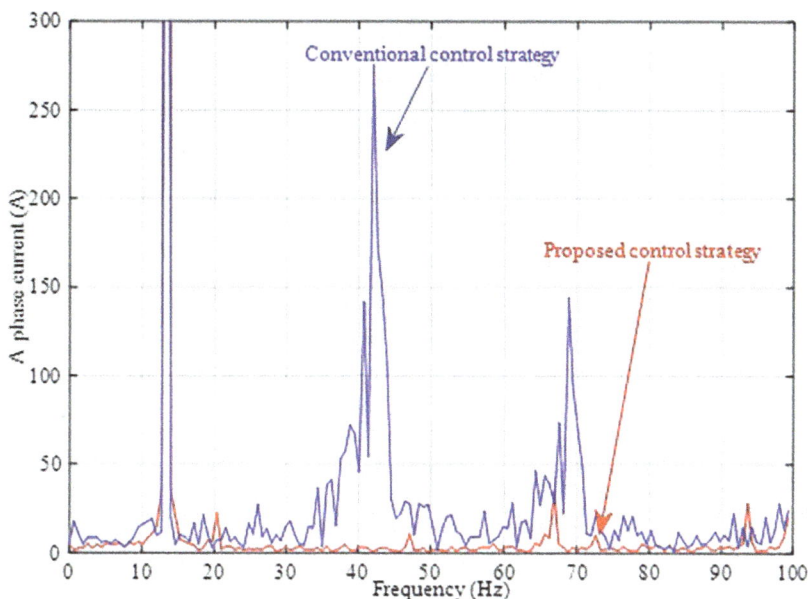

**FIGURE 4.50**  FFT analysis of phase-A current under control of both strategies for the case of a sudden drop in velocity.

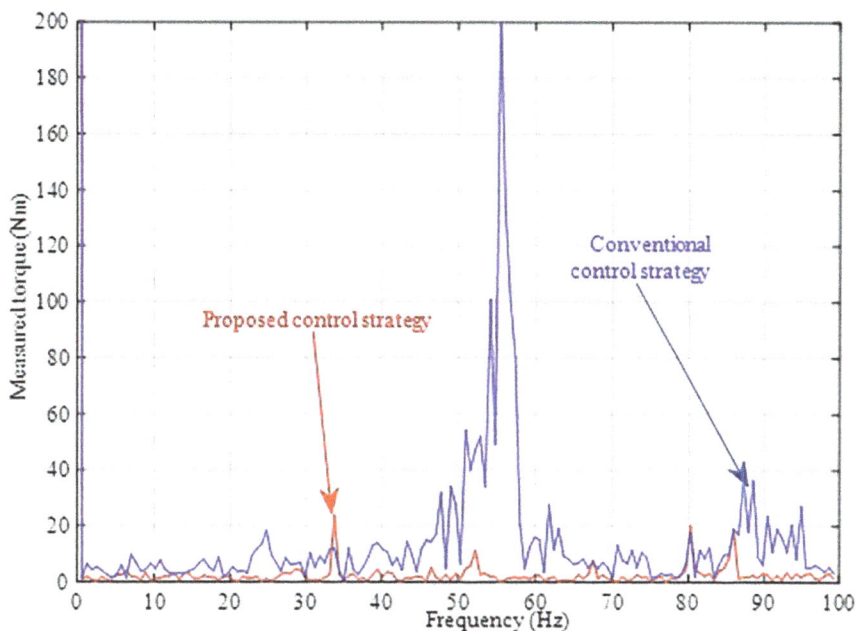

**FIGURE 4.51**  FFT analysis of the torque under control of both strategies for the case of a sudden drop in velocity.

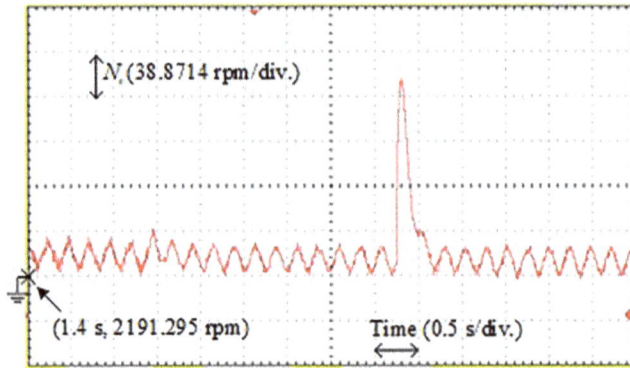

**FIGURE 4.52**  Rotational motor speed under control of the advanced strategy for the case of speed oscillation and load torque disruption.

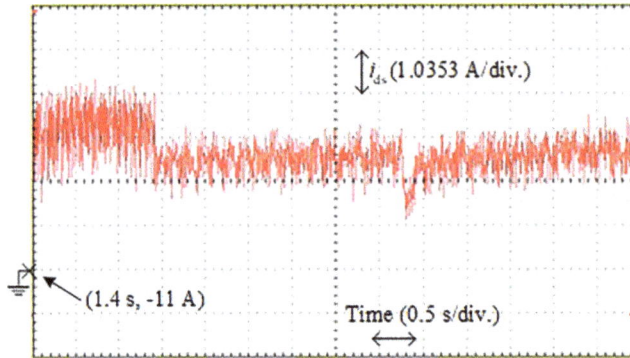

**FIGURE 4.53**  Stator current of the *d*-axis under control of the advanced strategy for the case of speed oscillation and load torque disruption.

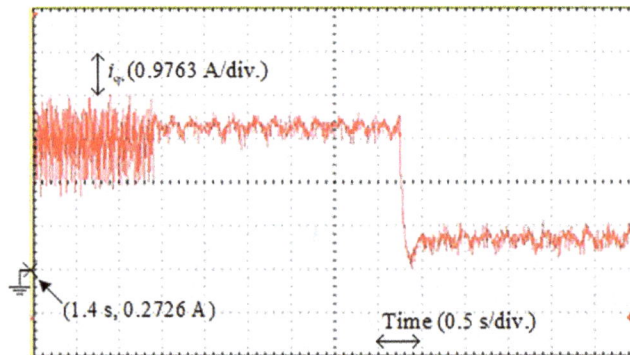

**FIGURE 4.54**  Stator current of the *q*-axis under control of the advanced strategy for the case of speed oscillation and load torque disruption.

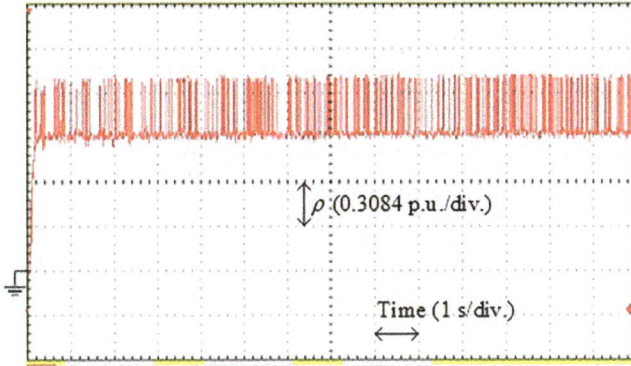

**FIGURE 4.55** Voltage vector under control of the advanced strategy for the case of speed oscillation and load torque disruption.

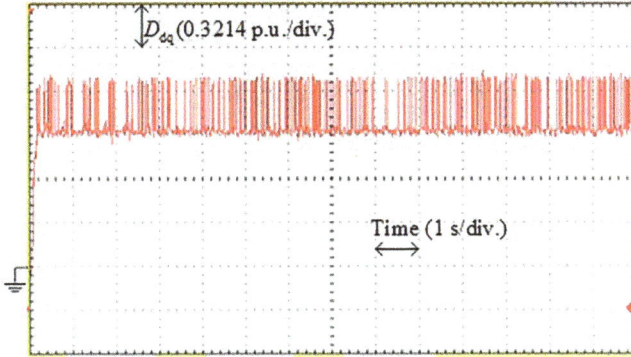

**FIGURE 4.56** $D_{dq}$ vector under control of the advanced strategy for the case of speed oscillation and load torque disruption.

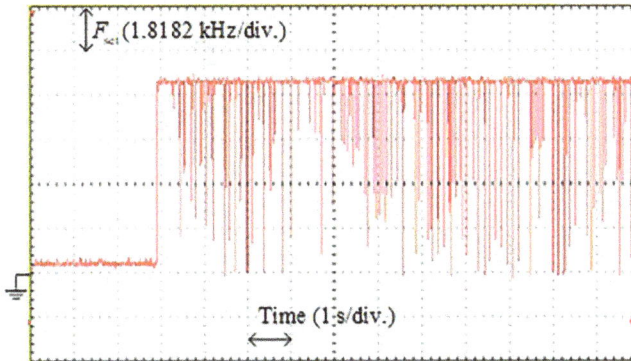

**FIGURE 4.57** Varying switching frequency under control of the advanced strategy for the case of speed oscillation and load torque disruption.

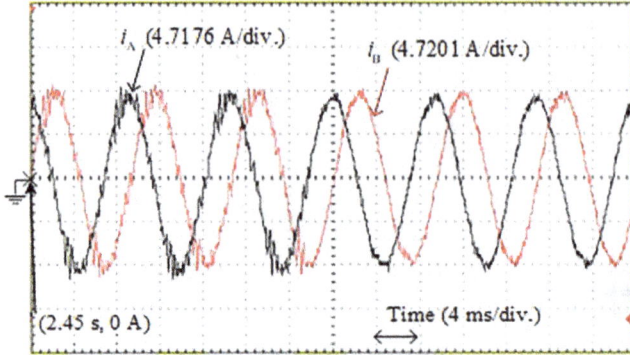

**FIGURE 4.58**   Effect of the FMPWM algorithm in stator phase currents under control of the advanced strategy for the case of speed oscillation and load torque disruption.

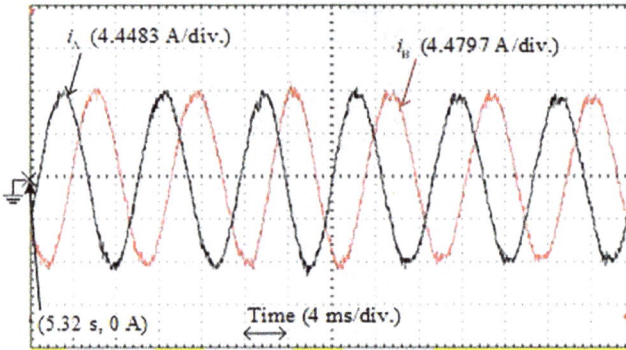

**FIGURE 4.59**   Effect of the load torque reduction in stator phase currents under control of the advanced strategy for the case of speed oscillation and load torque disruption.

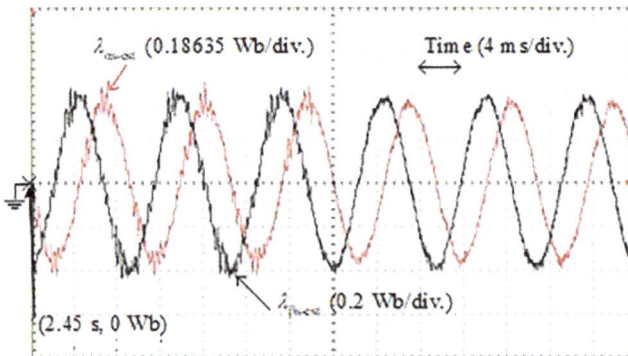

**FIGURE 4.60**   Estimated $\alpha$-$\beta$ axes fluxes based on the TOGI observer under control of the advanced strategy for the case of speed oscillation and load torque disruption.

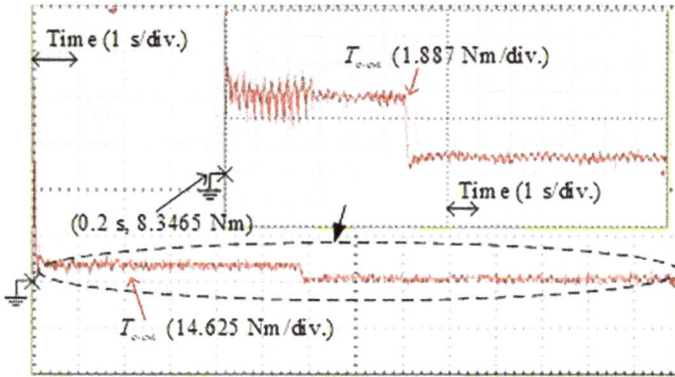

**FIGURE 4.61** Estimated torque based on the TOGI observer under control of the advanced strategy for the case of speed oscillation and load torque disruption.

Figures 4.28 and 4.38 show the measured motor velocity waves in the case of a sudden drop in reference velocity under this and the traditional control strategies separately. It can be observed from these figures that the advanced driving method can reduce the settling time compared to the traditional method; the times are 0.2892 s, and 3.15 s, respectively. In addition, the overrun above the reference velocity is reduced to about zero for the proposed drive and by 4.87% for the traditional drive system. Thus, the results index the robustness of this driving system in tracking the reference velocity.

The frequency modulation technique changes the switching frequency from 2.5 to 10 kHz, as illustrated in Figures 4.33 and 4.43, for the proposed and classical driving methods, respectively.

It can be seen from these figures that the mean value of the switching frequency is reduced to 6.25 kHz. The frequency modulation benefits are reflected in the stator currents of the $d$-$q$ axes and the estimated and measured electromagnetic waves as presented in Figures 4.29, 4.30, 4.37, and 4.48 separately. The effectiveness of this advanced control strategy can be verified relative to the traditional method. It can be observed from the stator currents of the $d$-axis that this driving system can significantly minimize the peak of the current ripple compared to the classical driving system from about 27% to 8%. The peak value of the $q$-axis stator current ripple is reduced from about 33.3% to 6.8% under the control of the proposed driving ripple. However, the peak values of the $d$-$q$ axes stator currents ripple under the control of the classical driving system are not affected because the PI gains of the inherent PI regulators make a difference in tracking the measured value with the reference value. The PI gains of the traditional driving system are adjusted randomly by trial and error. It can be observed from Figures 4.39 and 4.40 that the peak values of the stator current ripples of the $d$-$q$ axes are increased.

The average value of the $d$-axis stator current can be slightly reduced in the negative trajectory under the control of this driving system due to the sudden velocity step-down from about -9.27 A to -7.84 A, as presented in Figure 4.29. That is reduced to half of about -8.93 A, or -4.44 A, for the classical driving method, as presented in

Figure 4.39. It can be observed from Figures 4.30 and 4.40 that the mean values of the stator current of the $q$-axis under the control of this and traditional driving methods are constant during the step change of the reference velocity.

The electromagnetic torque is calculated to verify the ripple peak of the induced torque in terms of how it can be affected by the ripple peak of the stator $d$-$q$ axis currents. The calculated torque depends on the estimated stator fluxes in the stationary frame by an observer of a third-order generalized integrator. The estimated fluxes of the $\alpha$-$\beta$ axes for the driving systems are presented in Figures 4.36 and 4.46; the calculated induced torques are presented in Figures 4.37 and 4.47. It can be seen from these figures that the ripple peak can be minimized with the advanced driving method relative to the classical driving method.

The measured induced torques are calculated based on the measured stator current of the $q$-axis and the induced torque equation of the surface-mounted PMSM model. Figures 4.48 and 4.49 show the measured electromagnetic torque of this and traditional driving system, respectively.

The torque ripples caused by the $q$-axis stator current ripples can be shown in these figures to prove the ability of this driving method to reduce the ripple of the torque and the current compared to the classical method. The peak ripple of the measured torque can be mitigated from about 54.4% to about 10.4%, as presented in Figures 4.37 and 4.48. In contrast, the classical driving system has a ripple peak that is randomly increased because the stator current controller loses its controllability at the flux-weakening range.

Figures 4.32 and 4.42 show the voltage vector ($\rho$) of the inherent voltage inverter for this and classical driving systems, respectively. Recall that the maximum value of the voltage vector is 0.56 $V_{dc}$, which can be obtained from the modulation technique used. If the measured value is greater than or equal to 0.56 $V_{dc}$, the operation of the controlled PMSM is in the flux-weakening scope for a wide speed range. It can be seen from these figures that the voltage vector average value of the advanced drive command before the sudden speed variation is about 1.03, and that of the conventional drive command is about 1.01. The voltage vector average value of the advanced drive command after the sudden speed variation is about 1.1, and that of the conventional drive command is about 1.0.

Figures 4.31 and 4.41 show the duty-cycle vector of the inherent voltage inverter for the advanced and classical driving systems, respectively. It can be noted from these figures that the mean value of the duty-cycle vector of the advanced drive command before the sudden speed variation is about 1.03, and that of the conventional drive command is about 1.01. After the sudden speed variation, the mean value of that vector is about 1.1 and about 1.0 for the advanced and classical driving methods, respectively.

To calculate the efficiency, the input real power under the command of the advanced driving system is 5.72 kW at the operating point of 2300 rpm for the speed and 5.5 Nm for the load torque; the counterpart of the classical driving system is 5.54 kW. The output real power under the command of the advanced driving system is 4 kW at the same operating point, and the counterpart of the classical driving system is 3.87 kW. Recall that the real power consumed in the power inverter operation is considered the switching loss because it is difficult to measure. It can be concluded that the advanced driving method considered in this chapter can reduce switching losses from about 30% to 28.5% relative to the traditional driving method.

The mean values of the voltage vector and duty-cycle vector under the advanced driving system indicate that this advanced drive utilizes a DC-link voltage more than the average voltage utilized under the conventional driving system. As a result, the controlled PMSM has more input real power supplied under the advanced driving system than under the classical driving system. However, the output real power of the PMSM under the control of both driving systems is constant because the motor operates under the same operating point for the reference speed and load torque. The output real power of the controlled PMSM is about 1.33 kW. Motor efficiency is about 32.4% under the advanced driving method. However, it increases slightly to about 34.2% under the traditional driving method.

The output real power under the command of the advanced driving system is 3.79 kW at the operating point of 2100 rpm for the speed and 5.5 Nm for the load torque; the counterpart of the classical driving system is 2.25 kW. The efficiency of the motor is about 31.9% under the advanced driving method and about 53.7% under the traditional driving method [6].

An increase in the usage of DC-link voltage is beneficial for decreasing copper losses by decreasing the amplitude of the stator currents, as presented in Figures 4.35 and 4.45 for the advanced and traditional driving systems separately. As a result, the amplitude of the stator currents for the controlled PMSM under the advanced driving system is lower than when the motor is driven under the classical method. The amplitudes of the stator currents during the activated period of the FMPWM algorithm are about 10.18 A and 10.91 A, as shown in Figures 4.34 and 4.44, respectively. These are about 9.4 and 10.28 A for the period of reference velocity disorder, as shown in Figures 4.35 and 4.45, respectively.

Figure 4.50 shows the FFT analysis of the phase A stator current of the advanced and classical control strategies. Figure 4.51 shows the FFT analysis of the torque of the advanced and classical control strategies as well. It can be seen from these figures that the harmonic components can be significantly minimized under the control of the advanced drive system compared to the traditional drive system if the frequency modulation technique is employed on the flux-weakening range with adjusted PI controllers by the optimal parameters of the genetic optimization algorithm.

### 4.4.2 STUDYING DYNAMIC PERFORMANCE UNDER CONDITIONS OF SPEED OSCILLATION AND LOAD TORQUE DISRUPTION

Figures 4.62 to 4.75 display the dynamic performance comparison of the controlled motor using both advanced and traditional controllers. The fluctuations in the measured motor speed are attributed to the closed-loop control and machine structure.

These figures present the scenario of a sudden reference load torque disorder from 5.5 Nm to 1.5 Nm while the controlled PMSM runs in the flux-weakening region to reach a reference velocity of 2200 rpm. It can be seen from the figures that the step-down variation of the reference torque occurs at 5.6 s; the enabling time of the traditional frequency modulation technique is 2.8 s.

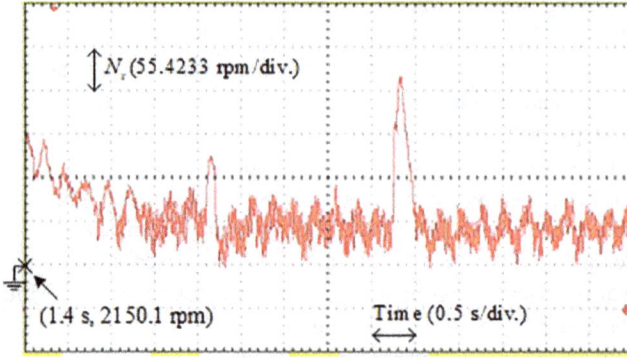

**FIGURE 4.62**  Rotational motor speed under control of the conventional strategy for the case of speed oscillation and load torque disruption.

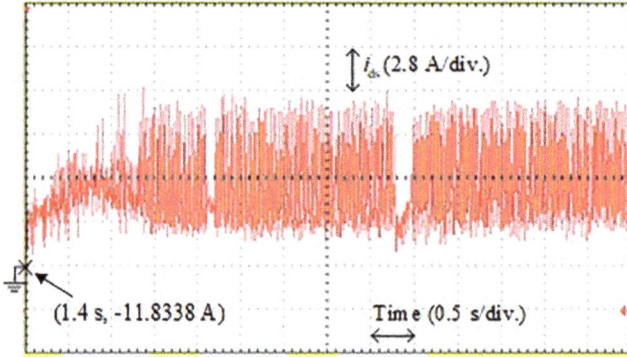

**FIGURE 4.63**  Stator current of the *d*-axis under control of the conventional strategy for the case of speed oscillation and load torque disruption.

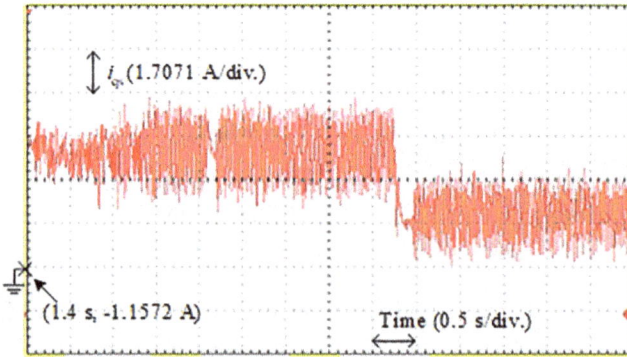

**FIGURE 4.64**  Stator current of the *q*-axis under control of the conventional strategy for the case of speed oscillation and load torque disruption.

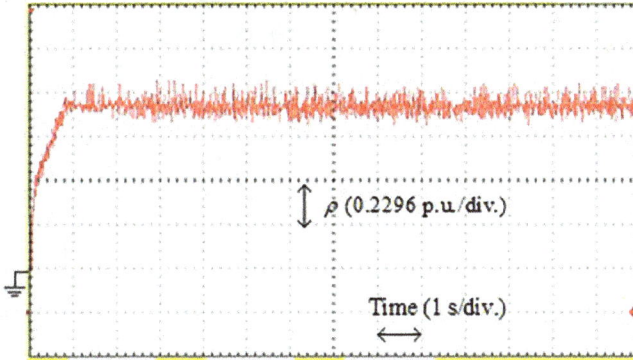

**FIGURE 4.65**  Voltage vector under control of the conventional strategy for the case of speed oscillation and load torque disruption.

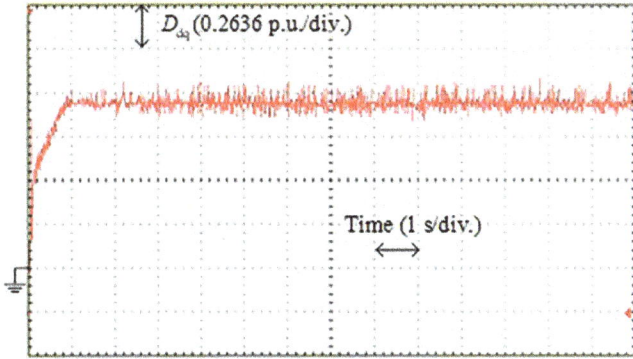

**FIGURE 4.66**  $D_{dq}$ vector under control of the conventional strategy for the case of speed oscillation and load torque disruption.

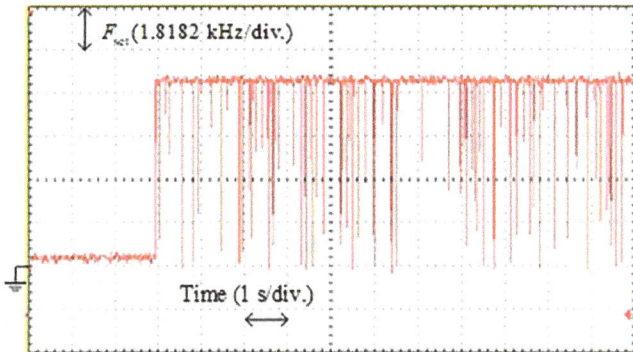

**FIGURE 4.67**  Varying switching frequency under control of the conventional strategy for the case of speed oscillation and load torque disruption.

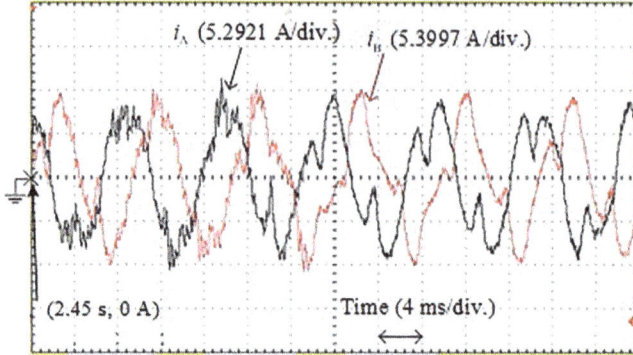

**FIGURE 4.68**  Effect of the FMPWM algorithm in stator phase currents under control of the conventional strategy for the case of speed oscillation and load torque disruption.

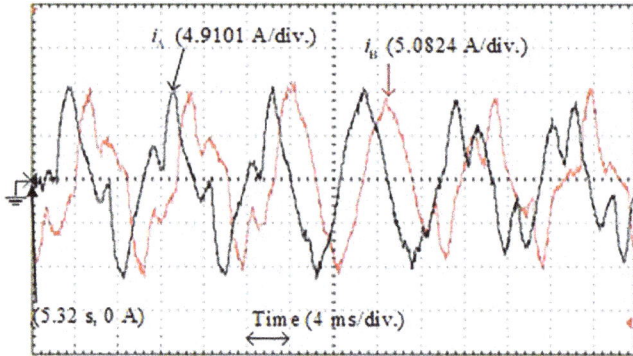

**FIGURE 4.69**  Effect of the load torque reduction in stator phase currents under control of the conventional strategy for the case of speed oscillation and load torque disruption.

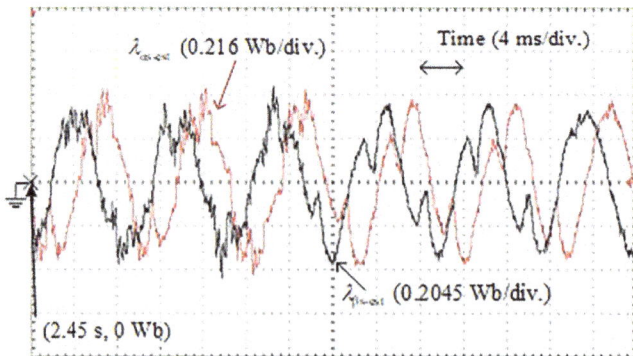

**FIGURE 4.70**  Estimated $\alpha$-$\beta$ axes fluxes based on the TOGI observer under control of the conventional strategy for the case of speed oscillation and load torque disruption.

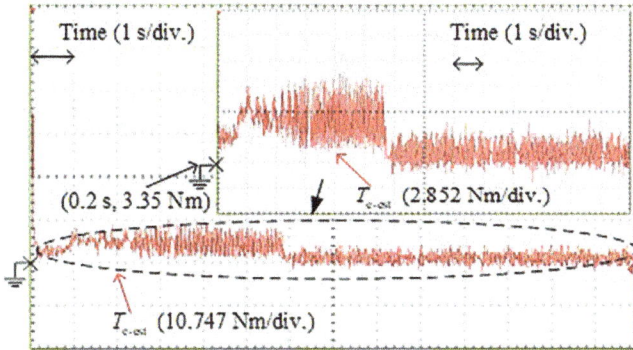

FIGURE 4.71 Estimated torque based on the TOGI observer under control of the conventional strategy for the case of speed oscillation and load torque disruption.

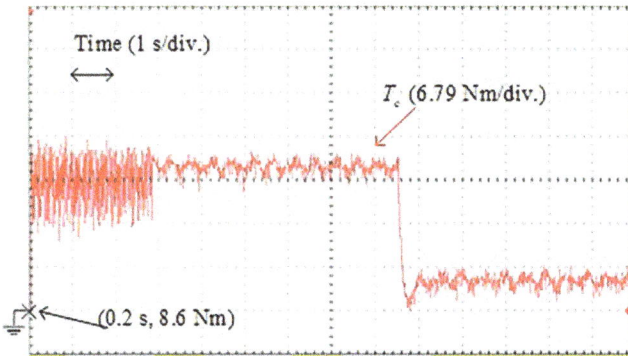

FIGURE 4.72 Measured torque under control of the advanced strategy for the case of speed oscillation and load torque disruption.

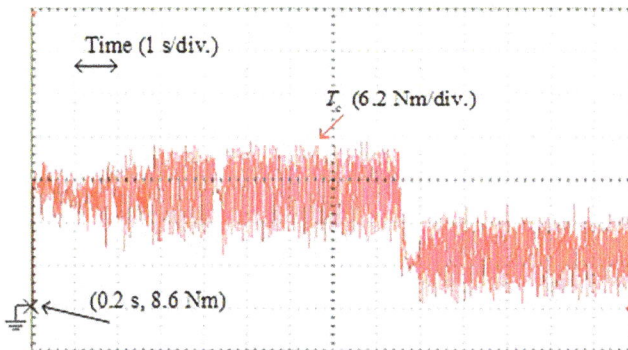

FIGURE 4.73 Measured torque under control of the conventional strategy for the case of speed oscillation and load torque disruption.

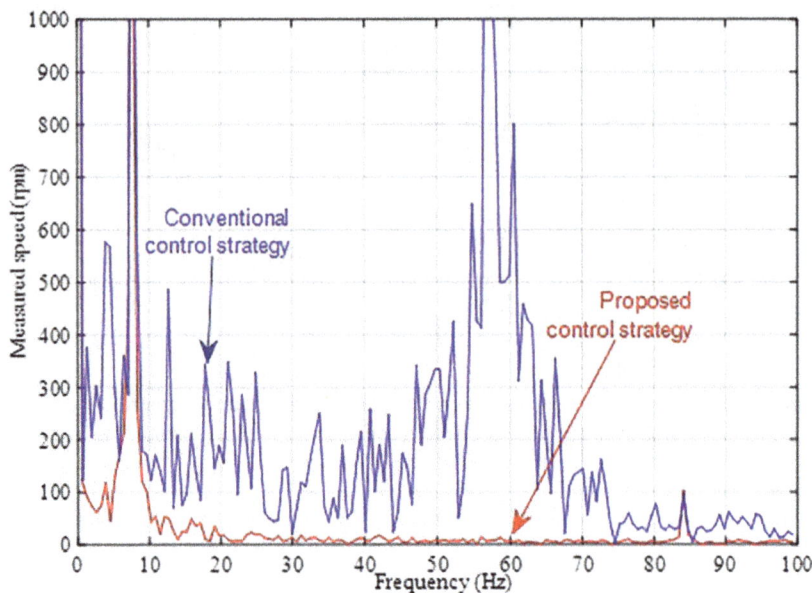

**FIGURE 4.74** FFT analysis of the measured motor speed under control of both strategies for the case of speed oscillation and load torque disruption.

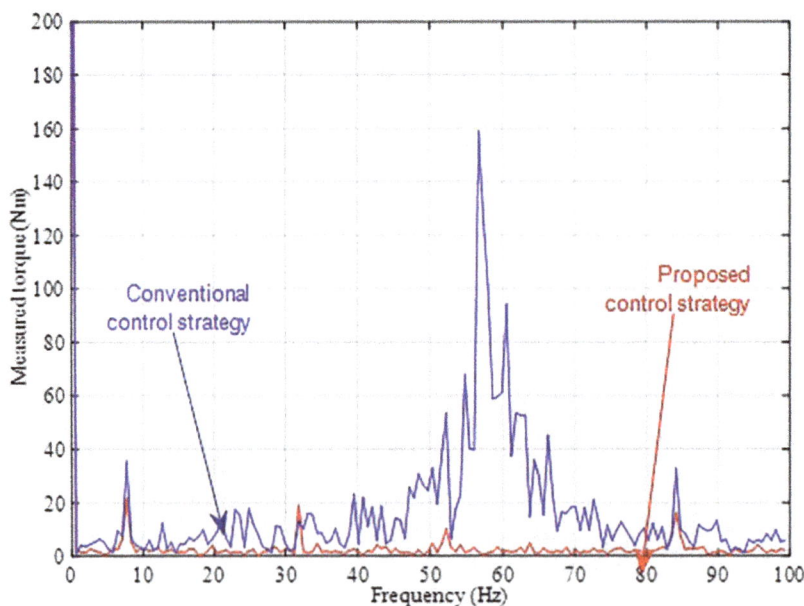

**FIGURE 4.75** FFT analysis of the torque under control of both strategies for the case of speed oscillation and load torque disruption.

The measured motor speeds of the advanced and traditional control systems can be seen in Figures 4.52 and 4.62, respectively. It can be noted that the advanced driving system can reduce the fluctuation peak of the measured speed and overrun above the reference speed compared to the traditional driving method. The values of the fluctuation peak in the steady-state period are about 0.86% and 2.6%, respectively, and the values of the speed overrun above the reference velocity of the transient period are 7.3% and 8.8%, respectively.

The frequency modulation technique of the advanced and classical driving methods changes the switching frequency from 2.5 to 10 kHz, as illustrated in Figures 4.57 and 4.67, respectively. It can be seen from these figures that the mean value of the switching frequency is reduced to 6.25 kHz. The frequency modulation benefits are reflected in the stator currents of the $d$-$q$ axes and the estimated and measured electromagnetic waves, as presented in Figures 4.53, 4.54, 4.61, and 4.72 separately. These figures demonstrate the superiority of the advanced driving method to the conventional method. The average values of the stator currents of the $d$-$q$ axes are step changed due to the reference torque disorder. The average $d$-axis stator current is slightly decreased in the negative directory from about -8.44 A to about -8.1 A. The counterpart of the traditional strategy is reduced in the negative trajectory from about -7 A to about -6 A.

The average value of the $q$-axis stator current under the advanced driving strategy is equal to that of the classic strategy and is reduced while the reference torque is reduced from about 3.5 A to 0.96 A. The ripples of the induced torque and stator current of the $q$-axis while the motor is controlled under the conventional driving system are caused by the velocity oscillation. In contrast, these ripples can be successfully reduced by the advanced system.

Figures 4.55 and 4.65 show the voltage vector ($\rho$) of the inherent voltage inverter for the advanced and classical driving systems, respectively. It can be seen from these figures that the voltage vector average value of the advanced driving system for the pre-period of reference torque disorder is about 1.07 and that of the conventional driving disorder is about 1.01.

The voltage vector average value of the advanced drive command for the post-period of reference torque disorder is about 1.08, and that of the conventional drive command is about 1.0. These average values mean that the controlled PMSM is working in the flux-weakening region. Figures 4.46 and 4.66 show the duty-cycle vector of the inherent voltage inverter for the advanced and classical driving systems, respectively. It can be noted from these figures that the mean value of the duty-cycle vector of the proposed drive command before the sudden torque disorder is about 1.074 and that of the conventional drive is about 1.013. After the sudden torque disorder, the mean value of that vector is about 1.095 and about 1.01 for the advanced and classical driving methods, respectively.

The input real power of the inherent voltage inverter under the command of the advanced driving system is 5.278 kW at the operating point of 2200 rpm for the speed and 5.5 Nm for the load torque; the counterpart of the classical driving system is 4.526 kW. The output real power is 3.927 kW, and the counterpart of the classical driving system is 3.173 kW. Thus, it can be concluded that the advanced driving method can reduce switching losses from about 74.4% to 70.1% relative to the traditional driving method.

The output real power of the motor under the command of the advanced driving system is 1.267 kW at the operating point of 2200 rpm for the speed and 5.5 Nm for the load torque. The counterpart of the classical driving system is the same. The efficiency of the motor is about 32.3% under the advanced driving method and about 39.9% under the traditional driving method.

This difference is due to the fact that the real power input of the motor under the advanced method is greater than that under the classical method [6]. An increase in the usage of DC-link voltage is beneficial for decreasing copper losses by decreasing the amplitude of the stator currents, as presented in Figures 4.58 and 4.68 for the advanced and traditional driving systems separately. The amplitude of the stator currents for the controlled PMSM under the advanced driving system is lower than when the motor is driven under the classical method. The amplitudes of the stator currents during the activated period of the FMPWM algorithm are about 9.3 A and 10.6 A and about 8.64 A and 10.0 A for the period after the reference load disorder, as shown in Figures 4.59 and 4.69, respectively.

Figure 4.74 shows the FFT analysis of the stator current of phase A for the advanced and classical control strategies. Figure 4.75 shows the FFT analysis of the torque of the advanced and classical control strategies. It can be seen from these figures that the harmonic components can be significantly minimized under the control of the advanced drive system compared to the traditional drive system if the frequency modulation technique is employed on the flux-weakening range with adjusted PI controllers by the optimal parameters of the genetic optimization algorithm.

Figures 4.61 and 4.71 show the calculated induced torque based on the TOGI observer for the advanced and classical driving systems, respectively. The induced fluxes in the stationary frame of the $\alpha$-$\beta$ axes are shown in Figures 4.60 and 4.70, respectively. Figures 4.72 and 4.73 present the measured induced torque based on the $q$-axis stator current under the command of the advanced and conventional driving systems, respectively. It can be seen from these figures that the advanced control drive can successfully minimize the ripple peak of the electromagnetic torque and stator current of the $q$-axis. The advanced driving strategy can minimize the ripple peak of the induced torque when the frequency modulation technique is activated from about 35.5% to 13.38%, as presented in Figures 4.61 and 4.72. On the other hand, the controlled PMSM continues to suffer from large ripples in the ripple peak of the induced torque under the control of the traditional driving system, as presented in Figures 4.71 and 4.73.

## 4.5 SUMMARY

The contribution of this chapter is the development of a strategy for reducing torque ripple while controlling a PMSM operating in the field-decreasing region. The improved torque ripple mitigation control strategy considered in this chapter has two parts that impact peak ripple reduction. The first part of the algorithm is based on the varying switching frequency of the voltage source power inverter, which is the basis of the torque ripple mitigation control strategy. In this chapter, a frequency linear equation is proposed to change the switching frequency function of the desired and predicted torque ripples. The calculation set precedes the equation to determine the predicted torque ripple based on the lookup table to determine the predicted ripple of the stator current phase. The second part of the algorithm aims to reduce the torque ripple based on the strong stator current control of the field-oriented control (FOC) strategy to reduce the stator current oscillation. In other words, the FOC strategy's parameters are adjusted based on the improved cost function. The cost function considers not only the error minimization of PI regulators but also the index of the system stability and maximizing the stability of the advanced drive system.

The following are the significant contributions of this improved control strategy:

(1) The cost function considers the error minimization of the PI regulators based on the linearized drive system model around the selected operating points of the flux-weakening region.
(2) The cost function aims to maximize the pole value of the drive system in the negative trajectory based on the calculation of the eigenvalues of the linear system model. As a result, system stability is increased to obtain strong stator current control.
(3) The technique of offline optimization of the cost function is based on the evolutionary GA optimization.
(4) The switching frequency of the space vector pulse-width modulation for the voltage source inverter is updated based on the predicted value of the stator current.
(5) The flux observer is used to validate the torque ripple reduction based on the third-order generalized integration by estimating the induced torque.

Theoretical study of the torque ripple control strategy is presented in this chapter. Bode diagrams are drawn to measure the speed and stator current of the controlled PMSM, and a performance comparison is conducted through in-depth and systematic simulation and experimentation to validate the effectiveness of the dynamic performance of the advanced control strategy. The advanced torque ripple control strategy operates more stably with lower torque and current ripple in the flux-weakening region. In other words, the comparator is the strategy that can use more DC-link voltage, lower copper losses, and lower switching losses. The criterion also involves reducing the root mean square of the harmonics. Finally, the advanced cost function of the torque ripple control strategy can be evaluated online in future work to supply the PI parameters in real-time.

## BIBLIOGRAPHY

[1]  A. A. Elbaset, M. Ismail, and A. H. K. Alaboudy. Particle swarm optimization for layout design of utility interconnected wind parks. 2018 IEEE Power & Energy Society Innovative Smart Grid Technologies Conference (ISGT), July 2018, pp. 1–6.
[2]  S. Sivanandam and S. Deepa. Genetic algorithm optimization problems. In Introduction to genetic algorithms, Springer, 2008, pp. 165–209.

[3]   Y. Jiang, W. Xu, C. Mu, J. Zhu, and R. Dian. An improved third-order generalized inte-
      gral flux observer for sensorless drive of PMSMs. IEEE Transactions on Industrial Elec-
      tronics, January 2019, 66(12):9149–9160.

[4]   W. Xu, Y. Jiang, C. Mu, and F. Blaabjerg. Improved nonlinear flux observer-based sec-
      ond-order SOIFO for PMSM sensorless control. IEEE Transactions on Power Electron-
      ics, April 2018, 34(1):565–579.

[5]   H. Klee and R. Allen. Simulation of dynamic systems with MATLAB and Simulink.
      CRC Press, 2016.

[6]   Jahns, Thomas M., Seok-Hee Han, Ayman M. El-Refaie, Jei-Hoon Baek, Metin Aydin,
      Mustafa K. Guven, and Wen L. Soong. Design and experimental verification of a 50
      kW interior permanent magnet synchronous machine. In Conference Record of the 2006
      IEEE Industry Applications Conference Forty-First IAS Annual Meeting, 4, 1941–1948.
      IEEE, 2006.

# 5 Advanced Flux-Weakening Control for Interior Permanent Magnet Synchronous Drives

## 5.1 DRIVING STRATEGY DESIGN FOR AN INTERIOR PMSM DRIVE CONSIDERING FLUX-WEAKENING RANGE

Figure 5.1 depicts the advanced drive system of an interior PMSM designed to achieve high performance across the flux-weakening region. The reference stator voltage is calculated using the field-oriented control strategy (as presented in Chapter 4), based on the outer loop of the duty-cycle vector. The switching gate pulses of the voltage source inverter are generated using a constant-frequency space-vector modulation technique. The six gate pulses are supplied to the six semiconductor switches of the two-level inverter. The inputs of the control strategy can be summarized as follows:

(1) The counted mechanical position of the PMSM rotor ($\Theta_m$)
(2) The three-phase stator currents of the controlled PMSM.
(3) The DC voltage of the inverter (Vdc).

The advanced control strategy presented in this chapter is mainly based on the previously known parameters of the PI regulators. This chapter presents an advanced

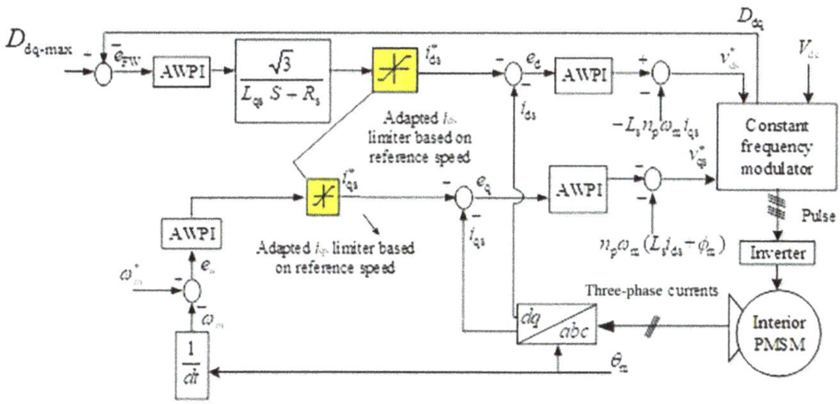

**FIGURE 5.1** Block diagram of the advanced method for the interior PMSM drive.

DOI: 10.1201/9781003320128-5

control strategy for a PMSM, building upon the field-oriented control strategy discussed in Chapter 3. The new control strategy optimizes the PI regulator parameters, which are finalized during an offline optimization procedure that relies on a genetic algorithm optimization technique, as described in Chapter 4. The objective function is modified to achieve the optimal values of the PI parameters. The duty-cycle vector's outer flux loop is used to calculate the reference value of the d-axis stator current, which adjusts the rotor flux's reduction level automatically. The limit reference value of the q-axis stator current can be determined to calculate the compensated stator voltages of the *d-q* axes. In the following sections, we discuss the design contributions of the advanced driving strategy relative to the control method presented in Chapter 3.

### 5.1.1 LIMITER MODIFICATION OF THE OUTER SPEED CONTROLLER

It is necessary for the inherent control strategy to cancel the speed overrun of the interior PMSM to obtain a high-performance drive system. The anti-windup technique discussed in Chapter 3 cannot completely keep the desired dynamic performance due to the saliency of this type of PMSM.

This section introduces the modification related to the limiter of the PI controllers of the outer control loops, as shown in Figure 5.1. The flux regulator is responsible for the output of the required *d*-axis current to adjust the motor flux level. The limit of the flux regulator can be defined in this chapter as [1]

$$i_{ds}^{*} \in \left[ 0 \quad i_{\text{ds-limit}}(\omega_{m}^{*}) \right] \tag{5.1}$$

where $i_{\text{ds-limit}}$ is the saturated value of the reference stator current of the *d*-axis, which can be computed as

$$i_{\text{ds-limit}}(\omega_{m}^{*}) = \frac{V_{dc}}{\sqrt{3}C_{e}L_{ds}\left|\omega_{m}^{*}\right|} - \frac{\lambda_{m}}{L_{ds}} \tag{5.2}$$

$$i_{\text{ds-limit}} \in \left[ 0 \quad -I_{s\text{-max.}} \right] \tag{5.3}$$

The speed regulator is responsible for the output of the required *q*-axis current to provide sufficient electromagnetic torque. The limit of the speed regulator can be calculated based on the fraction of the measured motor speed over the desired motor speed and can be defined in this chapter as [2, 3]

if

$$\left|\omega_{m} / \omega_{m}^{*}\right| < 0.98 \tag{5.4}$$

then,

$$i_{qs}^{*} \in \left[ -i_{\text{qs-limit1}}(i_{ds}^{*}) \quad i_{\text{qs-limit1}}(i_{ds}^{*}) \right] \tag{5.5}$$

where $i_{\text{qs-limit1}}$ is the saturated value of the reference stator current of the $q$-axis when the motor speed is under the reference speed, which can be computed by

$$i_{\text{qs-limit1}}\left(i_{\text{ds}}^{*}\right) = \sqrt{I_{\text{s-max.}}^{2} + i_{\text{ds}}^{*2}} \tag{5.6}$$

otherwise

$$i_{\text{qs}}^{*} \in \left[-i_{\text{qs-limit2}}\left(i_{\text{ds-limit}}, T_{\text{e-TOGI}}\right) \quad i_{\text{qs-limit2}}\left(i_{\text{ds-limit}}, T_{\text{e-TOGI}}\right)\right] \tag{5.7}$$

where $T_{\text{e-TOGI}}$ is the calculated induced torque based on the TOGI flux observer and $i_{\text{qs-limit2}}$ is the saturated value of the reference stator current of the $q$-axis when the motor speed is above the reference speed, which can be computed by

$$i_{\text{qs-limit2}} = T_{\text{e-TOGI}} / \left(n_{\text{p}}\left(\lambda_{\text{m}} + i_{\text{ds-limit}}\left(L_{\text{ds}} - L_{\text{qs}}\right)\right)\right) \tag{5.8}$$

## 5.1.2 Objective Function of the Parameters Optimization for Interior PMSM Drive

The linearized model of the drive system is the basis of this technique for optimizing the PI controller gains. The Taylor series is used for the linearization procedure to define the equilibrium in terms of the model of the drive system, as discussed in Chapter 4. The cost function can be obtained by setting the equation equal to the equilibrium fraction by zero. Lastly, the cost objective function based on the manipulated variables can be minimized for selecting the optimal PI parameters.

The state-space model of the control loops of the speed, flux, and $d$-$q$ axes currents for the control strategy neglecting the anti-windup technique can be represented as

$$\begin{bmatrix} \dot{i}_{\text{dso}} \\ \dot{i}_{\text{qso}} \\ \dot{\omega}_{\text{mo}} \\ \dot{e}_{\omega\text{o}} \\ \dot{e}_{\text{FWo}} \\ \dot{e}_{\text{do}} \\ \dot{e}_{\text{qo}} \end{bmatrix} = \begin{bmatrix} a_{11} & a_{12} & a_{13} & 0 & 0 & 0 & 0 \\ a_{21} & a_{22} & a_{23} & 0 & 0 & 0 & 0 \\ 0 & a_{32} & a_{33} & 0 & 0 & 0 & 0 \\ 0 & 0 & 0 & a_{44} & 0 & 0 & 0 \\ 0 & 0 & 0 & 0 & a_{55} & 0 & 0 \\ a_{61} & a_{62} & a_{63} & 0 & 0 & a_{66} & 0 \\ a_{71} & a_{72} & a_{73} & 0 & 0 & 0 & a_{77} \end{bmatrix} \begin{bmatrix} i_{\text{dso}} \\ i_{\text{qso}} \\ \omega_{\text{mo}} \\ e_{\omega\text{o}} \\ e_{\text{FWo}} \\ e_{\text{do}} \\ e_{\text{qo}} \end{bmatrix} \tag{5.9}$$

$$a_{11} = \frac{-R_{s}}{L_{\text{ds}}} \tag{5.10}$$

$$a_{21} = \frac{-L_{\text{ds}} n_{\text{p}} \omega_{\text{mo}}}{L_{\text{qs}}} \tag{5.11}$$

$$a_{61} = \frac{R_s}{K_{P\text{-}d}} \tag{5.12}$$

$$a_{71} = \frac{2n_p L_{ds} \omega_{mo}}{K_{P\text{-}q}} \tag{5.13}$$

$$a_{21} = \frac{L_{qs} n_p \omega_{mo}}{L_{ds}} \tag{5.14}$$

$$a_{22} = \frac{-R_s}{L_{qs}} \tag{5.15}$$

$$a_{32} = \frac{1.5 n_p \phi_m}{J} \tag{5.16}$$

$$a_{62} = \frac{-2n_p L_{qs} \omega_{mo}}{K_{P\text{-}d}} \tag{5.17}$$

$$a_{72} = \frac{R_s}{K_{P\text{-}q}} \tag{5.18}$$

$$a_{13} = \frac{L_{qs} n_p i_{qso}}{L_{ds}} \tag{5.19}$$

$$a_{23} = \frac{-n_p (L_{ds} i_{dso} + \lambda_m)}{L_{qs}} \tag{5.20}$$

$$a_{33} = \frac{-B}{J} \tag{5.21}$$

$$a_{63} = \frac{-2n_p L_{qs} i_{qso}}{K_{P\text{-}d}} \tag{5.22}$$

$$a_{73} = \frac{2n_p (L_{ds} i_{dso} + \lambda_m)}{K_{P\text{-}q}} \tag{5.23}$$

$$a_{44} = \frac{-K_{I\text{-}s}}{K_{P\text{-}s}} \tag{5.24}$$

$$a_{55} = \frac{-K_{\text{I-FW}}}{K_{\text{P-FW}}} \tag{5.25}$$

$$a_{66} = \frac{-K_{\text{I-d}}}{K_{\text{P-d}}} \tag{5.26}$$

$$a_{77} = \frac{-K_{\text{I-q}}}{K_{\text{P-q}}} . \tag{5.27}$$

The error matrix of the controllers shown in (5.9) includes the model of the four control loops shown in Figure 5.1. It can be seen from the error matrix that it depends on the input control variables of the PI regulator's gains. The new cost function can be determined as follow:

$$\text{Minimize} \left\{ \dot{e}_{\omega o} + \dot{e}_{\text{FWo}} + \dot{e}_{\text{do}} + \dot{e}_{\text{qo}} + \frac{1}{\left| \text{Max.( Real}(\lambda_j) \right) \right|} + Flag(\lambda_j) \right\}, \text{ at } j=1:7 \tag{5.28}$$

where

$$Flag(\lambda_j) = 0 \tag{5.29}$$

if the maximum of the real part of the penalty function of $\lambda_j$ is less than 0; otherwise,

$$Flag(\lambda_j) = \infty \tag{5.30}$$

The control stability parts are different for the objective function of (5.28) and the objective function presented in Chapter 3. It can be said that the selected optimal PI parameters cause a high real eigenvalue in the negative trajectory. The second further term for ensuring stability is the penalty function, which is designed to reject the selected PI parameters that cause the zero and positive eigenvalues.

## 5.2 SIMULATION RESULTS ANALYSIS

### 5.2.1 EVALUATION OF THE ADVANCED OBJECTIVE FUNCTION TO ADJUST THE PARAMETERS OF THE INTERIOR PMSM CONTROL STRATEGY

Recall that the optimum PI parameters of the control technique must be known in advance based on the offline genetic optimization algorithm discussed in Chapter 4. This section shows the evaluation result of the cost function of (5.28) with the help of MATLAB simulation software. The equilibrium points are selected from the flux weakening range. The machine parameters and information for the voltage source inverter are listed in Table 2.3 in Chapter 2. Figure 5.2 presents the fitness

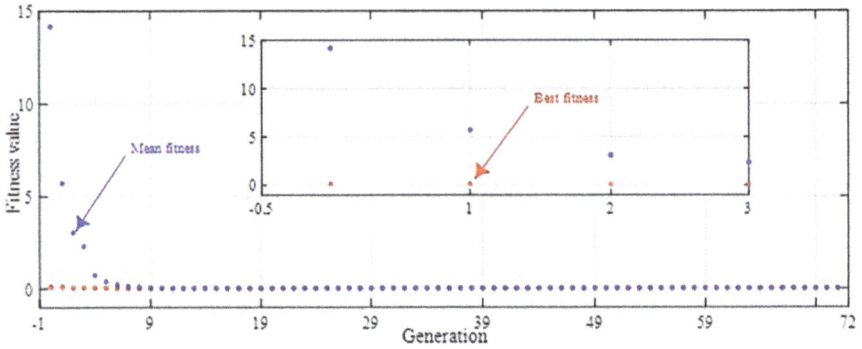

**FIGURE 5.2** Cost function evaluation of the inherent PMSM based on genetic optimization technique.

**TABLE 5.1**

**Optimal Selected PI Parameters of the Controlled Interior PMSM Based on Genetic Algorithm Optimization Procedure**

| | | |
|---|---|---|
| Outer Speed Controller | $K_{\text{I-}\omega}$ | 4.0 |
| | $K_{\text{P-}\omega}$ | 0.03 |
| $q$-axis Current Controller | $K_{\text{I-q}}$ | 940.5 |
| | $K_{\text{P-q}}$ | 1.4 |
| $d$-axis Current Controller | $K_{\text{I-d}}$ | 882 |
| | $K_{\text{P-d}}$ | 6.6 |
| Outer Flux Controller | $K_{\text{I-fw}}$ | 937.985 |
| | $K_{\text{P-fw}}$ | 7.0 |

evaluation and the corresponding iteration obtained in it. The minimum cost value is 0.00750084 and the total generation number is 71 iterations. The optimized PI parameters are listed in Table 5.1.

### 5.2.2 VERIFICATION OF DRIVE STABILITY

In order to verify the system stability, this section presents the closed-loop system responses of the $q$-axis stator current PI controller and speed controller as well as the drive system with bode diagrams. Figures 5.3 and 5.4 show the closed-loop system response of the speed controller and the $q$-axis stator current PI controller for the drive system to verify the stability of the control strategy. The bode diagrams describe robust stability thanks to the genetic optimization algorithm that sought the optimal PI parameters.

The system response shown in these figures is based on the transfer functions in which the measured motor speed and $q$-axis measure current are the output signals,

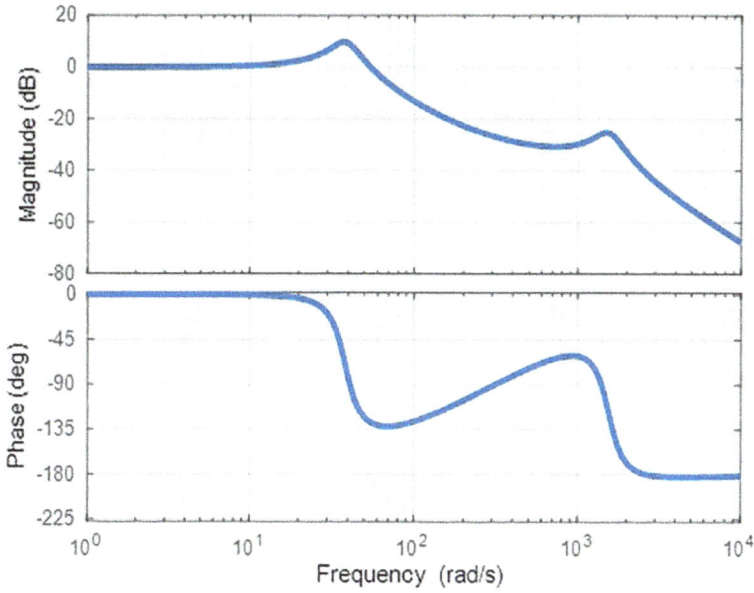

**FIGURE 5.3** Bode diagram of $\omega_m/\omega_m^*$ for the closed-loop response of the velocity regulator.

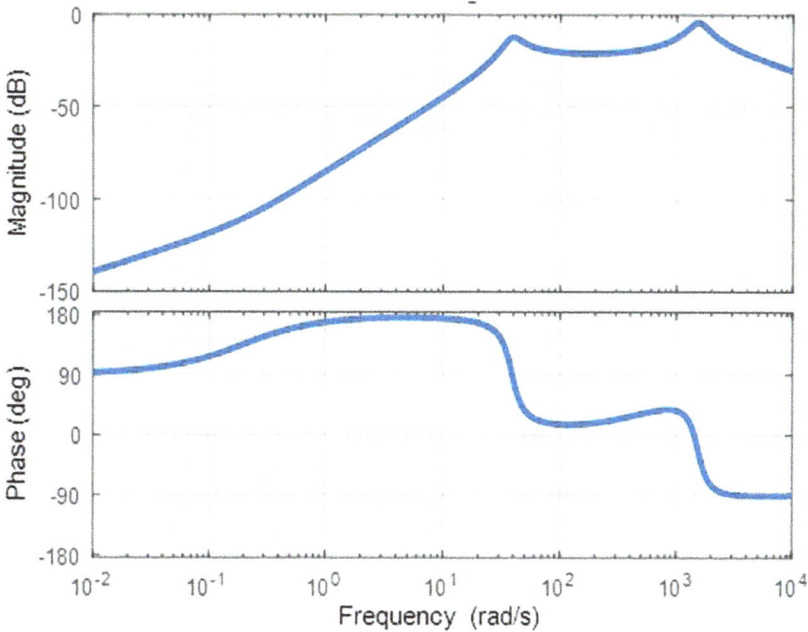

**FIGURE 5.4** Bode diagram of $i_{qs}/i_{qs}^*$ for the closed-loop response of the stator current regulator of the $q$-axis.

and the reference waves of the desired speed and $q$-axis stator current are the input signals. As a result, the gain margins are highly positive values of (118.919, 52.25 deg.) and (5560, infinity deg.), respectively. Achieving the robust stability of the drive system depends on the optimal selected PI parameters.

### 5.2.3 DYNAMIC PERFORMANCE ANALYSIS IN THE PRESENCE OF CHANGES IN THE REFERENCE TORQUE AND SPEED

The performance verification of the driving strategy considered in this chapter is presented in this section compared to the control technique discussed in Chapter 3. This section aims to validate the optimal obtained PI parameters during sudden changes in the reference speed and reference load torque. Figures 5.5 through 5.10 show the simulation case scenario in which the reference velocity is about 2200 rpm under 5.5 Nm of premier load torque. The sudden step-up change of reference velocity reaches about 2300 rpm at about 0.35 s. The load torque disturbance steps down to 0.5 Nm at about 0.7 s.

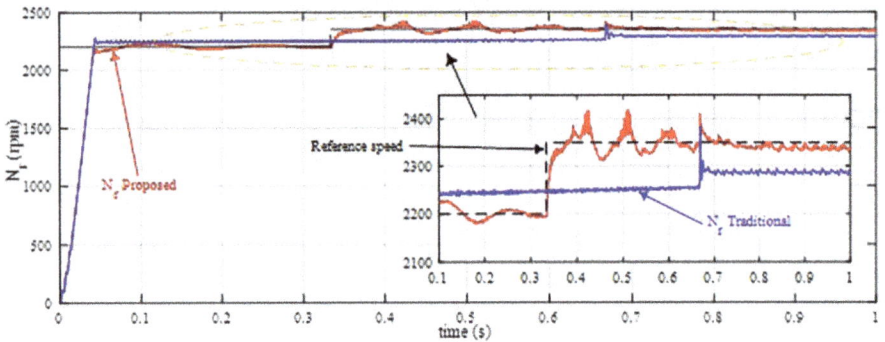

**FIGURE 5.5**   Rotational motor speed for the simulation case study of changes in the reference torque and speed.

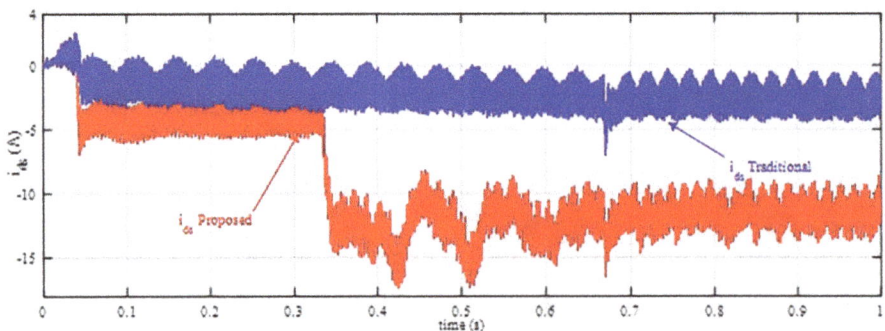

**FIGURE 5.6**   Stator currents of the $d$-axis for the simulation case study of changes in the reference torque and speed.

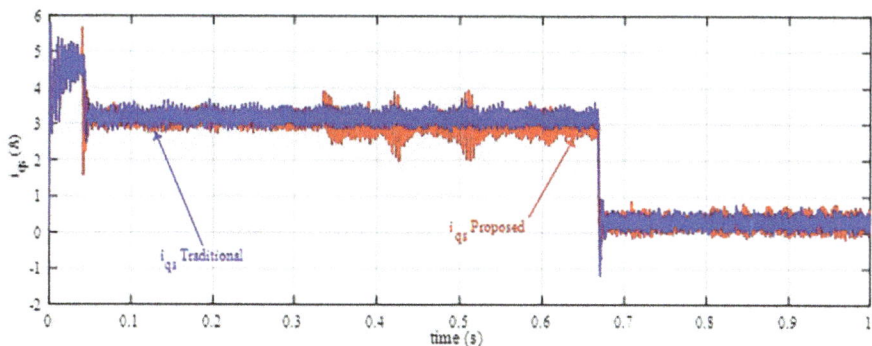

**FIGURE 5.7**   Stator currents of the $q$-axis for the simulation case study of changes in the reference torque and speed.

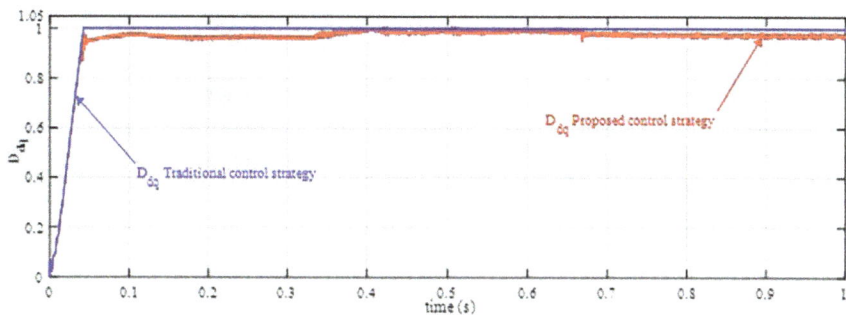

**FIGURE 5.8**   $D_{dq}$ vectors for the simulation case study of changes in the reference torque and speed.

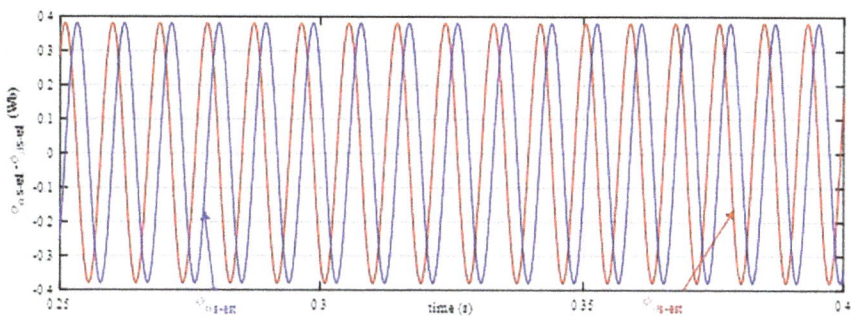

**FIGURE 5.9**   Estimated $\alpha$-$\beta$ axes fluxes based on the TOGI observer for the simulation case study of changes in the reference torque and speed.

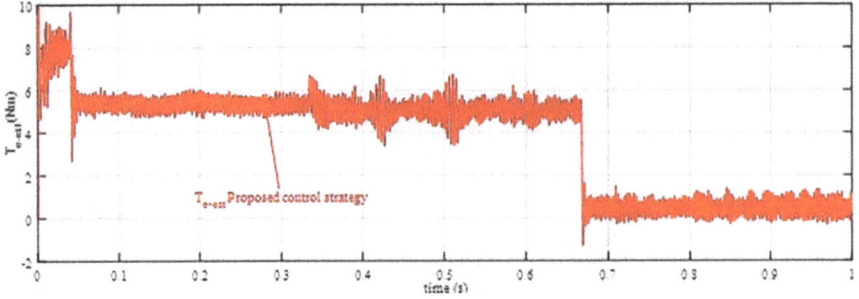

**FIGURE 5.10**   Estimated electromagnetic torque based on the TOGI observer for the simulation case study of changes in the reference torque and speed.

Figure 5.5 shows the measured motor speeds of the advanced and classical control strategies. As can be seen from this figure, the tracking ability during the flux-weakening range for the advanced driving method is better than the counterpart of the conventional control method; it cannot be affected by the disturbance of the reference torque or a change in the desired speed. There is no overshoot of the advanced motor speed. On the other hand, the dynamic performance of the conventional driving method is determined in terms of the steady-state error by about 42 rpm and 66 rpm of the pre-period and post-period of the reference torque disturbance, respectively.

Figures 5.6 and 5.7 show the measured stator currents of the $d$-$q$ axes of the advanced and classical control strategies, respectively. The average value of the stator currents of the $d$-axis is step changed due to the reference speed disorder. The average $d$-axis stator current is slightly increased in the negative trajectory in the negative directory from about -2.0 A to about -12.0 A. However, the counterpart of the traditional strategy is still constant at about -4.4 A. This is the reason behind the unsteady performance of the traditional method. The average values of the advanced and conventional drive systems are almost unaffected during the disturbance of the reference torque.

Meanwhile, the mean values of the $q$-axis stator currents under the control of the advanced and classical drive systems are still constant at about 3.2 A during the reference speed increase. These mean values are stepped down due to the reference torque disorder from about 3.2 A to about 0.5 A.

Figure 5.8 shows the duty-cycle vector of the controlled interior PMSM under the control of both advanced and conventional control strategies. The mean values during the pre-period of the torque disturbance are 0.99 and 0.98, respectively.

Lastly, Figures 5.9 and 5.10 present the estimated $\alpha$-$\beta$ axes fluxes based on the third-order generalized integrator (TOGI) observer under the advanced and classical drive methods and the estimated electromagnetic torque, respectively. It can be concluded from the methodology section of this chapter that the advanced driving method is dependent on these estimated figures.

## 5.3 EXPERIMENTAL DYNAMIC PERFORMANCE ANALYSIS IN CASE OF SUDDEN VARIANCE IN REFERENCE VELOCITY

In this section, the experimental case is presented to illustrate the performance validation of the driving method considered in this chapter compared to the control technique discussed in Chapter 3. This section aims to experimentally validate the optimal selected PI parameters during sudden changes in the reference speed in the flux-weakening section. Figures 5.11 through 5.18 show the experimental scenario in which the reference velocity of this case study is about 2150 rpm under 3.0 Nm of premier load torque. The reference speed increases suddenly to about 2300 rpm at the same initial reference load.

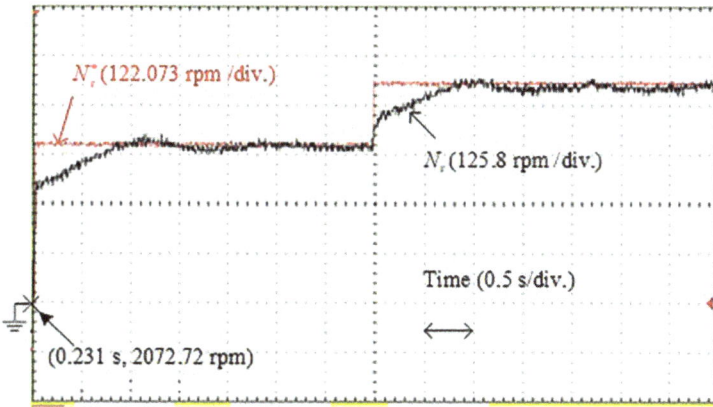

**FIGURE 5.11**    Rotational motor speed under control of the advanced strategy for the case of sudden variance in reference velocity.

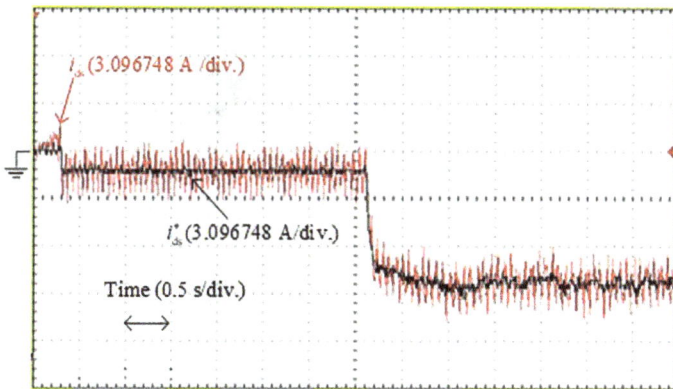

**FIGURE 5.12**    Stator current of the *d*-axis under control of the advanced strategy for the case of sudden variance in reference velocity.

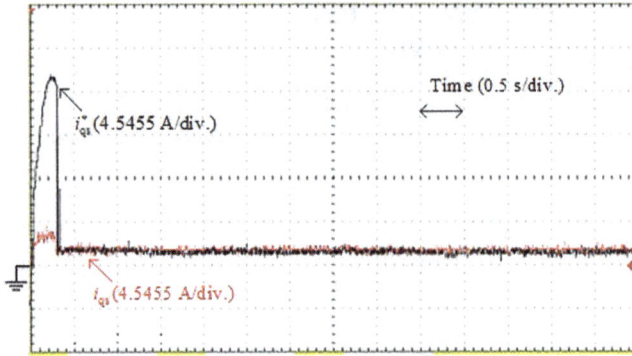

**FIGURE 5.13** Stator current of the *q*-axis under control of the advanced strategy for the case of sudden variance in reference velocity.

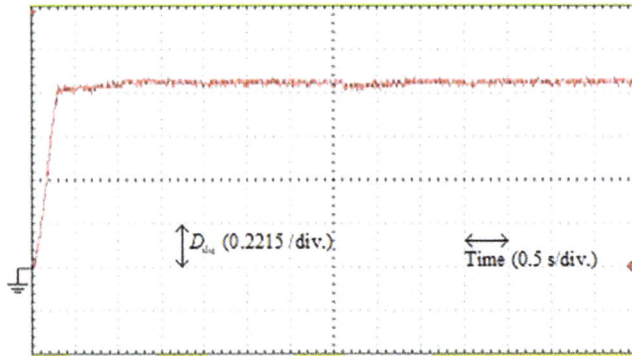

**FIGURE 5.14** $D_{dq}$ vector under control of the advanced strategy for the case of sudden variance in reference velocity.

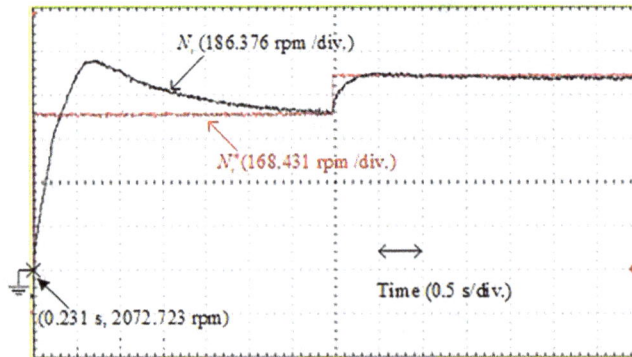

**FIGURE 5.15** Rotational motor speed under control of the conventional strategy for the case of sudden variance in reference velocity.

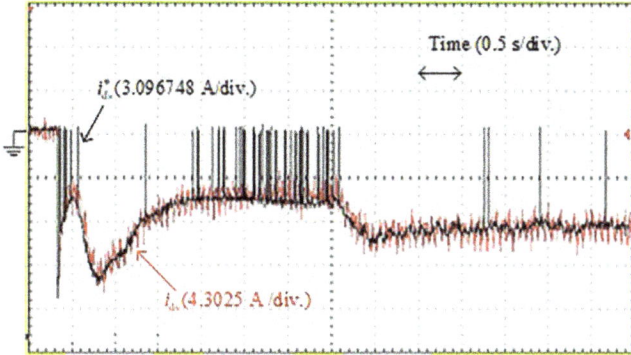

**FIGURE 5.16**   Stator current of the *d*-axis under control of the conventional strategy for the case of sudden variance in reference velocity.

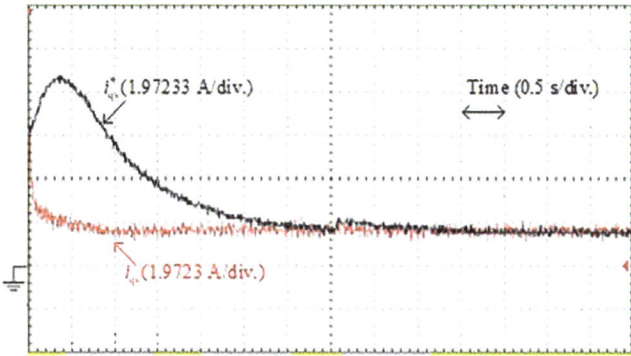

**FIGURE 5.17**   Stator current of the *q*-axis under control of the conventional strategy for the case of sudden variance in reference velocity.

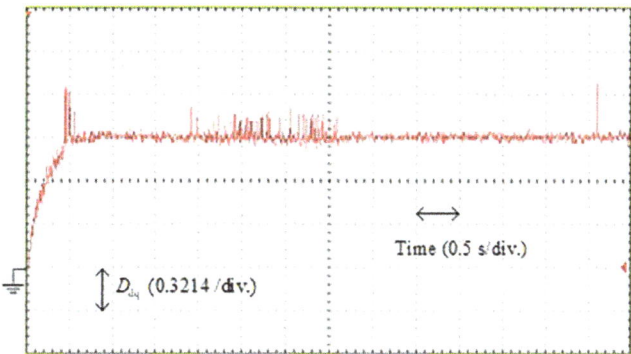

**FIGURE 5.18**   $D_{dq}$ vector under control of the conventional strategy for the case of sudden variance in reference velocity.

The advanced and traditional driving methods are implemented on the digital control unit of a TMS320F28335 float point digital signal processor (DSP) to control an interior type PMSM separately. This can study its dynamic performance to confirm the effectiveness of the advanced control technique. (The experimental laboratory used is shown in Figure 2.24.) The parameters of the controlled PMSM and power inverter are listed in Table 2.4 in Chapter 2. It should be mentioned that the DC-link voltage is saturated at 470 V, and the sample time of the control strategy is 100 μs. The hardware control unit provides a further ripple attached to the measured signal. Recall that it's difficult to measure the electromagnetic torque in this experiment because the inherent torque sensor is employed to display the electromagnetic torque on a monitor. The following sections present the performance comparison between the advanced and classical driving strategies.

Figures 5.11 and 5.15 show the measured motor velocity waves in the case of sudden variance in reference velocity under the advanced and traditional control strategies separately. It can be observed from these figures that the advanced driving method can reduce the settling time by about 85.8% compared to the traditional method. The overrun above the reference velocity is reduced to about zero for the advanced drive and about 10% for the traditional drive system. These results index the robustness of the advanced driving system in tracking the reference velocity.

Figures 5.12 and 5.16 show the measured $d$-axis stator current waves in the case of sudden variance in reference velocity under the advanced and traditional control strategies separately. Figures 5.13 and 5.17 show the measured $q$-axis stator current waves in the case of sudden variance in reference velocity under the advanced and traditional control strategies separately. It can be observed from these figures that the advanced driving method can reduce the overcurrent by about 91% for the $d$-axis stator current and 9.8% for the $q$-axis stator current compared to the traditional method. These mean values of the $q$-axis stator current of both advanced and conventional driving strategies are constant during the reference speed disorder, which equals about 1.7 A. The ripple peak of the $d$-axis stator current for the advanced driving method is lower than that of the conventional driving method.

Figures 5.14 and 5.18 shows the duty-cycle vector of the controlled interior PMSM under the control of both the advanced and conventional strategies. The mean values during the pre-period of the torque disturbance are about 1.0 and 0.99, respectively. The advanced driving method regulates the duty-cycle vector very smoothly, as opposed to the counterpart of the conventional driving method. It can be concluded from these figures that the operating points of the controlled interior PMSM are located in the flux-weakening scope.

## 5.4 SUMMARY

The main objective of this chapter is to develop an advanced control strategy that can be applied to the interior permanent magnet synchronous machine (PMSM) drive across all speed ranges. This strategy is based on the control strategies presented in Chapters 3 and 4. The reason for the development of this strategy is the differences in the values of the $d$-$q$ axes stator inductance of the interior PMSM compared to that of the surface-mounted type. The main objective of this chapter was to develop an advanced control strategy that can be applied to the interior permanent magnet synchronous machine (PMSM) drive across all speed ranges, building on the control strategies presented in Chapters 3 and 4. This is motivated by the differences in the values of the $d$-$q$ axis stator inductance of the interior PMSM compared to that of the surface-mounted type. These differences add further induced torque but complicate the control strategy implemented on the surface-mounted PMSM. Therefore, the modifications to the advanced field-oriented control method are divided into two parts:

(1) The flux observer of a third-order generalized integrator has been used to develop the limiter of the outer control loops of the velocity controller and flux controller. The induced torque is calculated based on the estimated flux to determine the maximum reference stator current of the q-axis.

(2) The PI parameters of the modified control method are optimized based on an objective function that considers the control stability terms. The control stability design increases the values of the system poles in the negative trajectory and controls the stability margin of the drive system using the penalty function.

The dynamic performance of this advanced method is compared with the performance of the traditional control method. The simulation and experimental results demonstrate that this advanced method performs effectively during sudden velocity changes and reference load torque disturbances. Consequently, the modified control strategy can reduce the settling time and cancel the overrun in the motor speed above the reference value while the motor is operating in the flux-weakening scope. The results show a stable stator current with less overshoot in the transient period.

## BIBLIOGRAPHY

[1]  J. Liu, C. Gong, Z. Han, and H. Yu. IPMSM model predictive control in flux-weakening operation using an improved algorithm. IEEE Transactions on Industrial Electronics, March 2018, 65(12):9378–9387.

[2]  Y. Jiang, W. Xu, C. Mu, J. Zhu and R. Dian. An improved third-order generalized integral flux observer for sensorless drive of PMSMs. IEEE Transactions on Industrial Electronics, December 2019, 66(12):9149–9160. doi: 10.1109/TIE.2018.2889627.

[3]  W. Xu, Y. Jiang, C. Mu and F. Blaabjerg. Improved nonlinear flux observer-based second-order SOIFO for PMSM sensorless control. IEEE Transactions on Power Electronics, January 2019, 34(1):565–579. doi: 10.1109/TPEL.2018.2822769.

# 6 Modified First-Order Flux Observer–Based Speed Predictive Control of Interior Permanent Magnet Drives

## 6.1 ANALYSIS AND DESIGN OF THE LINEAR FINITE SET MPC FOR INTERIOR PMSM DRIVE

Figure 6.1 displays the fundamental block diagram of the modified linear model predictive control (MPC) strategy. The controlled PMSM in the figure is of the interior magnet type. Typically, a two-level voltage source inverter is used as a voltage adaptor to link the control strategy with the controlled PMSM using a space-vector modulation technique. Nevertheless, in this study, the finite-set control evaluator technique [1–3] is employed in place of the space-vector modulation to determine the optimal reference voltage components of the $d$-$q$ axes.

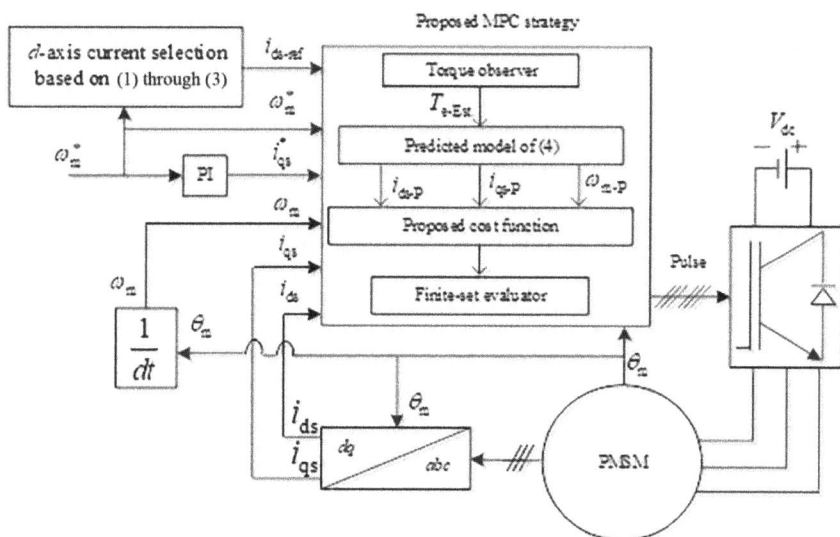

**FIGURE 6.1** Block diagram of the modified linear predictive speed control of interior PMSM drive.

DOI: 10.1201/9781003320128-6

The cost function considered in this chapter is based on the error values of the
$d$-$q$ axes' stator currents and motor speed. Therefore, a PI regulator is employed
to obtain the reference q-axis stator current from the reference motor velocity, while the required stator current of the $d$-axis component is obtained from
the model discussed in Chapter 5 based on the DC-link voltage source and the
machine parameters of direct stator inductance, permanent magnet flux, and
voltage constant, as shown in (5.2). The obtained $d$-axis reference stator current
is a demagnetizing current required for the PMSM during flux-weakening operation. Otherwise, the reference stator current of the $d$-axis component for the constant-torque region is set to zero. In other words, this advanced control method
first determines the maximum velocity in the constant-torque region to compare
it with the desired motor velocity. Second, it concludes the region required for
the PMSM to operate. Thus, the required stator current of the d-axis can be
defined as follows:

if
$$\omega_m^* < \omega_{\text{max-CT}} \tag{6.1}$$

then
$$i_{\text{ds-ref}} = 0$$

where
$$\omega_{\text{max-CT}} = \frac{V_{\text{dc}}}{\sqrt{3}C_e \lambda_m} \tag{6.2}$$

otherwise, at the range of
$$\omega_m^* > \omega_{\text{max-CT}} \tag{6.3}$$

$$i_{\text{ds-ref}} = \left(\frac{-\lambda_m}{L_{\text{ds}}}\right) + \left(\frac{V_{\text{dc}}}{\sqrt{3}L_{\text{ds}}C_e \left|\omega_m^*\right|}\right) \tag{6.4}$$

where $\omega_{\text{max-CT}}$ is the maximum velocity of the constant-torque region at the current operating condition of the PMSM. Finally, the design of the linear predictive speed control will be presented in the following sections to illustrate the
differences between the advanced method and the classical model predictive
method.

### 6.1.1 Interior PMSM Discrete Plant Model

Recall that the $d$-$q$ axes model of the PMSM is described in Chapter 2 by (2.10)
through (2.12), regardless of the eddy current, hysteresis losses, and iron saturation.
The discrete plant model of the PMSM can be expressed based on these equations in
terms of the Euler equation as shown in (6.5).

$$i_{\text{ds}}(k+1) = i_{\text{ds}}(k) + \frac{T_s}{L_{\text{ds}}}\left(v_{\text{ds}}(k) - R_s i_{\text{ds}}(k) + L_{\text{qs}} n_p \omega_m(k) i_{\text{qs}}(k)\right) \tag{6.5}$$

$$i_{qs}(k+1) = i_{qs}(k) + \frac{T_s}{L_{qs}}\left(v_{qs}(k) - L_{ds}n_p\omega_m i_{ds}(k) - R_s i_{qs}(k) - \lambda_m n_p\omega_m(k)\right) \quad (6.6)$$

$$\omega_m(k+1) = (1 - \frac{B_v T_s}{J})\omega_m(k) + \frac{T_s}{J}\left(1.5n_p i_{qs}(k)\left(\lambda_m + (L_{ds} - L_{qs})i_{ds}(k)\right) - T_L(k)\right) \quad (6.7)$$

where $k$ denotes that the time domain is converted to the discrete domain. The state variables of the PMSM plant can be written in the digital domain as $i_{ds}(k)$, $i_{qs}(k)$, $\omega_m(k)$, and $T_L(k)$.

### 6.1.2 Conventional Objective Function

Usually, the classical model predictive control strategies of the PMSM system control the $d$-$q$ axes stator current components [4]. The objective functions of these classical predictive strategies are based on the error values of the $d$-$q$ stator current components compared to their reference values.

A real-time optimization technique is implemented on this function to select the best reference voltage of the $d$-$q$ components or, as they are called, manipulated variables, which minimize the function evaluation to the nearest of zero. The conventional objective function can be expressed in the standard form as follows: [5]

$$\text{Minimize}\{(i_{ds\text{-P}} - i_{ds\text{-ref}})^2 + (i_{qs\text{-P}} - i_{qs}^*)^2\} \quad (6.8)$$

where $i_{ds\text{-P}}$ and $i_{qs\text{-P}}$ donate the predicted stator currents of the $d$-$q$ components that equal $i_{ds}(k+1)$ and $i_{qs}(k+1)$ from (6.5) and (6.6), respectively. Meanwhile, reference values of the stator currents can be calculated as discussed and shown in Figure 6.1.

### 6.1.3 Modified Objective Function

A cost function is one of the parameters that determine the performance of the model predictive control strategy [6]. The manipulated variables of the electric machine drive are usually the reference stator voltages of the $d$-$q$ axes components. In this chapter, three error value parts are considered for the objective function to reach the required motor performance in the flux-weakening region. The three parts are related to $d$-$q$ axes stator currents and motor velocity. The objective function can be written as:

$$\text{Minimize}\{W_1(i_{ds\text{-P}} - i_{ds\text{-ref}})^2 + W_1(i_{qs\text{-P}} - i_{qs}^*)^2 + W_2(\omega_{m\text{-P}} - \omega_m^*)^2\} \quad (6.9)$$

where $\omega_{m\text{-P}}$ is the next step value or the predicted rotor velocity, which can be calculated as:

$$\omega_{m\text{-P}} = (1 - \frac{BT_s}{J})\omega_m(k) + \frac{T_s}{J}\left(1.5n_p i_{qs\text{-P}}\left(\lambda_m + (L_{ds} - L_{qs})i_{ds\text{-P}}\right) - T_L(k)\right) \quad (6.10)$$

The predicted motor velocity is different relative to the predicted $d$-$q$ axes stator currents. Therefore, the weighting factors ($W_1$ and $W_2$) of the cost function are to increase and decrease the priority of minimizing the error values according to the motor working region, whether it is the constant-torque region or the flux-weakening region.

if
$$\omega_m^* < \omega_{max\text{-}CT} \tag{6.11}$$

then
$$W_1 = 0.9$$

$$W_2 = 0.1 \tag{6.12}$$

otherwise
$$W_1 = 0.05 \tag{6.13}$$

$$W_2 = 0.95 \tag{6.14}$$

It can be concluded from (6.11) and (6.14) that it is necessary to directly reduce the velocity error value in the flux-weakening range rather than those of the $d$-$q$ axes stator currents. However, the opposite is correct in the constant-torque region.

The weighting factors in the different motor range operations can be calculated based on the priority of decreasing the error value of the stator current. The following section will explain how to calculate the electromagnetic torque based on the modified low-pass filter observer to help predict motor velocity.

### 6.1.4 MODIFIED LOW-PASS FILTER OBSERVER

Recall that the modified linear finite control set predictive strategy as discussed earlier needs to determine the predictive motor velocity to evaluate the velocity error in the cost function stage. It can be seen in (6.7) and (6.10) that the calculation of the predictive velocity is based on electromagnetic torque, which can be calculated based on the estimated values of the $\alpha$-$\beta$ axes flux components. The modified low-pass first-order observer is discussed in this section. The advantages of this observer are lower DC offset and unwanted AC component waves that are generated from the mismatched parameters and the nonlinearity of the power converter plant model [7].

The connecting rotor flux in the interior PMSM can be defined in the steady-state period as:

$$\lambda_r(s) = \frac{1}{s}(v_s(s) - R_s i_s(s)) \tag{6.15}$$

where $s$ refers to the Laplace operator, $\lambda_r$ is rotor fluxes vector, $i_s$ is the stator currents vector, $v_s$ is the reference stator voltages vector, and $L_s$ is the stator inductances vector. The rotor fluxes vector, reference stator voltages vector, and stator currents vector can be expressed as

$$\lambda_r = \begin{bmatrix} \lambda_{r\alpha} \\ \lambda_{r\beta} \end{bmatrix} \qquad (6.16)$$

$$v_s = \begin{vmatrix} V_{\alpha s}^* \\ V_{\beta s}^* \end{vmatrix} \qquad (6.17)$$

$$i_s = \begin{bmatrix} i_{s\alpha} \\ i_{s\beta} \end{bmatrix} \qquad (6.18)$$

where $\lambda_{r\alpha}$, $\lambda_{r\beta}$ refer to stationary frame components rotor fluxes and $V_{s\alpha}^*$ and $V_{s\beta}^*$ are the stationary frame components of the reference stator voltages. The phase and gain of the modified low-pass observer can be calculated as

$$\theta = \tan^{-1}(\frac{\omega_e}{\omega_c}) \qquad (6.19)$$

$$G = \frac{1}{\sqrt{\omega_e^2 + \omega_c^2}} \qquad (6.20)$$

where $\vartheta$ is a phase of the low-pass filter, $G$ is the gain, $\omega_e$ refers to the operating frequency in radian per second of the interior PMSM, and $\omega_c$ is the low-pass filter cut-off frequency. The connecting rotor flux linkage can be calculated as

$$\lambda_{r\alpha\text{-}F}(s) = \lambda_{r\alpha}(s) + \frac{\omega_c}{\omega_e}\lambda_{r\beta}(s) \qquad (6.21)$$

$$\lambda_{r\beta\text{-}F}(s) = \lambda_{r\beta}(s) - \frac{\omega_c}{\omega_e}\lambda_{r\alpha}(s) \qquad (6.22)$$

Meanwhile, the rotor flux in (6.21) and (6.22) can be expressed by a block diagram as shown in Figure 6.2. The electromagnetic torque ($T_{e\text{-Est.}}$) of the modified low-pass filter observer can also be calculated based on (4.60).

The modification of this observer is to correct the reference voltage components to approximate the actual value of the stator voltage. The difference between the values of the reference voltages and the actual voltages is caused by the dead time of the space vector modulation technique and the nonlinearity of the power converter model. The corrected stator voltages in the stationary frame of $\alpha$-$\beta$ axes can be calculated as

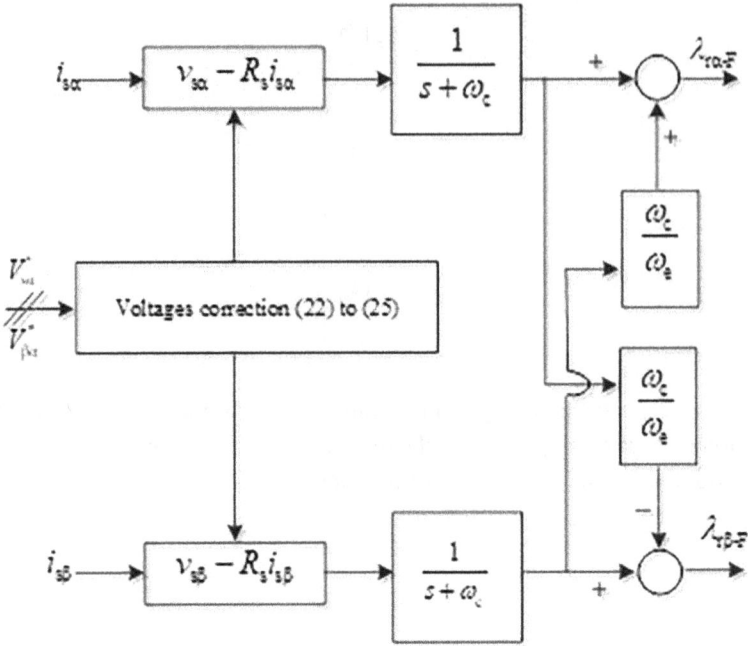

**FIGURE 6.2** Block diagram of the stationary frame flux observer based on the modified low-pass filter.

$$v_{s\alpha} = v_{s\alpha\text{-Ref.}} - \frac{1}{2}(r_{ce} + r_d)i_{s\alpha} + V_F\left(2\,\text{sign}(i_{sa}) - \text{sign}(i_{sb}) - \text{sign}(i_{sc})\right) \qquad (6.23)$$

$$v_{s\beta} = v_{s\beta\text{-Ref.}} - \frac{1}{2}(r_{ce} + r_d)i_{s\beta} + \sqrt{3}V_F\left(\text{sign}(i_{sb}) - \text{sign}(i_{sc})\right) \qquad (6.24)$$

where
$$\text{sign}(x) = \begin{cases} 1, & \text{when } x > 0 \\ -1, & \text{when } x < 0 \end{cases} \qquad (6.25)$$

$$V_F = \frac{1}{6}\left(\frac{(V_{dc} - V_{ce} - V_d)(T_{off} - T_{on} - T_d)}{T_s} - V_{ce} - V_d\right) \qquad (6.26)$$

where $r_{ce}$ is on-state active switch slope resistance, $r_d$ is the on-state freewheeling diode slope resistance, $V_{ce}$ is the active switch threshold voltage, and $V_d$ is the freewheeling diode threshold voltage. Additionally, $T_d$ is PWM dead time, $T_{on}$ is the power device turn-on times, and $T_{off}$ is the power device turn-off times. The objective function is ready now for evaluation. The following section explains the finite control set evaluation technique.

### 6.1.5 FINITE-SET EVALUATION TECHNIQUE

The finite control set technique is used instead of a real-time optimization technique to speed up the process of a predictive control technique. This evaluation technique has eight iterations for each time the strategy runs. Therefore, the number of the eight-voltage sectors relates to the two-level voltage source power converter that can be tested sector by sector in seven iterations. The voltage vector that has the minimum evaluation will be chosen as the optimal manipulated variable; thus, the objective function in (6.9) would be evaluated by the finite control set to select the optimal reference $d$-$q$ axes stator voltages. Figure 6.3 presents the on-off states of the two-level power inverter switches during each voltage vector. It can be seen from this figure that there are six working vectors, and the remaining two vectors are zero vectors.

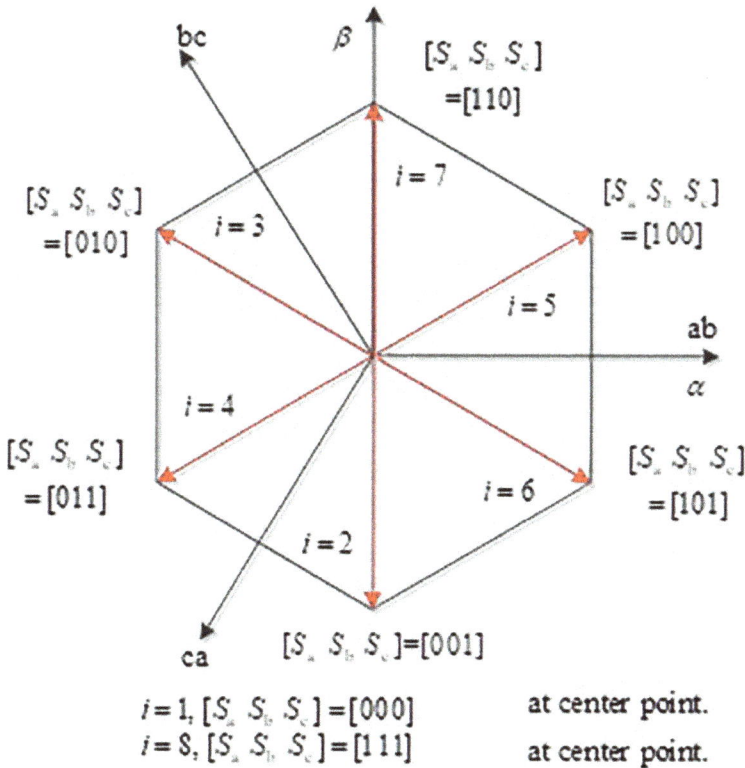

FIGURE 6.3  Two-level voltage vectors.

To explain the voltage vector computing in detail, there are $d$-$q$ axes components ($v_{ds}$ and $v_{qs}$) of each voltage vector. These components can be calculated based on the DC-link voltage, the velocity position ($\Theta_e$), and the states of the upper switches of about one or zero ($S_a$, $S_b$, and $S_c$). The park transformed from three phases to two phases is employed to achieve the voltage vector as shown in (6.27).

$$v_{ds} = \frac{2}{3}\left( v_A \cos(\theta_e) + v_B \cos(\theta_e + \frac{4\pi}{3}) + v_C \cos(\theta_e + \frac{2\pi}{3}) \right) \qquad (6.27)$$

$$v_{qs} = \frac{2}{3}\left( v_A \sin(\theta_e) + v_B \sin(\theta_e + \frac{4\pi}{3}) + v_C \sin(\theta_e + \frac{2\pi}{3}) \right) \qquad (6.28)$$

where

$$\theta_e = n_p \theta_m$$

$$v_A = \frac{V_{dc}}{3}(S_a - S_b - S_c)$$

$$v_B = \frac{V_{dc}}{3}(S_b - S_a - S_c)$$

$$v_C = \frac{V_{dc}}{3}(S_c - S_b - S_a)$$

The voltage vectors can be evaluated to select the best sector concerning the operating condition. Figure 6.4 is a flow chart showing how the finite control set is implemented in the modified linear predictive control to obtain high-performance interior PMSM drive.

## 6.2 SIMULATION RESULTS ANALYSIS

In this section, the modified predictive control strategy is verified by means of MATLAB simulations compared to the conventional model predictive control (MPC) strategy based on the classical cost function discussed earlier. The difference between the predictive and traditional strategies is that the traditional predictive control strategy has no flux observer and no direct speed control.

Meanwhile, the traditional strategy has the same finite control set evaluator technique. The major parameters of the controlled motor drive in this simulation scenario of the interior-type PMSM are shown in Table 2.3 in Chapter 2. The pulse-width modulation (PWM) technique of the inherent power inverter has a constant switching frequency of 10 kHz, and the motor voltage constant ($C_e$) of the controlled interior PMSM is 3.203. The following sections present three scenarios to validate the modified control strategy compared to the classical predictive control.

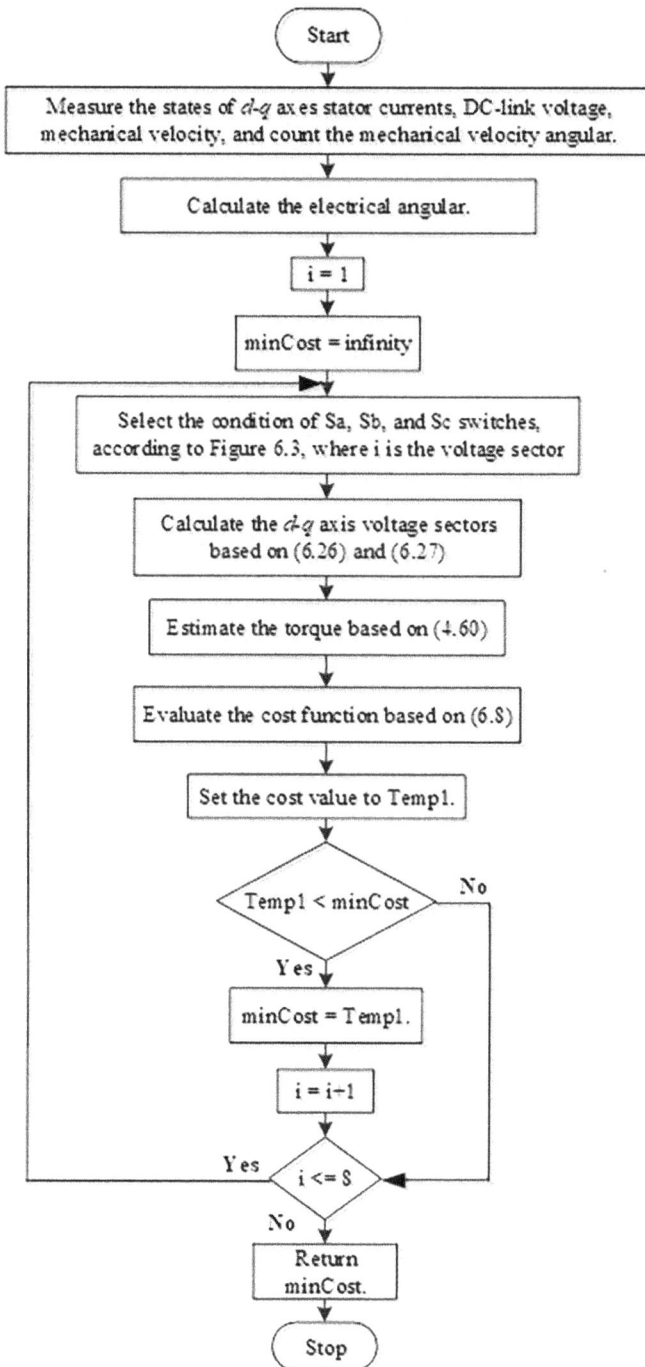

**FIGURE 6.4**   Flow chart of the finite control set procedure for the modified linear predictive control.

### 6.2.1 DYNAMIC PERFORMANCE VALIDATING OF THE LINEAR PREDICTIVE CONTROL DRIVE IN A SCENARIO OF FOUR QUADRANTS CONTROL

In this section, the dynamic performance of the controlled interior PMSM of the modified predictive control strategy is compared to the traditional predictive control strategy in the case of driving the reference speed of about 193.732 rad/s at the flux-weakening region in the forward direction and about 209.44 rad/s covering the four-quadrant regions of the drive system. The initial load torque is 3.0 Nm. Figures 6.5 through 6.9 show the four-quadrant control to track the reference speed acting as a step model.

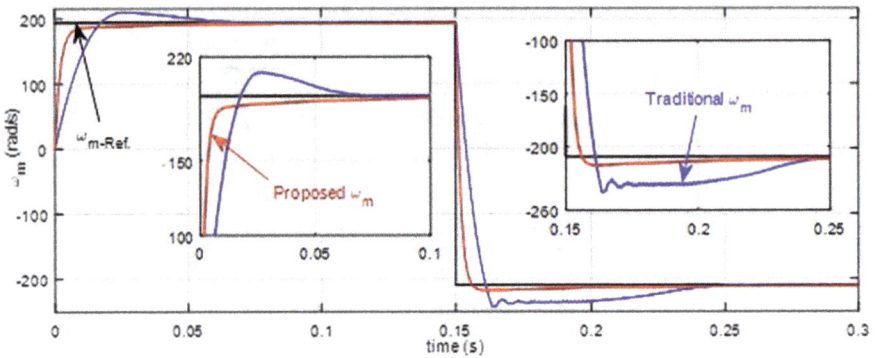

**FIGURE 6.5** Measured velocity under the control of both modified and traditional strategies for the case of four-quadrant control.

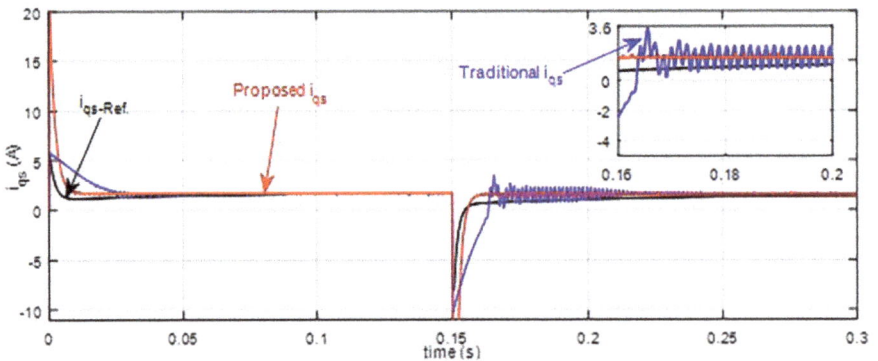

**FIGURE 6.6** Measured $q$-axes stator currents under the control of both modified and traditional strategies and the reference $q$-axis stator current (under modified strategy) for the case of four-quadrant control.

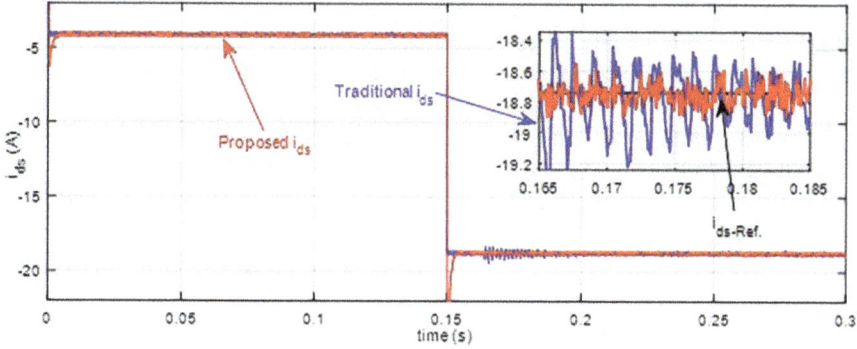

**FIGURE 6.7** Measured $d$-axis stator currents under the control of both modified and traditional strategies for the case of four-quadrant control.

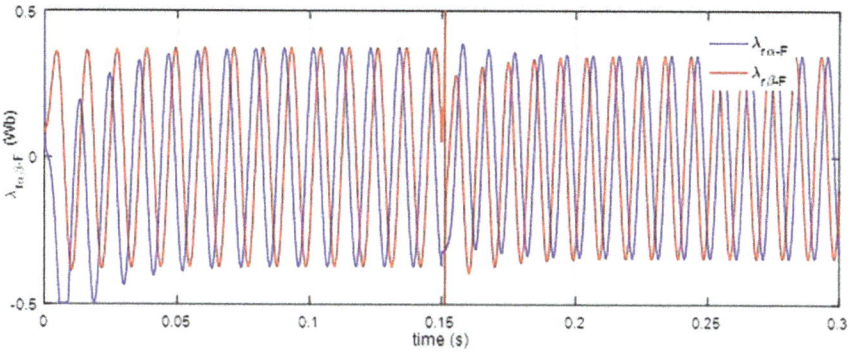

**FIGURE 6.8** Estimated $\alpha$-$\beta$ axes fluxes under the control of both modified and traditional strategies for the case of four-quadrant control.

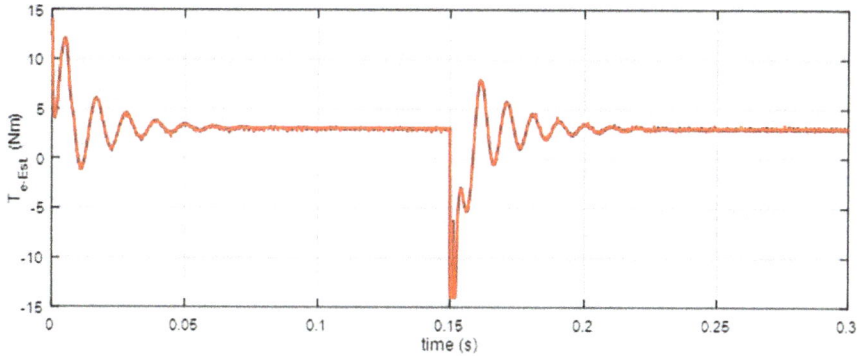

**FIGURE 6.9** Calculated electromagnetic torque under the control of both modified and traditional strategies for the case of four-quadrant control.

The modified low-pass observer has an essential role in this predictive control strategy. The predictive procedure of this method is based on the calculated induced torque, as shown in Figure 6.9, which can be determined from the estimated flux, as shown in Figure 6.8.

Figure 6.5 shows the measured speeds of the controlled PMSM under the control of both the modified and classical predictive strategies. It can be seen from this figure that the speed overshoot reaches 3.8% in the reverse direction under the control of the modified predictive control. However, this modified control method cancels the speed overshoot in the forward direction. In contrast, the counterpart for the classical predictive control is 8.2% in the forward direction and 15.5% in the reverse direction. There are no steady-state errors during the steady-state period under the control of either the modified or the classical predictive strategy. The settling time of the transient period can be reduced to 0.034 s under the control of the modified predictive model relative to the conventional predictive model. Thus, the modified predictive model can reduce the speed overshoot and speed settling time of the transient period relative to the conventional predictive control model. Figures 6.6 and 6.7 present the measured stator currents of the $q$-$d$ axes under both the modified and conventional predictive control strategies, respectively. Figure 6.6 presents the reference stator current of the $q$-axis as well. It can be seen from these figures that the modified predictive control can reduce the peak ripple of the stator currents of the $d$-$q$ axes during the section in the reverse direction.

## 6.2.2 Dynamic Performance Validating of the Linear Predictive Control Drive in a Load Torque Disorder Case

The performance verification of the modified predictive control strategy is presented in this section compared to the traditional predictive control technique during sudden changes in the reference load torque at the flux-weakening region. Figures 6.10 through 6.14 show the simulation case scenario in which the reference velocity is about 225.148 rad/s under 7 Nm of premier load torque. The load torque disturbance will step down and reach 2 Nm.

The modified low-pass observer has an essential role in this modified predictive strategy. The predictive procedure of the modified method is based on the calculated induced torque, as shown in Figure 6.14, which can be determined from the estimated flux, as shown in Figure 6.13.

Figure 6.10 shows the measured speeds of the PMSM under the control of both the modified and classical predictive strategies. It can be seen from this figure that the speed overshoot reaches 6.5% during the sudden torque disturbance under the control of the modified predictive control. In the contrast, the counterpart for the classical predictive control is 7.4% during the sudden torque disturbance. There are no steady-state errors during the steady-state period under the control of either the modified or the classical predictive strategy. The modified predictive model can reduce the speed overshoot relative to the conventional predictive control model.

Figures 6.11 and 6.12 present the measured stator currents of the $q$-$d$ axes under both the modified and conventional predictive control strategies, respectively. Figure 6.11 presents the reference stator current of the $q$-axis as well. It can be seen from these figures that the modified predictive control can reduce the peak ripple of

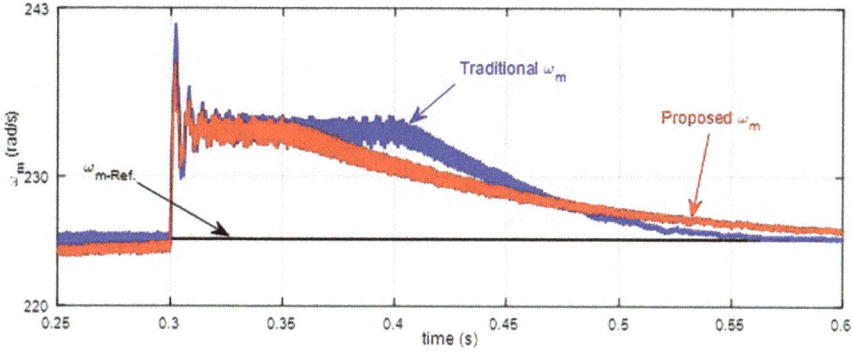

**FIGURE 6.10** Measured velocity under the control of both modified and traditional strategies for the case of a load torque disturbance.

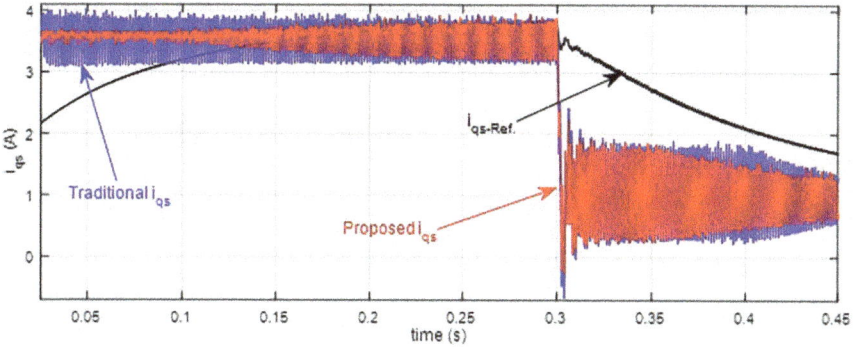

**FIGURE 6.11** Measured $q$-axes stator currents under the control of both modified and traditional strategies and the proposed reference $q$-axis stator current for the case of a load torque disturbance.

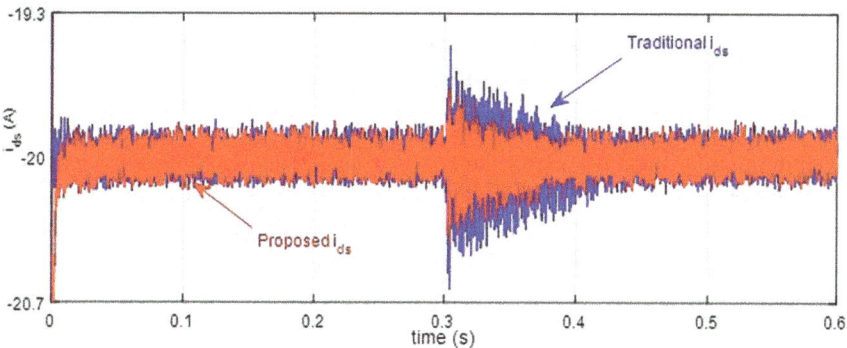

**FIGURE 6.12** Measured $d$-axis stator currents under the control of both modified and traditional strategies for the case of a load torque disturbance.

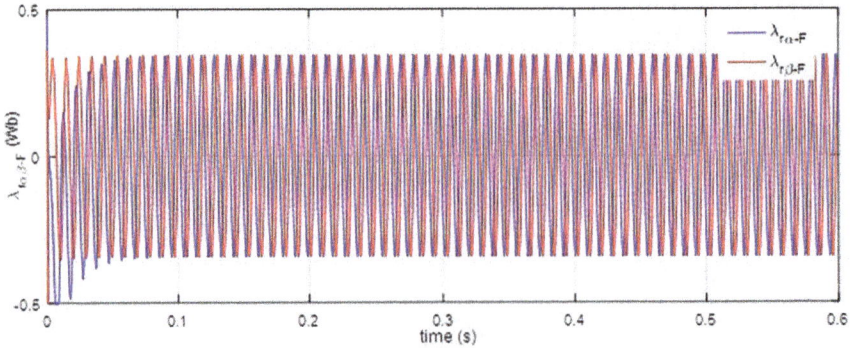

**FIGURE 6.13**   Estimated $\alpha$-$\beta$ axes fluxes under the control of both modified and traditional strategies for the case of a load torque disturbance.

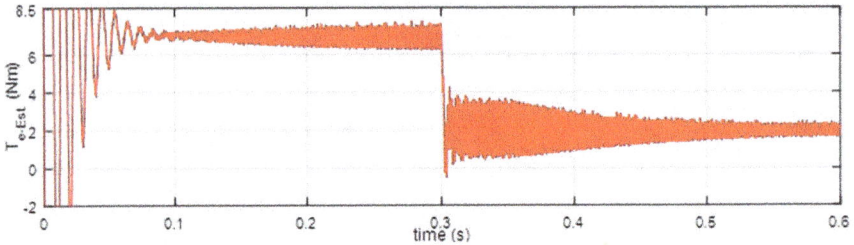

**FIGURE 6.14**   Calculated electromagnetic torque under the control of both modified and traditional strategies for the case of a load torque disturbance.

the $q$-axis stator current from 4 to 3.6 A compared to the classical predictive method under a load torque of about 7 Nm during the operation of the interior PMSM at the flux-weakening region. The peak ripple of the $d$-axis stator current is reduced as well compared to the classical method under a load torque of about 2 Nm.

## 6.2.3 DYNAMIC PERFORMANCE VALIDATING OF THE LINEAR PREDICTIVE CONTROL DRIVE IN A REFERENCE SPEED DISORDER CASE

The dynamic performance validation of the modified predictive control model is presented in this section compared to the conventional predictive model technique during sudden changes in the reference speed at the flux-weakening region. Figures 6.15 through 6.19 show the simulation case scenario in which the reference velocity is about 52.36 rad/s under 3 Nm as the initial load torque.

The desired speed disturbance will step up from the constant speed region and reach about 225.148 rad/s in the flux-weakening region. The modified low-pass observer has an essential role in the modified predictive strategy. The predictive procedure of the modified method is based on obtaining the calculated induced torque, as shown in Figure 6.18, which can be determined from the estimated flux shown in Figure 6.19.

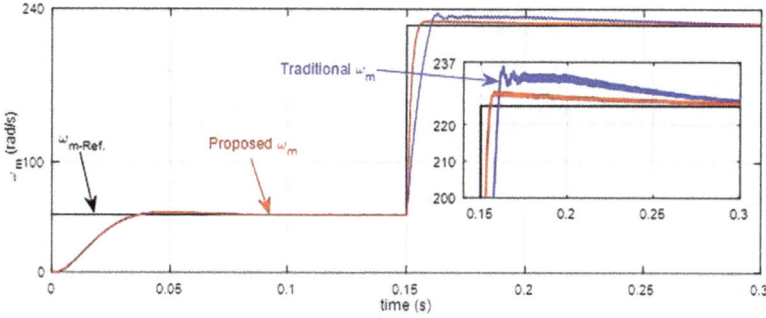

**FIGURE 6.15** Measured velocity under the control of both modified and traditional strategies for the case of a reference speed disorder.

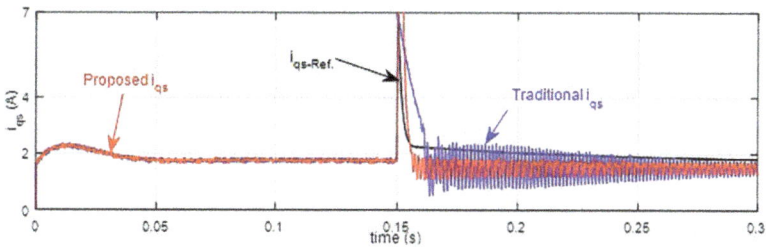

**FIGURE 6.16** Measured $q$-axes stator currents under the control of both modified and traditional strategies and the reference $q$-axis stator current (under modified strategy) for the case of a reference speed disorder.

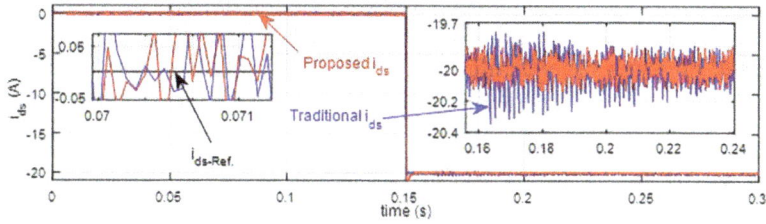

**FIGURE 6.17** Measured $d$-axis stator currents under the control of both modified and traditional strategies for the case of a reference speed disorder.

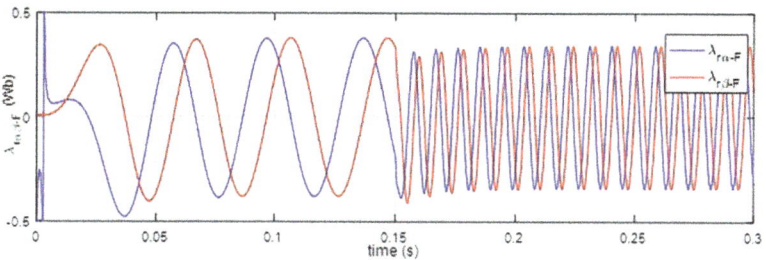

**FIGURE 6.18** Estimated $\alpha$-$\beta$ axes fluxes under the control of both modified and traditional strategies for the case of a reference speed disorder.

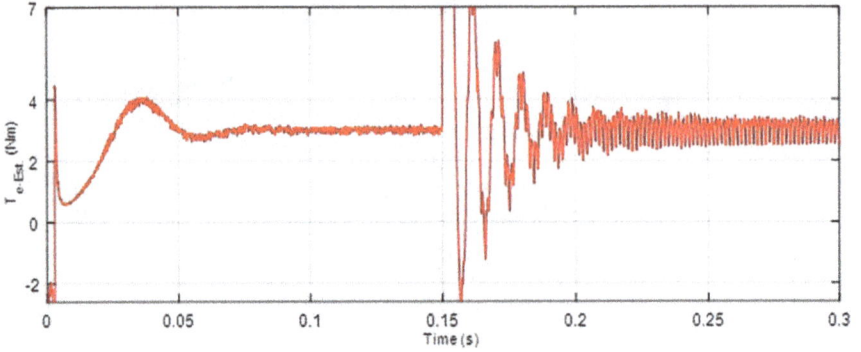

**FIGURE 6.19**  Calculated electromagnetic torque under the control of both modified and traditional strategies for the case of a reference speed disorder.

Figure 6.15 shows the measured speeds of the PMSM under the control of both the modified and the classical predictive strategies. It can be seen from this figure that the drive can easily shift from the constant-torque region to the flux-weakening region.

The speed overshoot reaches 1.6% during the sudden speed disturbance under the control of the modified predictive control. In the contrast, the counterpart for the classical predictive control is 4.8% during the sudden speed disturbance. There are no steady-state errors during the steady-state period under the control of either the modified or the classical predictive strategy.

Figures 6.16 and 6.17 present the measured stator currents of the $q$-$d$ axes under both modified and conventional predictive control strategies, respectively. Figure 6.16 presents the reference stator current of the $q$-axis as well. It can be seen from these figures that the modified predictive control can reduce the peak stator current ripples during the operation of the interior PMSM at the flux-weakening region compared to the classical predictive method.

## 6.3  SUMMARY

This chapter presents a linear finite control set model predictive control strategy to improve the dynamic performance of the interior PMSM drive. The conventional model predictive control scheme evaluates the error value of the $q$-axis stator current to indicate the motor speed and induced torque proportionally, which can result in excess speed and stator current. To address this issue, the following contributions were made:

(1) The discrete linear predictive model includes the induced torque terms, making the modified predictive model capable of representing all speed regions.
(2) The modified first-order generalized integrator is used to calculate the electromagnetic torque needed for the speed prediction calculation.
(3) The objective function directly evaluates the speed error value and the error value of the d-axis stator current to improve the control strategy's

performance. The objective function evaluates the speed error value and the error value of the d-axis stator current directly.

As a result, the modified predictive control strategy reduces the overrun during operation in the flux-weakening region. Additionally,

(4) the *d*-axis stator current model is formulated based on the machine parameter to calculate the required *d*-axis stator current for setting the flux level during the flux-weakening operation.

Simulation scenarios were conducted to validate the dynamic performance of the modified linear model predictive control, including sudden speed disorder, reference torque disturbance, and four-quadrant speed drive operation. Based on the results of the performance comparison with the conventional current predictive control strategy, it can be concluded that the modified model predictive technique presented in this chapter reduces the overrun of the motor speed and stator current and controls the ripple peak of the stator current.

## BIBLIOGRAPHY

[1]   Y. Tang, W. Xu, D. Dong, Y. Liu and M. M. Ismail. Low-Complexity Multistep Sequential Model Predictive Current Control for Three-Level Inverter-Fed Linear Induction Machines. IEEE Transactions on Industrial Electronics, 2022.
[2]   J. Liu, C. Gong, Z. Han, and H. Yu. IPMSM model predictive control in flux-weakening operation using an improved algorithm. IEEE Transactions on Industrial Electronics, March 2018, 65(12):9378–9387.
[3]   J. Liu, C. Gong, Z. Han, and H. Yu. IPMSM model predictive control in flux-weakening operation using an improved algorithm. IEEE Transactions on Industrial Electronics, March 2018, 65(12):9378–9387.
[4]   T. Geyer. Model predictive control of high-power converters and industrial drives, John Wiley & Sons, November 2016.
[5]   Z. Zheng and D. Sun. Model predictive flux control with cost function-based field weakening strategy for permanent magnet synchronous motor. IEEE Transactions on Power Electronics, February 2020, 35(2):2151–2159.
[6]   J. Zou, W. Xu, J. Zhu, and Y. Liu. Low-complexity finite control set model predictive control with current limit for linear induction machines. IEEE Transactions on Industrial Electronics, February 2018, 65(12):9243–9254.
[7]   Z. Wu, K. Lu, and Y. Zhu. A practical torque estimation method for interior permanent magnet synchronous machine in electric vehicles. PLoS One, June 2015, 10(6):e0130923.

# 7 Adaptive Linear Model Predictive Control for Flux-Weakening Control Based on Particle Swarm Optimization

## 7.1 INTRODUCTION

The intersection of state variables for the controlled plant renders the model predictive control (MPC) strategy a reliable algorithm that can handle constraints. Although the control strategy discussed in Chapter 4 demonstrates excellent dynamic performance in the flux weakening region, the control loops operate independently, precluding the algorithms from simultaneously controlling all state variables. Consequently, the conventional field-oriented control (FOC) strategy is unable to achieve the best dynamic performance because the accurate nonlinear constraints cannot be jointly applied. Therefore, in this chapter, a new adaptive model for a linear permanent magnet synchronous motor (PMSM) plant is considered to investigate motor performance. This adaptive model regulates the motor velocity in the flux-weakening region using a modulator based on a space vector pulse-width modulation (SVPWM) technique. The adaptability of the MPC strategy presented in this chapter lies in its ability to update the operating points of the plant model to more accurately express all operating regions of the controlled motor. Specifically, at each time step, the operating points of the discrete linear plant model (DLPM) are updated using mathematical formulas based on the measured speed and $d$-$q$ axes stator currents. The third-order generalized integral (TOGI) flux observer strengthens the DLPM by estimating the load torque. To reduce the overshoot of motor speed and currents, the required demagnetization current is calculated using the discussed algorithm for the persuasive reference speed. The maximum stator current is calculated as a limitation for the prediction procedure to reduce the overshoot current, and a novel real-time implementation of the adaptive velocity particle swarm optimization (AVPSO) algorithm is considered to determine the reference voltages of the $d$-$q$ axes. Simulation results and comprehensive experiments are presented to efficiently verify the control strategy. The control system considered in this chapter is compared with the conventional plant model (CPM) presented in [1], demonstrating the advantages of the new method.

The performance of the new adaptive MPC strategy is compared to that of the conventional FOC strategy discussed in Chapter 3. In this chapter, the interior permanent magnet synchronous motor (PMSM) is chosen as the controlled motor instead of the

DOI: 10.1201/9781003320128-7

surface-mounted type because of the saliency issue that causes control difficulties compared to the surface-mounted type. The following sections provide more details and explanations about the main principle of model predictive control (MPC), the new adaptive MPC approach aimed at eliminating speed overshoot and overcurrent, the real-time implementation of the adaptive velocity particle swarm optimization (AVPSO) algorithm, and the new discrete linear plant model of the interior PMSM.

## 7.2 ANALYSIS AND DESIGN OF THE ADAPTIVE MODEL PREDICTIVE CONTROL

Generally, an MPC is one of the most important advanced strategies to efficiently control the dynamic performance of electric machinery drives, as explained in Chapter 1. Figure 7.1 shows a new adaptive MPC drive system. As can be seen from the figure, the adaptive MPC controls the interior PMSM using a space vector modulation technique to generate six get pulses for the voltage source inverter (VSI). At each time step of the prediction procedure, the embedded digital controller calculates the mechanical angle and the rotor shaft speed. It also measures the $d$-$q$ axes stator currents and DC-link voltage to adjust the MPC algorithm with the current state variables. The main steps of the adaptive MPC algorithm are summarized as follows:

(1) The load torque is estimated using the TOGI flux observer mentioned in Chapter 4.
(2) The current state variables ($i_{ds}$, $i_{qs}$, and $\omega_m$) are sampled, and then the operating points of the DLPM are updated.
(3) The performance control algorithm manipulates the reference speed ($\omega_{ref}$) to reduce the speed overshoot.

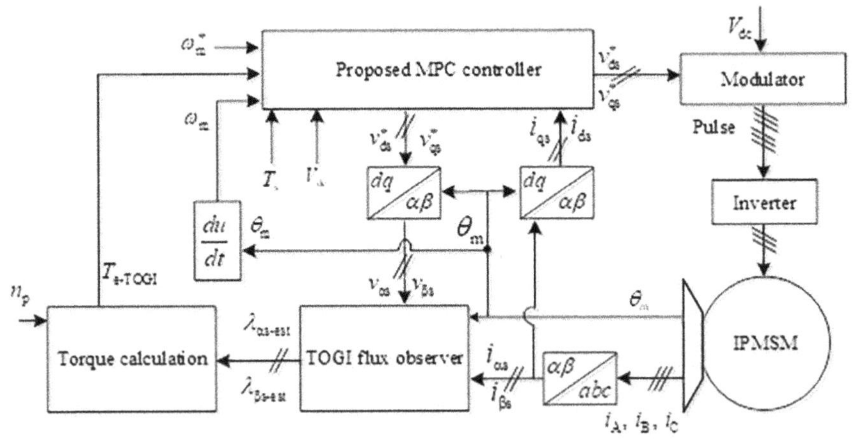

**FIGURE 7.1**   Block diagram of the adaptive MPC drive system.

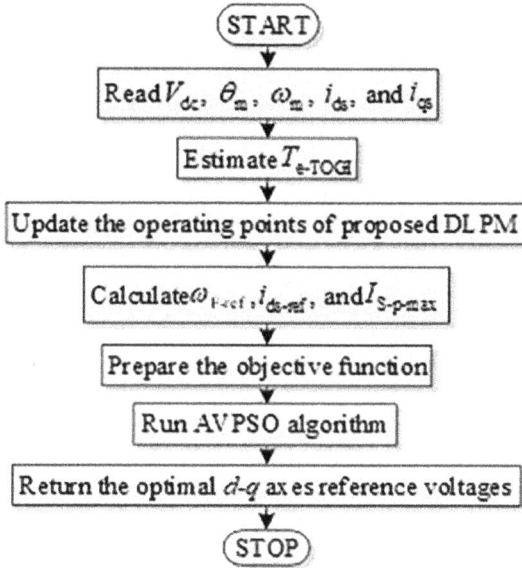

FIGURE 7.2    Flow chart of the adaptive MPC at each time step.

(4) The required reference demagnetizing current ($i_{ds\text{-}ref}$) and the maximum limit prediction of the stator current ($I_{S\text{-}P\text{-}max}$) are calculated based on the manipulated reference speed ($\omega_{F\text{-}ref}$).

(5) Finally, for interior PMSM operation, the AVPSO algorithm optimizes the cost function to find the optimal $d$-$q$ axes reference voltages ($v^*_{ds}$, $v^*_{qs}$). Figure 7.2 presents a flow chart showing the adaptive MPC steps. The following sections explain these five steps in detail.

## 7.2.1 DISCRETE LINEAR PLANT MODEL

In general, the differential equations of interior PMSM can be given as (2.10) through (2.12). The load torque is considered equivalent to $T_{e\text{-}TOGI}$ during the steady-state period. Typically, a Taylor series expansion is used to provide a small linear signal of the plant model based on the state-space representation. The following time-variant linear plant model can be implemented by the state-space:

$$\frac{dx(t)}{dt} = A(t)\, x(t) + B\, u(t) + E\, \delta T_{L}(t) \tag{7.1}$$

$$y(t) = C\, x(t) + D\, u(t) \tag{7.2}$$

$$x(t) = [\delta i_{ds}(t),\ \delta i_{qs}(t),\ \delta \omega_m(t)]^{\mathrm{T}} \tag{7.3}$$

$$u(t) = [\delta v_{ds}(t),\ \delta v_{qs}(t)]^{\mathrm{T}} \tag{7.4}$$

$$y(t) = x(t) \tag{7.5}$$

where $\delta$ indicates a small signal of the state variable, $x(t)$ is the state vector, $u(t)$ is the input vector, and $y(t)$ is the output vector. The discretization matrices of the state-space representation can be given as

$$A(3,3) = \begin{bmatrix} \dfrac{-R_s}{L_{ds}} & \dfrac{L_{qs}n_p\omega_{m0}}{L_{ds}} & \dfrac{L_{qs}n_p i_{qs0}}{L_{ds}} \\[3mm] \dfrac{-L_{ds}n_p\omega_{m0}}{L_{qs}} & \dfrac{-R_s}{L_{qs}} & \dfrac{n_p((-L_d i_{ds0})-\lambda_m)}{L_{qs}} \\[3mm] \dfrac{1.5n_p i_{qs0}(L_{ds}-L_{qs})}{J} & \dfrac{1.5n_p(\lambda_m + i_{ds0}(L_{ds}-L_{qs}))}{J} & \dfrac{-B_v}{J} \end{bmatrix} \tag{7.6}$$

$$B(3,2) = \begin{bmatrix} \dfrac{1}{L_{ds}} & 0 & 0 \\[3mm] 0 & \dfrac{1}{L_{qs}} & 0 \end{bmatrix}^{\mathrm{T}} \tag{7.7}$$

$$C(3,3) = \begin{bmatrix} 1 & 0 & 0 \\ 0 & 1 & 0 \\ 0 & 0 & 1 \end{bmatrix} \tag{7.8}$$

$$E(3,1) = \begin{bmatrix} 0 & 0 & \dfrac{-1}{J} \end{bmatrix}^{\mathrm{T}} \tag{7.9}$$

where $E$ is load torque matrix, and $D$ is the feedthrough matrix equal to zero. To clearly describe the interior PMSM plant based on the DLPM, the operating points are updated by the measured $i_{ds}$, $i_{qs}$, and $\omega_m$. To implement the time-variant linear plant model on a digital microprocessor, it should be transformed into a linear plant model with time-invariant discrete variables. Herein lies the principle of adaptability, which is the adaptation of operating points in the plant model used in the adaptive MPC control strategy. The operating points are updated in the state matrix, and then

(7.1) and (7.2) are discretized with a time step using a zero-order hold method [2], as illustrated by

$$x[k+1] = A_d[k] \, x[k] + B_d \, u[k] + E_d \, \delta T_L[k] \tag{7.10}$$

$$y[k] = C \, x[k] + D \, u[k] \tag{7.11}$$

where $k$ is the sampling time, and $A_d$, $B_d$, and $E_d$ are the discretization matrices of $A$, $B$, and $E$, respectively, as defined as

$$A_d = e^{AT_s} \tag{7.12}$$

$$B_d = A^{-1}(A_d - diag([1 \quad 1 \quad 1]))B \tag{7.13}$$

$$E_d = A^{-1}(A_d - diag([1 \quad 1 \quad 1]))E \tag{7.14}$$

The challenge from (7.12) through (7.14) is to compute $e^{AT_s}$ accurately and timely because the dimension of the state matrix is 3 by 3. Taking the Laplace inverse of (7.12) through (7.14) will get

$$A_d(i,j)(S) = \cfrac{1}{diag\!\left(\begin{vmatrix} S & 0 & 0 \\ 0 & S & 0 \\ 0 & 0 & S \end{vmatrix}\right) - A(i,j)}, \quad i=1{:}3, \; j=1{:}3 \tag{7.15}$$

The main diagonal elements of a matrix $A_d(S)$ contain the numerator of the second order and the denominator of the third order. The other elements contain the first-degree numerator and third-order denominator. After computing the inverse Laplace with the help of the fraction decomposition and then replacing each $t$ of the time domain with $T_s$, the main diagonal elements of $A_d$ can be defined as

$$A_d(i,i) = q_1 e^{-q_2 T_s} + q_3 e^{-q_4 T_s} \cos(q_0 T_s) + q_5 e^{-q_6 T_s} \sin(q_0 T_s), \quad i=1{:}3 \tag{7.16}$$

Other elements of a matrix $A_d$ can be defined as

$$A_d(i,j) = q_7 e^{-q_8 T_s} + q_9 e^{-q_{10} T_s} + q_{11} e^{-q_{12} T_s}, \quad i=1{:}3, \, j=1{:}3, \, i \neq j \tag{7.17}$$

where $q_0$, $q_1$, $q_2$, $q_3$, $q_4$, $q_5$, $q_6$, $q_7$, $q_8$, $q_9$, $q_{10}$, $q_{11}$, and $q_{12}$ are variables resulting from calculations that are closely determined by the motor parameters and the updated operating points. Finally, the DLPM is ready to be evaluated as a fitness function using the AVPSO algorithm.

## 7.2.2 ALGORITHM TO AVOID OVER SPEED AND HIGH TRANSIENT CURRENT IN FLUX-WEAKENING REGION

The high-efficiency PMSM drive system features a fast settling time, no overdrive, and no speed fluctuations. This chapter presents an algorithm for regulating rotor shaft acceleration. Focusing on the problem, the reason the speed transgresses is that the sampled speed, at $k$ sample time, is far from the reference speed. For example, the reference speed ($\omega^*_m$) is 1000 rpm, and the measured speed ($\omega_m$) is 0 rpm, -1000 rpm, or 2000 rpm. This difference in the speeds forces the MPC controller to use a large reference voltage value in an attempt to reduce this difference. But this high reference voltage causes the rotor shaft to accelerate much faster than required, making it difficult to stop at the reference speed without overrunning and speed fluctuating. Therefore, the new algorithm presented in this section controls the acceleration of the rotor shaft by two processes that can be presented as follows:

(1) At every prediction time, the control algorithm processes the reference speed ($\omega^*m$) to reduce the difference in the speeds depending on the measured and reference speeds, which can be defined as

$$\omega_{F\text{-ref}} = (\omega^*_m + \omega_m)/2 \tag{7.18}$$

where $\omega_{F\text{-ref}}$ is a persuasive reference speed. Thus, $\omega_{F\text{-ref}}$ is set in the cost function as reference speed instead of $\omega^*_m$. The aim of this step is for the MPC optimizer to extract a suitable $d$-$q$ axes reference voltage to stop the rotor shaft acceleration if necessary.

The solution of (7.18) remains enabled each time for the prediction until the difference in velocities is less than or equal to $z_1$ by about 0.2095. Besides this condition, the reference speed scale should be larger than the measured speed scale, because if (7.18) is not applied in this case, the MPC optimizer will choose large reference voltages, which should be avoided. Meanwhile, the condition of $z_1$ ensures that the solution of (7.18) is applied in case of a large difference between the speeds, but if the solution of (7.18) continues with a small difference between the speeds, the velocity will fluctuate.

(2) When the difference between the speeds is small, a persuasive reference speed will be set by a value dependent on the measured speed and the maximum speed in the region of constant torque ($\omega_{\text{max-CT}}$), which can be defined as

$$\omega_{F\text{-ref}} = \frac{\left|\omega_{F\text{-ref}}\right|}{\omega_{F\text{-ref}}}\left|\omega_m\right| - z_4\omega_{\text{max-CT}} \tag{7.19}$$

where $z_4$ is 0.020944; the goal of $z_4$ is to obtain a portion of this maximum speed to accurately process the manipulated reference speed. And $\omega_{\text{max-CT}}$ can be represented against the DC-link voltage and motor parameters, as illustrated by

$$\omega_{\text{max-CT}} = \frac{V_{dc}}{\sqrt{3}\lambda_m C_e} \tag{7.20}$$

where $C_e$ is the motor voltage constant. The implementation of the process in (7.19) has a condition to ensure that the difference between the velocities is small, which is that the difference between the reference and measured velocity scales is greater than $z_2$ and less than $z_3$. Thus, $z_2$ and $z_3$ are specified to define the difference in the velocities whether they are small or not and can be defined as 0.020944 and 1.6755161, respectively. The idea of the process (7.19) is to give $\omega_{\text{F-ref}}$ a value close to $\omega_m$, so the value of $z_4$ is small. Moreover, the process (7.19) aims to cancel the steady-state error of the speed.

Finally, Figure 7.3 illustrates a flow chart of how to calculate $\omega_{\text{F-ref}}$ at each prediction time step. For flux-weakening operation, the required demagnetizing current $(i_{\text{d-ref}})$ is calculated as

$$\text{if } |\omega_{\text{max-CT}}| < |\omega_{\text{F-ref}}| \text{ then } i_{\text{ds-ref}} = (\frac{-\lambda_m}{L_{ds}}) + (\frac{V_{dc}}{\sqrt{3}L_{ds}C_e|\omega_{\text{F-ref}}|}) \tag{7.21}$$

However, if $\omega_{\text{F-ref}}$ is predicted in the constant-torque region, $i_{\text{d-ref}}$ can be defined as

$$\text{if } |\omega_{\text{F-ref}}| \leq |\omega_{\text{max-CT}}| \text{ then } i_{\text{ds-ref}} = 0 \tag{7.22}$$

The calculating method of $i_{\text{d-ref}}$ in (7.21) is not always effective because of its dependence on motor parameters that are usually variable. To reduce the high current, the MPC control algorithm proposes a new constraint of the cost function to limit the stator current and thus defines a variable that is the maximum stator current vector $(I_{\text{S-p-max}})$. That is considered the indirect constraint on the rotor shaft acceleration that limits the stator current excited from the VSI at the transition period, having a maximum inverter current $(I_{\text{s-max}})$. The maximum limit of the stator current vector can be defined as

if

$$|\omega_{\text{max-CT}}| < |\omega_{\text{F-ref}}| \tag{7.23}$$

then

$$I_{\text{S-p-max}} = |i_{\text{ds-ref}}| + I_L$$

if

$$|\omega_{\text{F-ref}}| \leq |\omega_{\text{max-CT}}| \tag{7.24}$$

then

$$I_{\text{S-p-max}} = I_{\text{rated}}$$

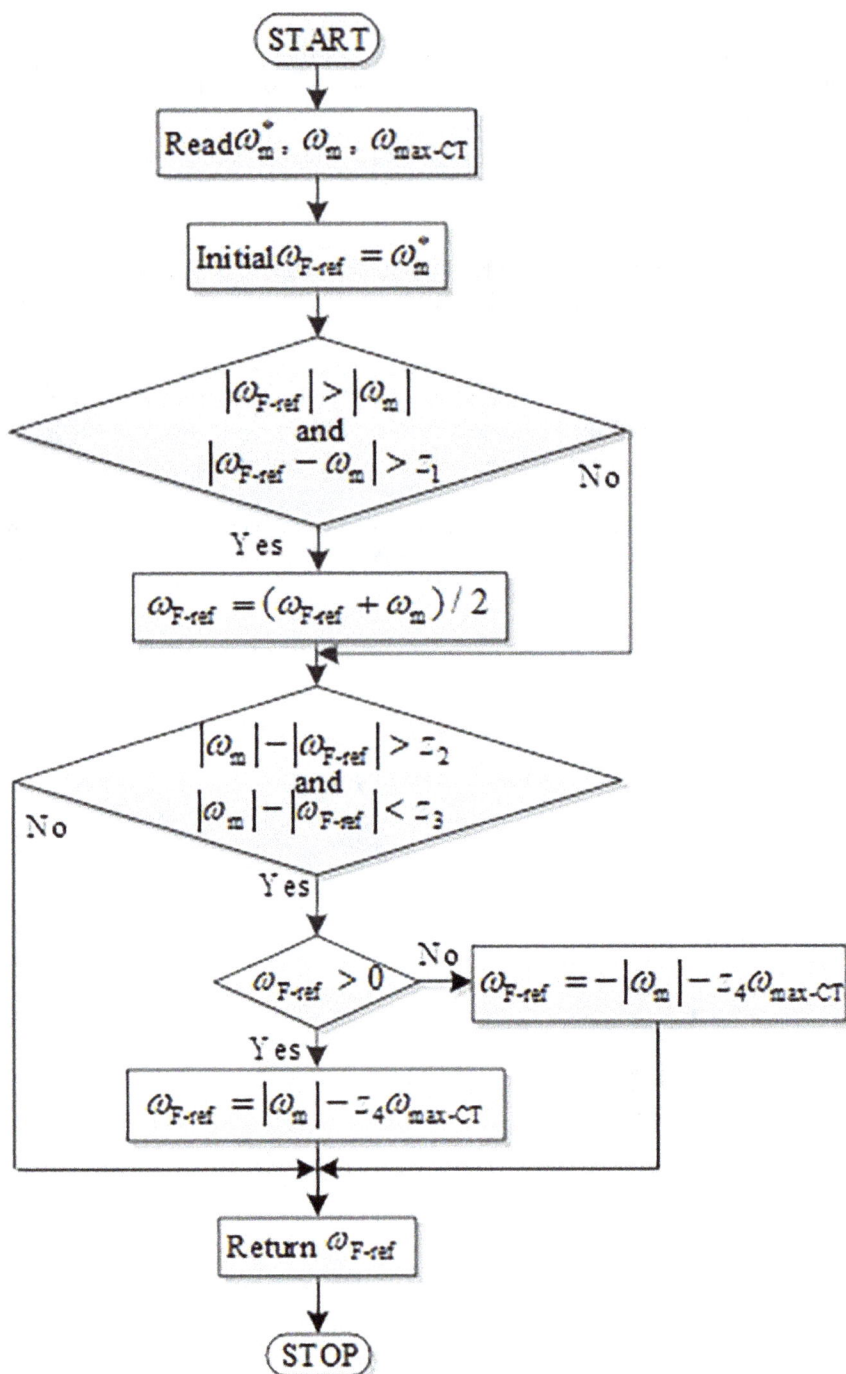

**FIGURE 7.3** Algorithm to get rid of over speed.

where $I_{rated}$ is the rated current of IPMSM, $I_L$ can be defined as

if

$$I_L < 1 \tag{7.25}$$

then

$$I_L = 1$$

otherwise,

$$I_L = \frac{T_{e\text{-}TOGI}}{1.5 n_p \lambda_m} \tag{7.26}$$

### 7.2.3 ADAPTIVE MPC COST FUNCTION

The cost function considered in this chapter can be presented by

$$\text{Minimize} \left\{ W_{id} \left| i_{ds\text{-}ref} - i_{ds\text{-}p} \right| + W_{wm} \left| \omega_{F\text{-}ref} - \omega_{m\text{-}p} \right| \right\} \tag{7.27}$$

As seen from (7.27), the objective function includes two summed components, $d$-axis stator current and motor speed, which may optimize the relevant coefficients, $W_{id}$ and $W_{wm}$.

To obtain effective solutions, the following nonlinear constraints are applied to this optimization process:

$$\sqrt{\left( \delta v_{ds} \right)^2 + \left( \delta v_{qs} \right)^2} \leq \left( \frac{V_{dc}}{\sqrt{3}} \right) \tag{7.28}$$

$$\sqrt{\left( \delta i_{ds\text{-}p} \right)^2 + \left( \delta i_{qs\text{-}p} \right)^2} \leq I_{S\text{-}p\text{-}max} \tag{7.29}$$

where $\omega_{m\text{-}p}$, $i_{ds\text{-}p}$, and $i_{qs\text{-}p}$ are the predicted motor velocity and predicted $d$-$q$ axes components of stator current during the next sample time $(k+1)$, respectively. Meanwhile, the discrete state vector can be written as

$$x[k] = \begin{bmatrix} \delta i_{ds\text{-}p} & \delta i_{qs\text{-}p} & \delta \omega_{m\text{-}p} \end{bmatrix}^T \tag{7.30}$$

The discrete input and output vectors can be written as

$$u[k] = \begin{bmatrix} \delta v_{ds} & \delta v_{qs} \end{bmatrix}^T \tag{7.31}$$

$$y[k] = \begin{bmatrix} \delta i_{ds\text{-}p} & \delta i_{qs\text{-}p} & \delta \omega_{m\text{-}p} \end{bmatrix}^T \tag{7.32}$$

Regarding the weighting factors of $W_{id}$ and $W_{\omega m}$, in the flux-weakening region, minimizing the error of the $d$-axis reference stator current is necessary to achieve the required speed. Therefore, $W_{id}$ should be larger than $W_{\omega m}$ using this method, as illustrated by

$$\text{if } |\omega_{max-CT}| < |\omega_{F-ref}| \text{ then } W_{id} = 1, W_{\omega m} = 0.5 \qquad (7.33)$$

$$\text{if } |\omega_{max-CT}| \geq |\omega_{F-ref}| \text{ then } W_{id} = 0.2, W_{\omega m} = 1 \qquad (7.34)$$

### 7.2.4 REAL-TIME IMPLEMENTATION OF THE ADAPTIVE-VELOCITY PARTICLE SWARM OPTIMIZATION ALGORITHM

The particle swarm optimization (PSO) algorithm is simple to program and contains less programming syntax [3–6]. Therefore, the adaptive velocity PSO (AVPSO) algorithm is chosen to correct the modulator's $d$-$q$ axes voltages by iteratively trying to improve a local particle concerning the given adaptive velocity [7]. It can solve the objective function of (7.27) to obtain a global minimum solution ($p_{global}$) by population size in this chapter of ten candidate solutions: i.e., local particles. The AVPSO algorithm moves the local particles in the research place and updates particle velocity and the particle position according to simple mathematical formulas that can be defined as (3.16) through (3.18), mentioned in Chapter 3, where $g$ is the iteration counter of solving the optimization problem that starts in this chapter from 1 to 50.

For real-time implementation, the population size and iterations are set to a low number, allowing the ability of the DSP28335 as a digital controller to accommodate the processing of the PSO algorithm. Initial solutions for any algorithm optimizer are among the factors controlling the access speed of the best optimal solution [8–12]. Therefore, the ten initial solutions are equivalent to the reference $d$-$q$ voltages obtained from the FOC strategy in Figure 7.4, as shown in Figure 3.1 in Chapter 3.

In Figure 7.4, the gains of four PI controllers are optimized using the PI parameters tuning method, as mentioned in Chapter 3. The proposed AVPSO algorithm is convenient for real-time control since the initial solutions are provided near optimal

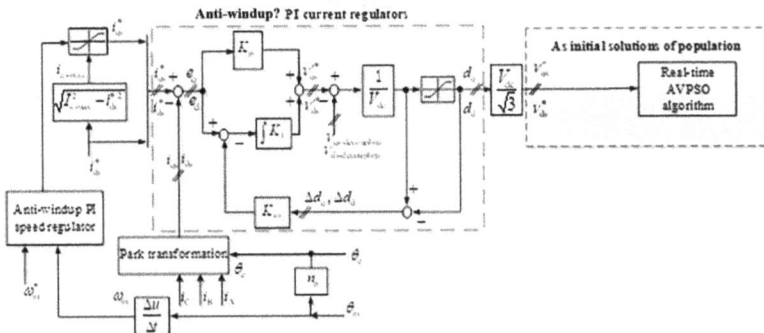

**FIGURE 7.4** Field-oriented control (FOC) strategy to adjust the initial solutions.

solutions by narrowing the research place. Afterwards, the lower and upper bounds of the manipulating variables can be defined as

$$v_{ds\_limit} = v_{ds}^* + v_{ds}^*[-0.1 \quad 0.1], \ v_{ds}^* = 0 : \frac{V_{dc}}{\sqrt{3}} \tag{7.35}$$

$$v_{qs\_limit} = v_{qs}^* + v_{qs}^*[-0.2 \quad 0.2], \ v_{qs}^* = 0 : \frac{V_{dc}}{\sqrt{3}} \tag{7.36}$$

where $v_{ds\_limit}$ and $v_{qs\_limit}$ are lower and upper bounds of $d$-$q$ axes voltages. Finally, the steps of the AVPSO implementation procedure are shown in the flow chart in Figures 7.5 and 7.6. In addition, Table 7.1 lists the optimum PI gain used in Figure 3.1 to initialize the initial solutions.

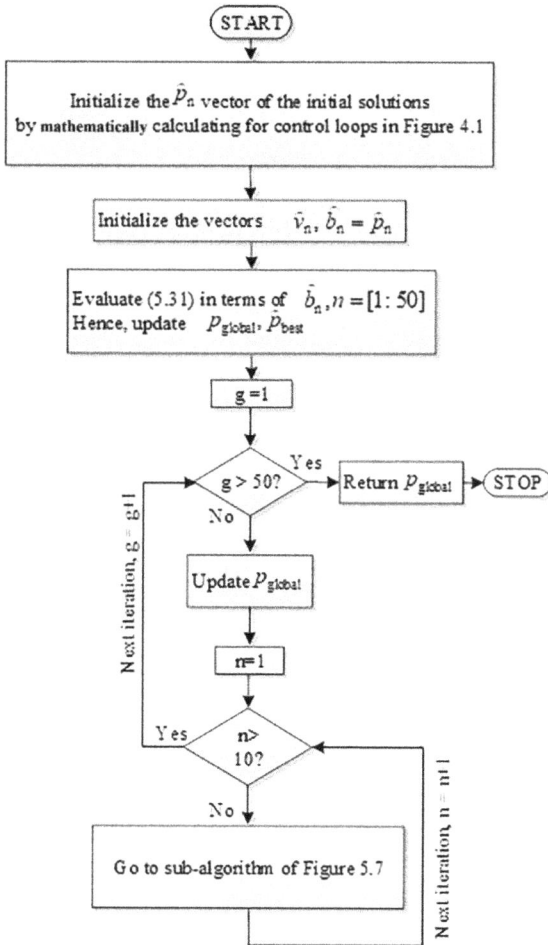

**FIGURE 7.5** MPC optimizer of AVPSO algorithm.

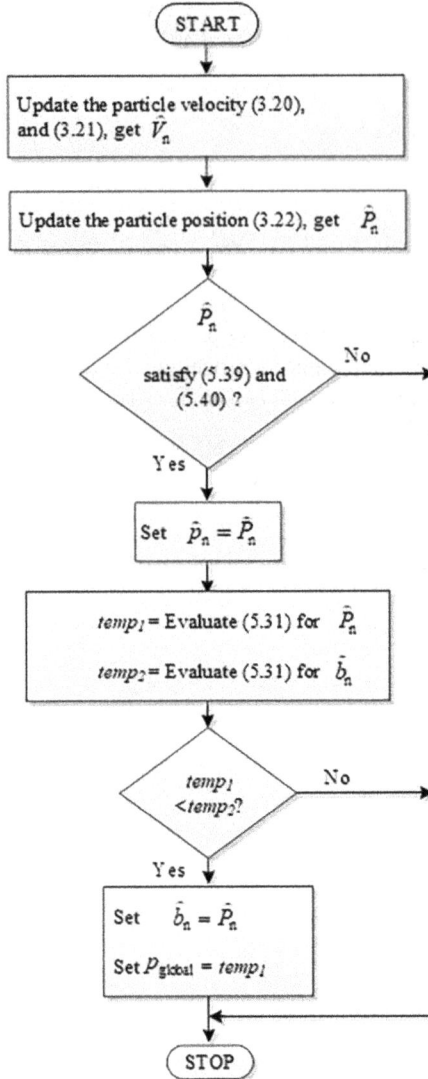

**FIGURE 7.6** A sub-algorithm for implementing the AVPSO algorithm of population size evaluation.

**TABLE 7.1**

**Optimum PI Gains of FOC Strategy to Process of Initial Solutions Adjustment**

| Speed controller | $K_{P\text{-}\omega}$ | 0.1 | $K_{I\text{-}\omega}$ | 97.67 |
|---|---|---|---|---|
| $q$-axis current controller | $K_{P\text{-}q}$ | 2.9 | $K_{I\text{-}q}$ | 98.72 |
| $d$-axis current controller | $K_{P\text{-}d}$ | 0.85 | $K_{I\text{-}d}$ | 85.94 |
| Flux weakening controller | $K_{P\text{-}FW}$ | 5.36 | $K_{I\text{-}FW}$ | 99 |

## 7.2.5 CONVENTIONAL MODEL PREDICTIVE CONTROL

The different points between the adaptive MPC and the conventional MPC (CMPC) control strategy shown in Figure 7.7 can be explained as follows [13]:

(1) The load torque must be pre-known as a constant before depending on the CLPM.
(2) The standard design of the MPC involves a linear plant model based on the Euler method can be described as follows:

$$X(k) = [i_{ds}(k), i_{qs}(k), \omega_m(k), i_{ds}i_{qs}(k), \omega_m i_{ds}(k), \omega_m i_{qs}(k), 1]^T \tag{7.37}$$

$$U(k) = [v_{ds}(k), v_{qs}(k)]^T \tag{7.38}$$

$$Y(k) = [i_{ds}(k), i_{qs}(k), \omega_m(k)]^T \tag{7.39}$$

where $X(k)$ is the state vector, $U(k)$ is the input vector, and $Y(k)$ is the output vector. The discretization matrices of the state-space representation can be given as [2]

$$B(7,2) = \begin{bmatrix} \dfrac{T_s}{L_d} & 0 & 0 & 0 & 0 & 0 & 0 \\[2mm] 0 & \dfrac{T_s}{L_q} & 0 & 0 & 0 & 0 & 0 \end{bmatrix}^T \tag{7.40}$$

$$C(7,7) = \begin{bmatrix} 1 & 0 & 0 & 0 & 0 & 0 & 0 \\ 0 & 1 & 0 & 0 & 0 & 0 & 0 \\ 0 & 0 & 1 & 0 & 0 & 0 & 0 \\ 0 & 0 & 0 & 0 & 0 & 0 & 0 \\ 0 & 0 & 0 & 0 & 0 & 0 & 0 \\ 0 & 0 & 0 & 0 & 0 & 0 & 0 \\ 0 & 0 & 0 & 0 & 0 & 0 & 0 \end{bmatrix} \tag{7.41}$$

$$A(7,7) = \begin{bmatrix} a_{11} & 0 & 0 & 0 & 0 & a_{16} & 0 \\ 0 & a_{22} & a_{23} & 0 & a_{25} & 0 & 0 \\ 0 & a_{32} & a_{33} & a_{34} & 0 & 0 & a_{37} \\ 0 & 0 & 0 & 1 & 0 & 0 & 0 \\ 0 & 0 & 0 & 0 & 1 & 0 & 0 \\ 0 & 0 & 0 & 0 & 0 & 1 & 0 \\ 0 & 0 & 0 & 0 & 0 & 0 & 1 \end{bmatrix} \tag{7.42}$$

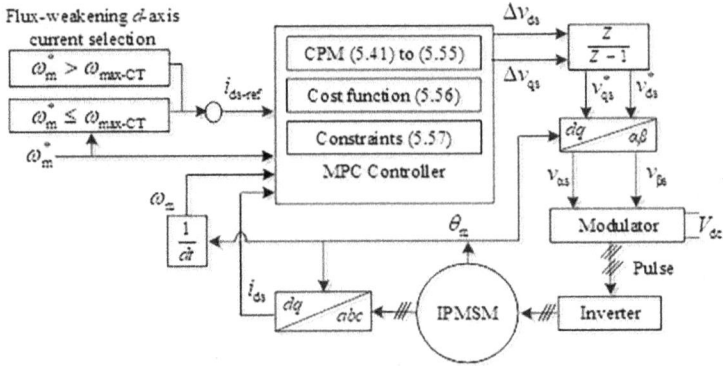

**FIGURE 7.7**   Conventional MPC block diagram [13].

$$a_{11} = 1 - (\frac{T_s R_s}{L_{ds}})$$ (7.43)

$$a_{16} = \frac{T_s n_p L_{qs}}{L_{ds}}$$ (7.44)

$$a_{22} = 1 - (\frac{T_s R_s}{L_{qs}})$$ (7.45)

$$a_{23} = \frac{-T_s n_p \lambda_m}{L_{qs}}$$ (7.46)

$$a_{25} = \frac{-T_s n_p L_{ds}}{L_{qs}}$$ (7.47)

$$a_{32} = \frac{1.5 T_s n_p \lambda_m}{J}$$ (7.48)

$$a_{33} = \frac{B_v T_s}{J} + 1$$ (7.49)

$$a_{34} = \frac{1.5 T_s n_p (L_{ds} - L_{qs})}{J}$$ (7.50)

$$a_{37} = \frac{-T_s T_L}{J}$$ (7.51)

The objective function of CMPC can be determined as

$$\text{Minimize}\left\{\left|i_{\text{ds-ref}} - i_{\text{ds-p}}\right| + \left|\omega_{\text{ref}} - \omega_{\text{m-p}}\right|\right\} \tag{7.52}$$

In order to reduce the speed and current overshoot, additional constraints can be defined as

$$\left|\Delta v_{\text{ds}}\right| \leq \Delta u_{\text{d-max}}, \quad \left|\Delta v_{\text{qs}}\right| \leq \Delta u_{\text{q-max}} \tag{7.53}$$

$$\Delta u_{\text{d-max}} = \Delta u_{\text{q-max}} = \frac{T_s V_{\text{dc}}}{\sqrt{3}\, t_s} \tag{7.54}$$

where $\Delta u_{\text{d-max}}$ and $\Delta u_{\text{q-max}}$ are approximate change rates of $d$-$q$ axes stator voltage components, and $t_s$ is the required settling time of the system, about 0.5 s. Meanwhile, the $d$-axis flux-weakening reference current ($i_{\text{ds-ref}}$) is calculated by (7.21) and (7.22) in terms of the reference speed ($\omega^*_m$). More details about CMPC can be found in [13].

## 7.3 SIMULATION RESULTS AND DISCUSSION

The adaptive MPC (PMPC) strategy is verified by means of MATLAB simulations compared to the conventional MPC strategy [14]. The given parameters and the specifications of the IPMSM motor and VSI used for simulation are listed in Table 2.3 in Chapter 2. The inverter has a maximum current limit of 20 A, the pulse-width modulation (PWM) has a constant switching frequency of 2.5 kHz, and the regulated DC-link voltage is 500 V. Figures 7.8 to 7.22 present a scenario of a sudden change in motor speed to half the value. The motor velocity works at 261.7994 rad/s (2500 rpm) as reference speed with an initial load of 3 Nm.

Figure 7.8 shows the adaptive MPC and CMPC speeds and the persuasive reference speed ($\omega_{\text{F-ref}}$). In this figure, the overshoot of the speed is canceled, and

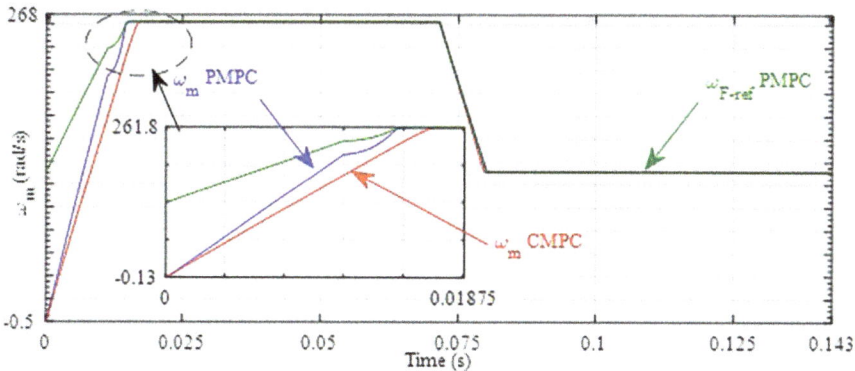

**FIGURE 7.8** Motor speeds of simulation performance for MPC strategies.

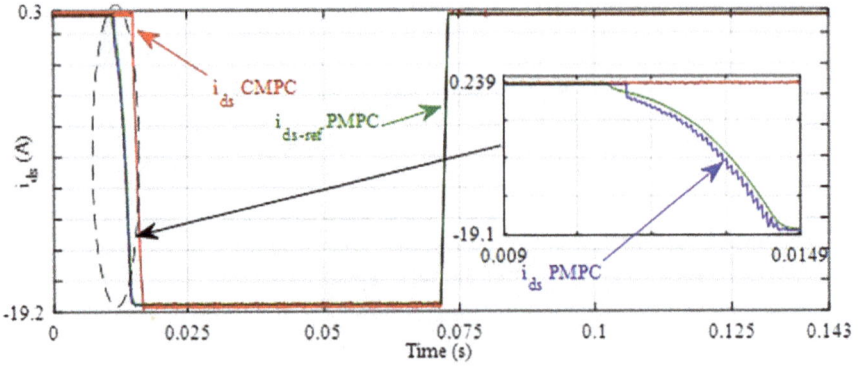

**FIGURE 7.9**   Measured $d$-axis stator current of simulation performance for MPC strategies.

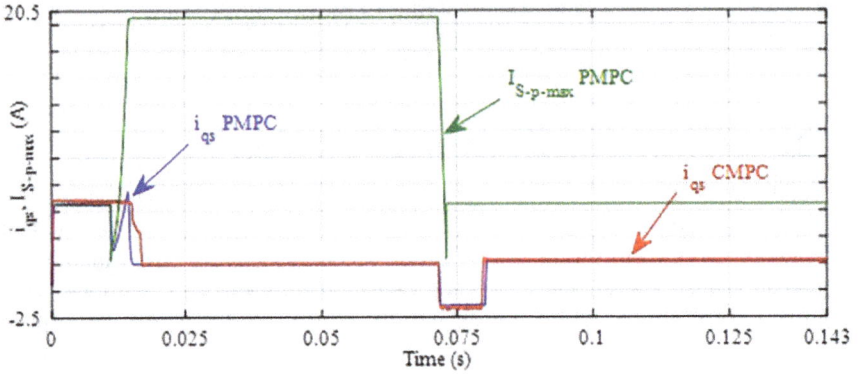

**FIGURE 7.10**   Measured $q$-axis stator current of simulation performance for MPC strategies.

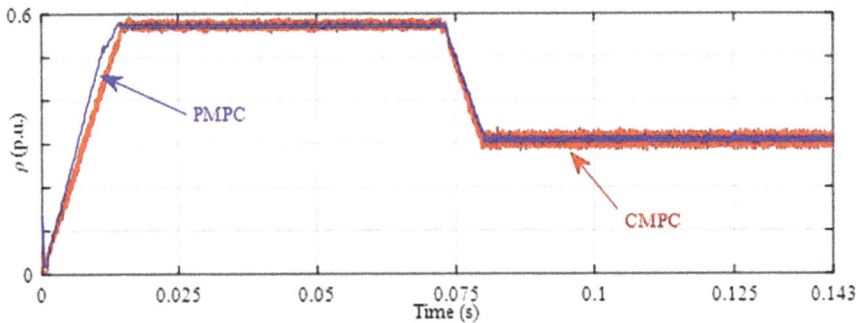

**FIGURE 7.11**   Voltage vector rated value 500 V of simulation performance for MPC strategies.

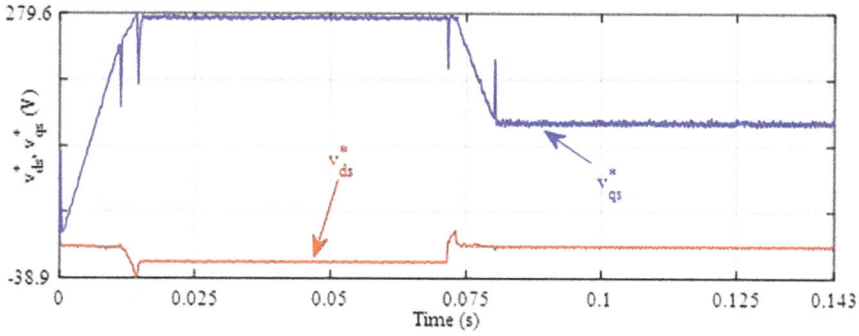

**FIGURE 7.12** Optimum manipulating variables of simulation performance for MPC strategies.

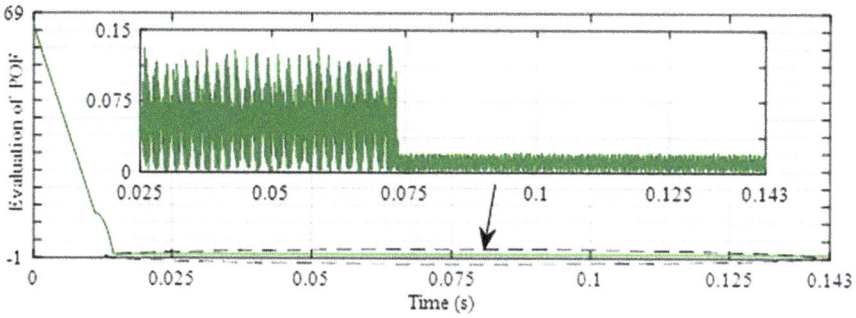

**FIGURE 7.13** Cost function evaluation of simulation performance for MPC strategies.

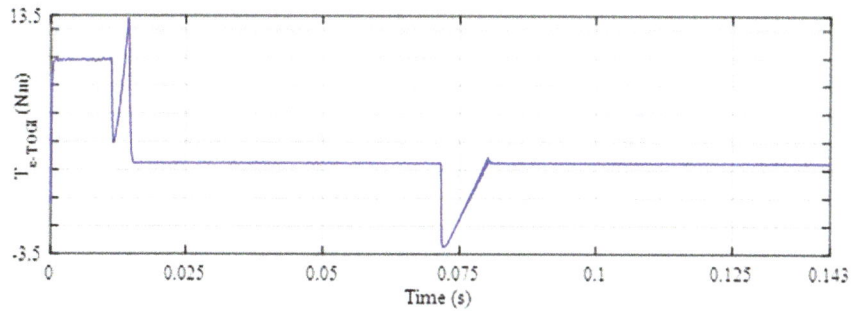

**FIGURE 7.14** Estimation of electromagnetic torque of simulation performance for MPC strategies.

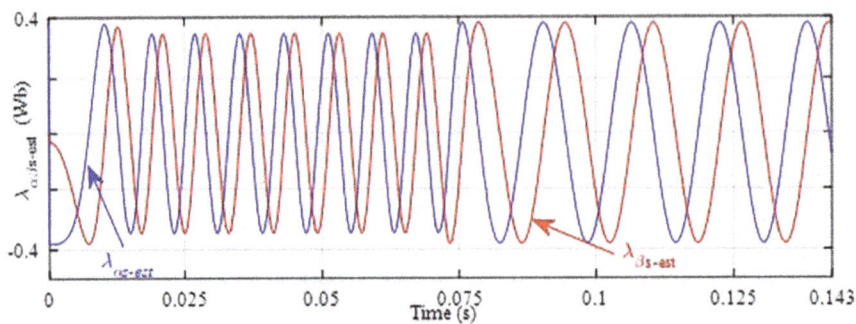

**FIGURE 7.15**  Estimation fluxes of simulation performance for MPC strategies.

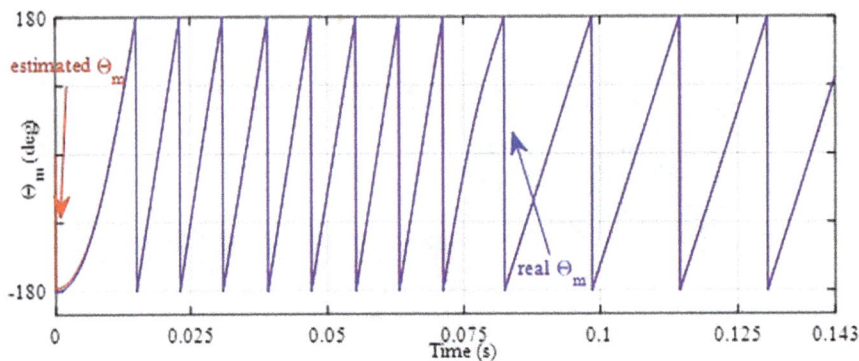

**FIGURE 7.16**  Rotor position of simulation performance for MPC strategies.

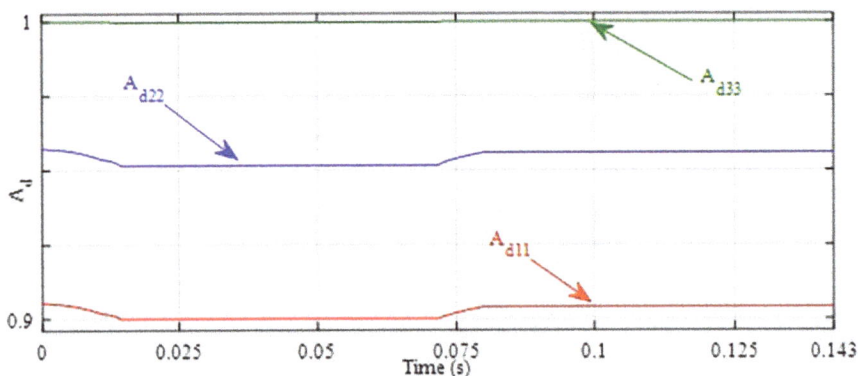

**FIGURE 7.17**  Discretization matrices of $A_{11}$, $A_{22}$, and $A_{33}$ of simulation cases for MPC strategies.

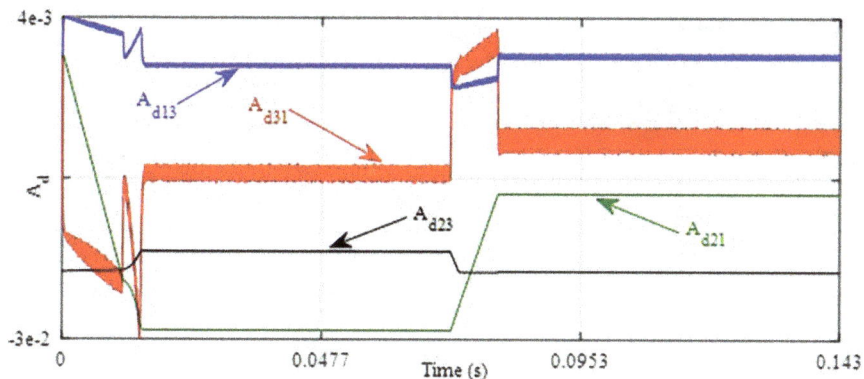

**FIGURE 7.18** Discretization matrices of $A_{13}, A_{31}, A_{23}$, and $A_{21}$ of simulation cases for MPC strategies.

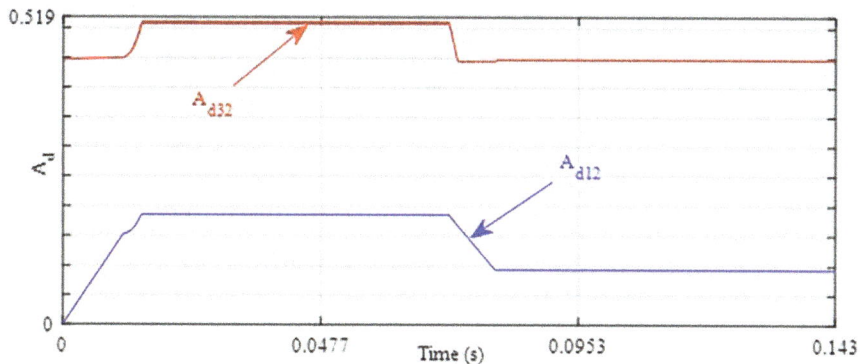

**FIGURE 7.19** Discretization matrices of $A_{32}$ and $A_{12}$ of simulation case for MPC strategies.

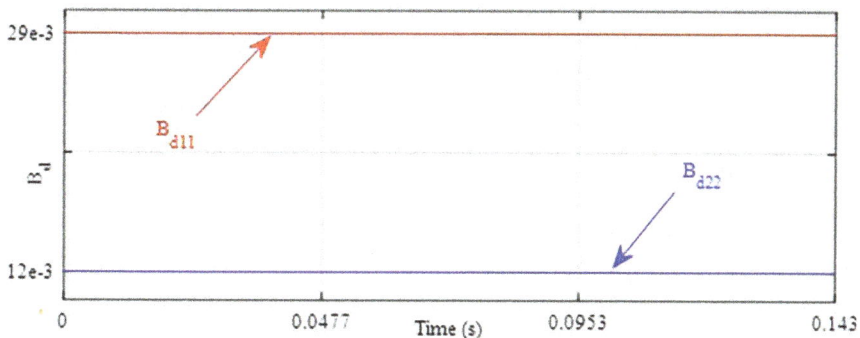

**FIGURE 7.20** Discretization matrices of $B_{11}$ and $B_{22}$ of simulation case for MPC strategies.

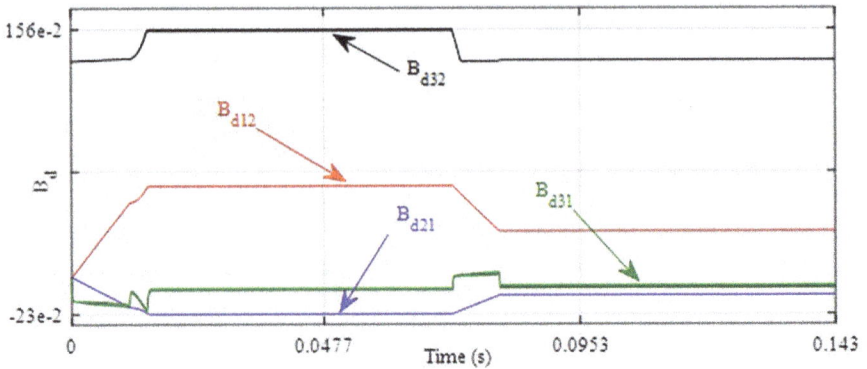

**FIGURE 7.21**  Discretization matrices of $B_{12}$, $B_{32}$, $B_{21}$, and $B_{31}$ of simulation case for MPC strategies.

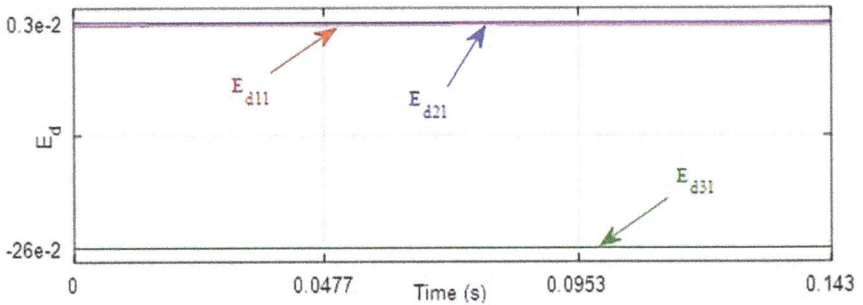

**FIGURE 7.22**  Discretization matrices of $E_{11}$, $E_{21}$, and $E_{31}$ of simulation case for MPC strategies.

the credit goes to the adaptive algorithm for calculating $\omega_{F\text{-ref}}$, which controls the shaft acceleration. The adaptive MPC speed reaches the reference faster than that of CMPC.

As can be seen from Figure 7.9, the calculated $i_{ds\text{-ref}}$ and $d$-axis currents are displayed under both strategies. The $d$-axis demagnetizing current of the adaptive MPC tracks the calculated $i_{ds\text{-ref}}$ having a mean value at a steady state of -18.74 A. Meanwhile, the CMPC current has the same mean value, but the breakout time is delayed compared to the adaptive MPC due to the operation delay in the flux-weakening region. When reducing the velocity, the $d$-axis demagnetizing currents for both strategies fade to zero because the reference speed is below the maximum speed in the region of the constant torque ($\omega_{\text{max-CT}}$) of about 71.14 rad/s (679.35 rpm).

Figure 7.10 presents the $q$-axis currents under both control strategies and calculates $I_{S\text{-p-max}}$. In this figure, the $q$-axis current of the adaptive MPC strategy does not contain a high transient current. The new strategy does not limit the rate of the $d$-$q$ axes reference voltages but depends on the definition of the maximum prediction stator current. The performance of the $q$-axis current in the transient period reflects the significance of the constraint of $I_{\text{S-p-max}}$. Moreover, the mean value in the steady-state period is the same for both MPC controllers, which increases slightly from about 1.5 to 1.7 A.

Figure 7.11 shows the magnitude of the voltage vector for both strategies per unit, with the base voltage being 500 V. The new adaptive calculation algorithm for $\omega_{\text{F-ref}}$ and $I_{\text{S-p-max}}$ controls the rate of the reference voltage variation better than the limitation defined by CMPC of the voltage rate.

Figures 7.12 and 7.13 present the optimum value of the manipulating variables and the evaluation of the cost function, respectively. In Figure 7.12, the voltage rate of the adaptive MPC is controlled by the new algorithm for avoiding speed overshoot and high transient currents.

Figure 7.14 shows the estimated torque relied on by the TOGI observer to support the adaptive IPMSM linear model. The estimated torque is calculated in every instance based on the $\alpha$-$\beta$ axes fluxes in the stationary frame, as shown in Figure 7.15. In Figure 7.15, the estimated rotor position compared to the real position testifies to the efficiency of the TOGI observer, where the estimated rotor position can be calculated as $tan^{-1}(\lambda_{\beta s}/\lambda_{\alpha s})$.

Figures 7.17 through 7.22 present the instant time value of the $A$, $B$, and $E$ discretization matrices. As noted in these figures, their instant values differ from each time step, which makes the cost function like the real conditions of IPMSM.

## 7.4 EXPERIMENTAL RESULTS AND DISCUSSION

In this section, to verify the effectiveness of the new algorithm, three experimental scenarios for the IPMSM were carried out in the laboratory using the controller unit of the floating-point digital signal processor (DSP) TMS320F28335. The laboratory prototype is shown in Figure 2.24 in Chapter 2. The measured signals have an additional ripple due to the hardware noises. The sampling frequencies of VSI and MPC are set to 10 kHz and 2 kHz, respectively. The motor parameters are listed in Table 2.3 in Chapter 2 as well as the control parameters. The proposed AVPSO algorithm is used to optimize the CPM. In the experimental scenarios, the DC-link voltage is changed to demonstrate that the DLPM efficiently represents the IPMSM at any DC-link voltage. According to (7.20) and the voltage-limiting ellipse explained in Chapter 2, the maximum velocity in the region of constant torque ($\omega_{\text{max-CT}}$) is proportional to the DC-link voltage.

Thus, the reference velocities in the flux-weakening region are set differently in each scenario. Table 7.2 also lists the inverter voltage and the required high operating speed, considering that the inverter current ($I_{\text{s-max}}$) is constant.

**TABLE 7.2**

**Reference Velocities in Each Case**

| | $V_{\text{dc}}$ **(V)** | $\omega_m^*$ **(rad/s)** | $\omega_{\text{max-CT}}$ **(rad/s)** |
|---|---|---|---|
| Case 1 | 150 | 75.4 | 71.14 |
| Case 2 | 500 | 261.8 | 237.1379 |
| Case 3 | 300 | 143 | 142.2828 |

### 7.4.1 Dynamic Performance of Four-Quadrant Control Study

Figures 7.23 to 7.30 (Case 1) present the dynamic performance of IPMSM under PMPC and CMPC with a load torque of about 2.5 Nm. In this case, a four-quadrant operation scenario is provided with a reference speed of 75.4 rad/s (720 rpm), and the DC bus voltage of the two-level VSI is 150 V. Meanwhile, the performance of the TOGI observer is rated by measured rotor position and matched with estimated position, which can be calculated as $tan^{-1}(\lambda_{\beta s}/\lambda_{\alpha s})$.

The calculation of $\omega_{F\text{-ref}}$, as shown in Figure 7.23, increases the acceleration of the suggested MPC velocity without speed overpeering. As shown in this figure, the adaptive MPC has a settling time less than CMPC and equal to about 0.454 s, while the counterpart of CMPC is 0.53 s. Figure 7.23 also shows that the adaptive MPC is rapidly and smoothly shifted between the four quadrants. Meanwhile, the reason the adaptive MPC behaves faster than the classical MPC is that the authors of the conventional MPC strategy in Reference [1] used an approximate change rate of $d$-$q$ axes reference stator voltage to cancel the speed overshoot and overcurrent. They did not mention how the rate of change was calculated. They even have a fast transient time of motor speed in their study because they chose an acceptable rate of change with their application.

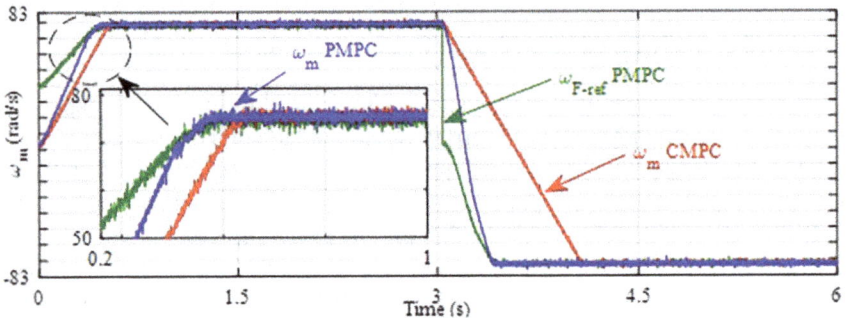

**FIGURE 7.23**  Motor velocity (Case 1).

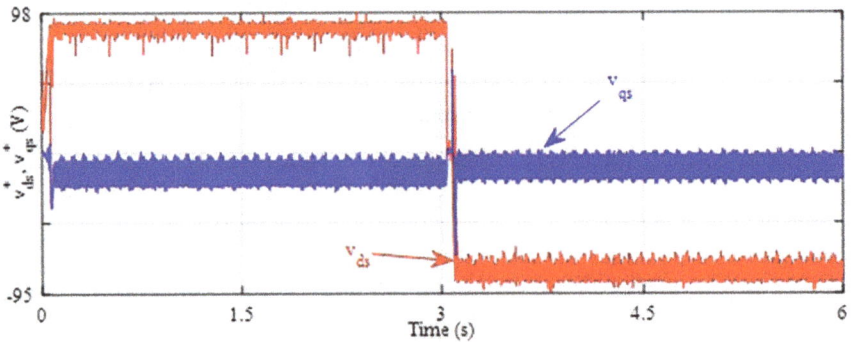

**FIGURE 7.24**  Optimized $d$-$q$ axes stator voltages of the adaptive MPC (Case 1).

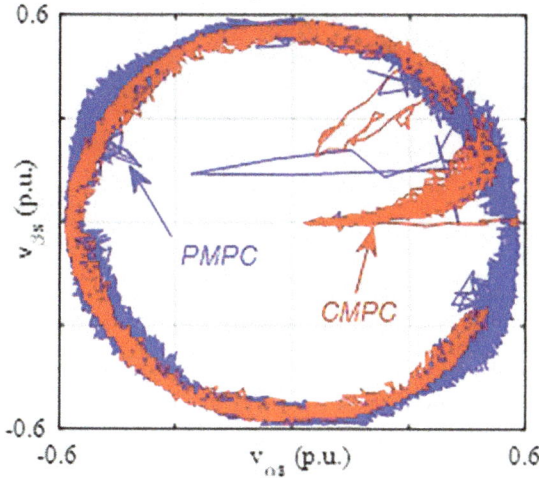

**FIGURE 7.25**   Voltage trajectory, base value 150 V (Case 1).

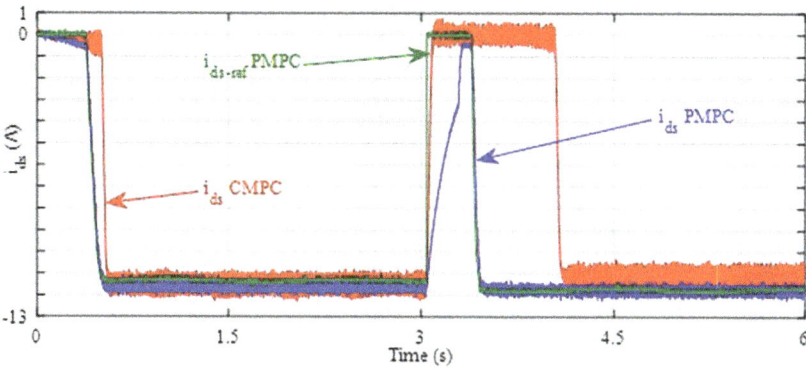

**FIGURE 7.26**   Measured stator $d$-axis currents (Case 1).

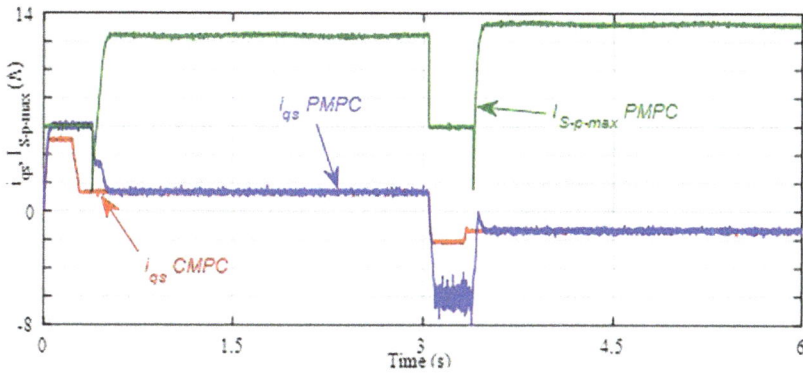

**FIGURE 7.27**   Measured $q$-axis stator currents and magnitude of stator current (Case 1).

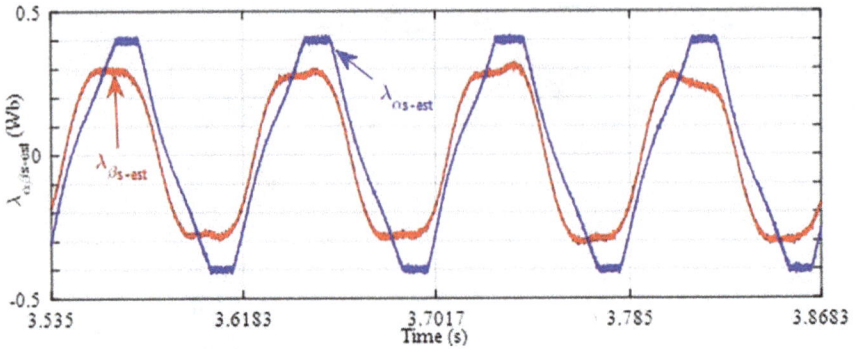

**FIGURE 7.28** Estimation fluxes in the stationary frame based on TOGI observer (Case 1).

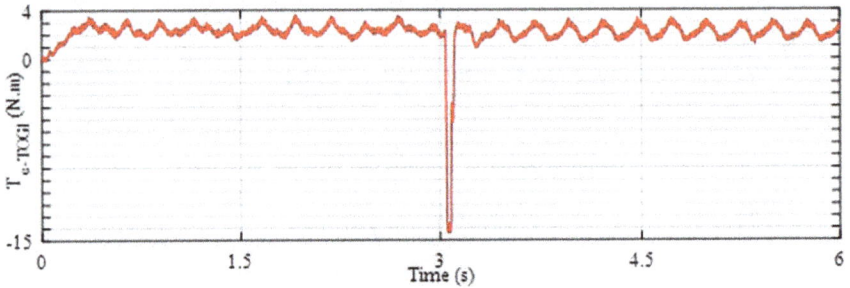

**FIGURE 7.29** Estimated $T_{e\text{-TOGI}}$ (Case 1).

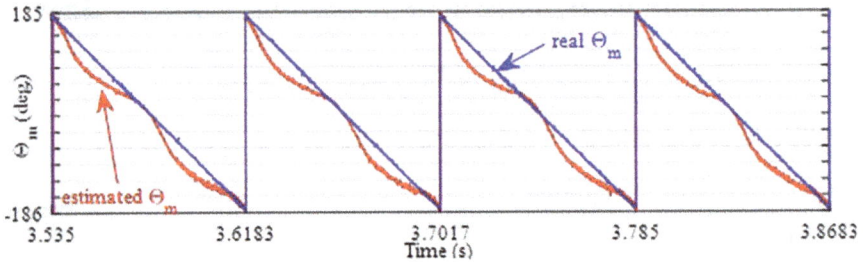

**FIGURE 7.30** Estimated and real rotor positions (Case 1).

Figure 7.24 specifies the manipulated variables, which are the reference voltages of the $d$-$q$ axes for the adaptive MPC, which is optimized by the AVPSO technique. The search areas for these manipulated variables range between -86.6 and 86.6 V. The voltage trajectory shown in Figure 7.25 illustrates that the $d$-$q$ axes' voltage vectors in the flux-weakening region are on the boundary of the maximum voltage that the inverter can supply to the motor of 86.6 V.

Figures 7.26 and 7.27 show the $d$-$q$ axes current of both MPC control strategies. The increase in the angle ($\Theta_{\text{Ids-Is}}$) between the $d$-axis component current vector and the stator current magnitude vector proportionally indicates the amount of the negative $d$-axis current value in the flux-weakening region, as calculated by

$$\theta_{\text{Ids-IS}} = (\pi - cos^{-1}(L_{ds}i_{ds} + \lambda_m + \sqrt{\lambda_{ds}^2 + \lambda_{qs}^2}))(\frac{180}{\pi}) \qquad (7.55)$$

The $\theta_{\text{Ids-Is}}$ of Figure 7.26 under both strategies is about 135.698°. As seen in Figure 7.26, the calculated $i_{ds\text{-ref}}$ in steady state under both strategies is -11.275 A. Both strategies produce a similar average of $i_{ds}$ in the clockwise (CW) rotational direction that are about -11.535 A. The average CMPC is changed to about -11.09 A in the counterclockwise (CCW) rotational direction.

Although the adaptive MPC cannot limit the $d$-$q$ axes reference voltages rate, the $q$-axis current has no high transient current depending on the definition of the maximum prediction stator current, as shown in Figure 7.27. In this figure, the transient current performance indicates the importance of calculation $I_{\text{S-p-max}}$ as a limitation of solving the objective function. As shown in this figure, the average $q$-axis current for the adaptive MPC and CMPC is changed from about 1.48 to 1.37 A.

The estimated rotor position in Figure 7.30 has proven that the power of the TOGI flux observer can provide the DLPM with estimated load torque because the error between the estimated and real positions is small. However, the error between the estimated and real rotor positions in this figure is more than others, in the direction of the reverse rotor for the flux-weakening region, due to the DC offset of estimated flux that has not been eliminated.

Figures 7.28 and 7.29 present the estimated $\alpha$-$\beta$ axes fluxes and estimated electromagnetic torque equal to the load torque in the steady-state period, respectively. The overall efficiency of about 11.3% is low as the motor is running in the flux-weakening region and under unrated torque. The output power of both MPC strategies is 0.1885 kW.

Figures 7.31 through 7.34 show the performance of the conventional FOC strategy of the Case 1 scenario. From these figures, the superiority of the adaptive MPC over

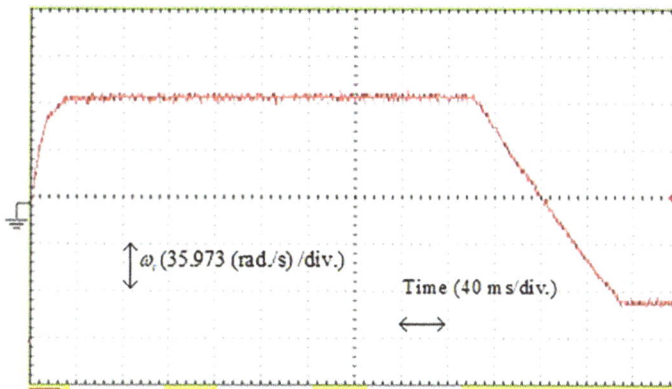

**FIGURE 7.31**　Motor speed of FOC strategy (Case 1).

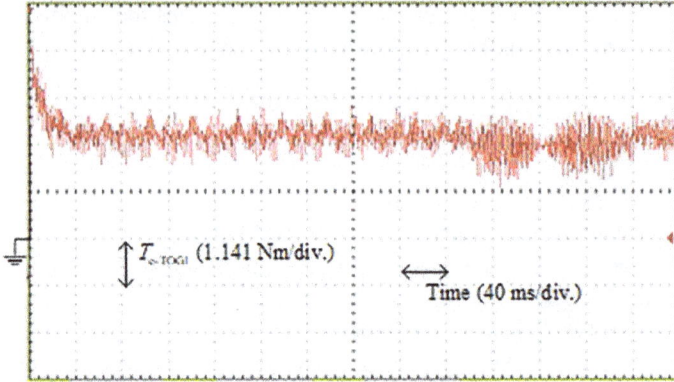

**FIGURE 7.32**   Estimated torque of FOC strategy (Case 1).

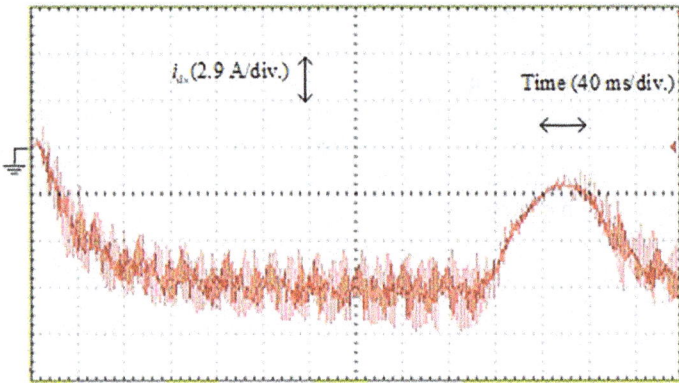

**FIGURE 7.33**   *d*-axis current of FOC strategy (Case 1).

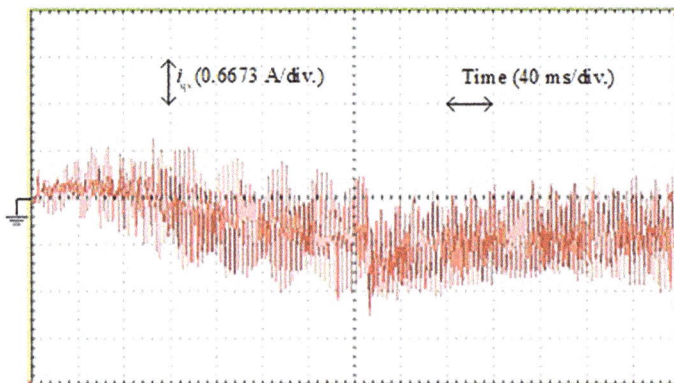

**FIGURE 7.34**   *q*-axis current of FOC strategy (Case 1).

conventional FOC is evident. Figure 7.31 shows the motor speed according to the FOC strategy, which has a long settling time of about 2.28 s in the reverse rotation direction. It has a high torque ripple of about 28% compared to 20% and low-order harmonics for the adaptive MPC strategy, as shown in Figures 7.32 and 7.29. Figures 7.33 and 7.34 present the $d$-$q$ axes stator currents having mean values of about -8.8 A and 1.37 A, respectively.

### 7.4.2 Dynamic Performance Study in Case of Speed Step-Down Condition

Verifying the ability of the drive to shift from the constant-torque region and the flux-weakening region is shown in Figures 7.30 to 7.37 (Case 2). These figures present the motor speed driven from standstill to flux-weakening region of 261.8 rad/s (2500 rpm) under a load torque of 3 Nm and a VSI voltage of 500 V, and then the speed is halved.

The calculation of $\omega_{F\text{-ref}}$, as presented in Figure 7.30, increases the speed acceleration of the adaptive MPC without speed overshoot. As shown in this figure, the adaptive MPC has a settling time less than that of CMPC equal to about 0.658 s while the counterpart of CMPC is 0.697 s. Figure 7.35 confirms the smooth transition between regions of constant torque and flux weakening.

Figure 7.36 describes the manipulated variables, which are the reference voltages of $d$-$q$ axes for the adaptive MPC, optimized by the new AVPSO algorithm. The range of search areas for these manipulated variables varies from -288.6 and 288.6 V. The voltage trajectory shown in Figure 7.32 illustrates the $d$-$q$ axes voltage vectors in the flux-weakening region placed at the limits of the maximum voltage that the inverter can provide to the motor of 288.6 V.

Figures 7.38 and 7.39 show the $d$-$q$ axes current of both MPC control strategies. The $\Theta_{\text{Ids-Is}}$ of Figure 7.38 under both strategies is about 139.31°. In Figure 7.38, during the case of steady state, the calculated $i_{ds\text{-ref}}$ is about -18.145 A. The same average $i_{ds}$ of about -17.94 A is produced under both strategies.

The stator current maximum algorithm for the prediction does not cause a high transient current of the $q$-axis, as shown in Figure 7.39. The mean value of the $q$-axis

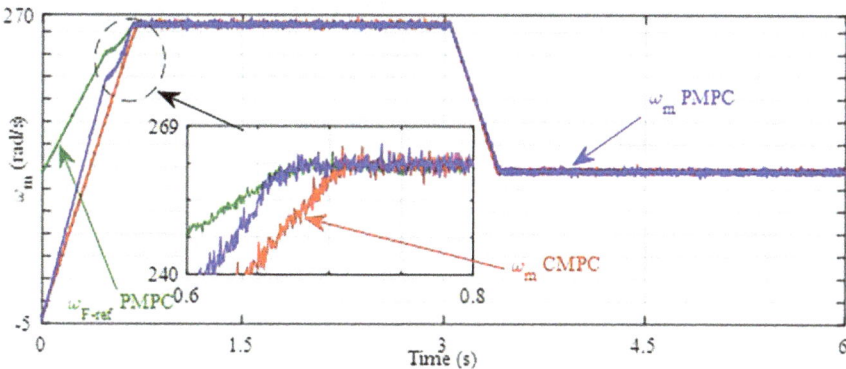

**FIGURE 7.35**  Motor velocity (Case 2).

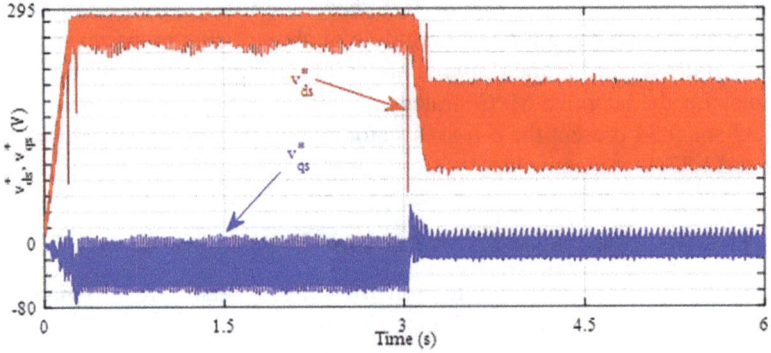

**FIGURE 7.36** Optimized *d-q* axes stator voltages of the proposed MPC (Case 2).

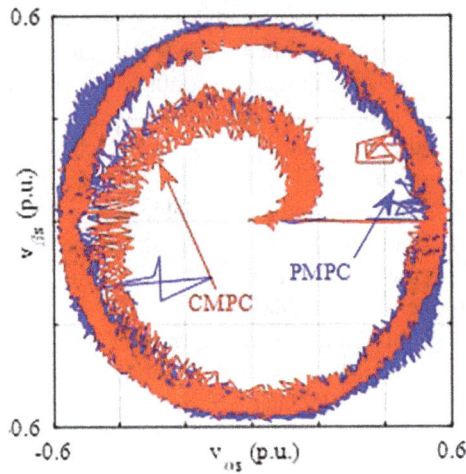

**FIGURE 7.37** Voltage trajectory, base value 500 V (Case 2).

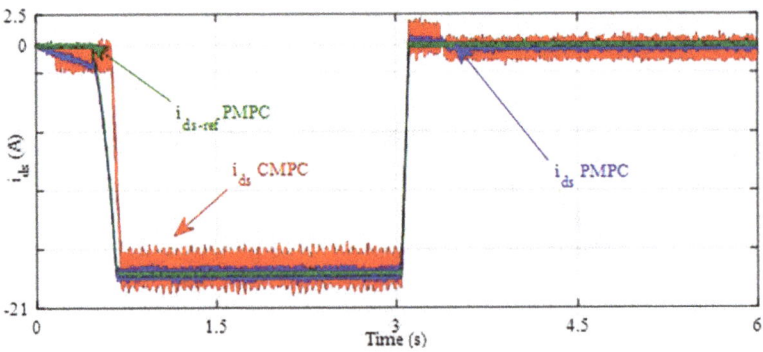

**FIGURE 7.38** The *d*-axis currents (Case 2).

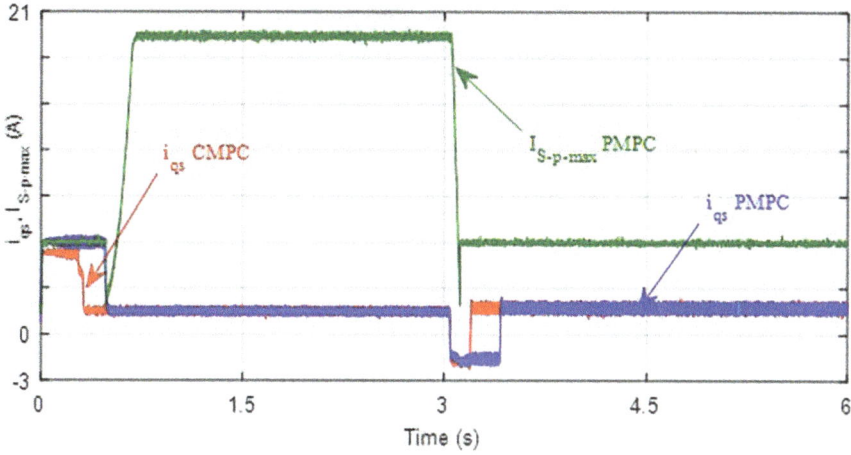

**FIGURE 7.39** The *q*-axis currents and magnitude of stator current (Case 2).

current of both strategies is constant at about 5.64 A. The estimated rotor position in Figure 7.42 has proven the ability of the TOGI flux observer to provide the DLPM with estimated load torque because the error between the estimated and real position is small. Figures 7.40 and 7.41 present the estimated $\alpha$-$\beta$ axes fluxes and estimated electromagnetic torque equal to the load torque in the steady-state period, respectively.

The overall efficiency is low due to the motor running in the flux-weakening region and under unrated torque. The efficiency of both control strategies increases as the speed changes from about 8.62% to 14.4%. The output power of both MPC strategies is 0.7854 kW.

Figures 7.43 through 7.46 show the performance of the conventional FOC strategy of the Case 2 scenario. It can be seen from these figures that the adaptive MPC exceeds the FOC strategy in reducing the torque ripple and settling the motor speed

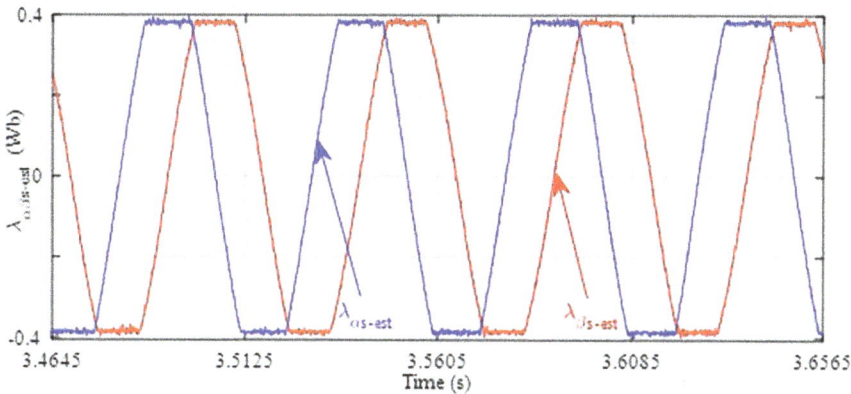

**FIGURE 7.40** Estimation fluxes in the stationary frame based on TOGI observer (Case 2).

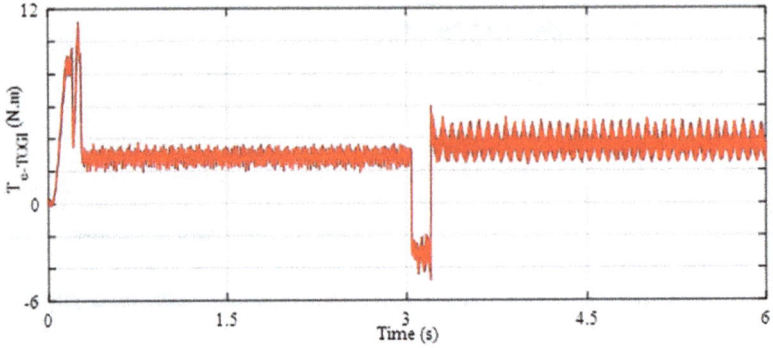

**FIGURE 7.41**    Estimated $T_{e\text{-TOGI}}$ (Case 2).

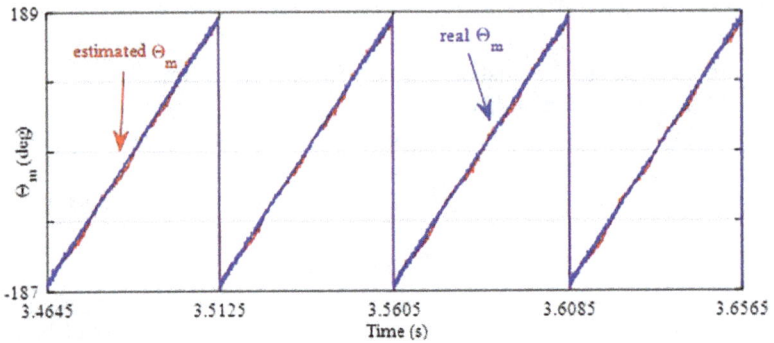

**FIGURE 7.42**    Estimated and real rotor positions (Case 2).

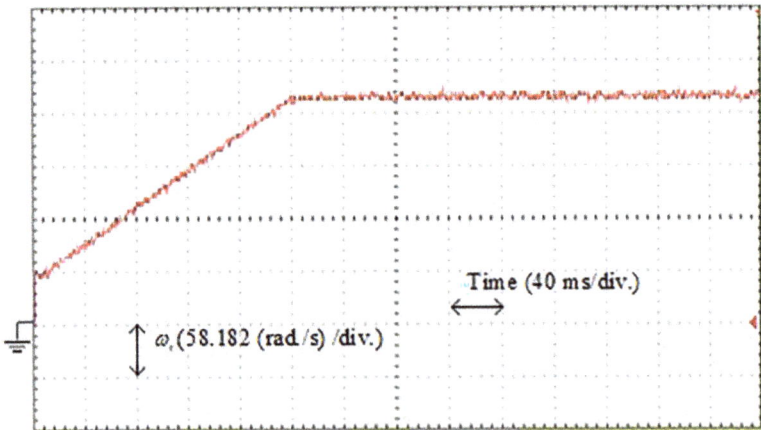

**FIGURE 7.43**    Motor speed of FOC strategy (Case 2).

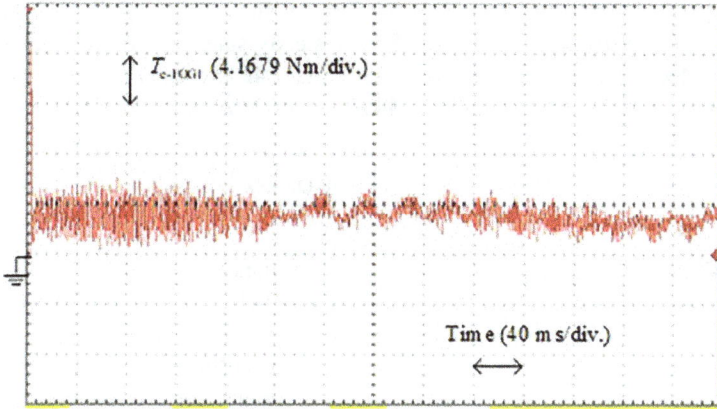

**FIGURE 7.44**   Estimated torque of FOC strategy (Case 2).

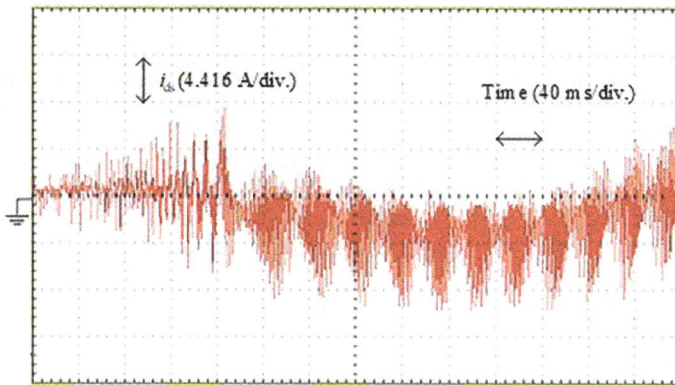

**FIGURE 7.45**   The $d$-axis current of FOC strategy (Case 2).

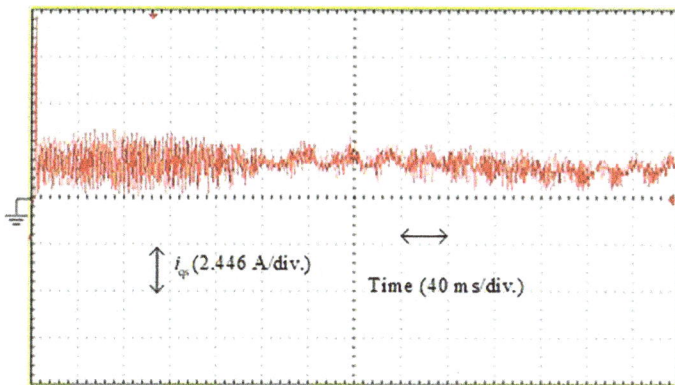

**FIGURE 7.46**   The $q$-axis current of FOC strategy (Case 2).

time as the attribute ripple reduction is 13.74% versus 69% for the FOC strategy, which has a settling time of more than 2.5 s and steady-state error of 4.1%, as shown in Figures 7.43, 7.44, and 7.41. Figures 7.45 and 7.46 present the $d$-$q$ axes stator currents having mean values of about -1.44 A and 1.823 A, respectively.

### 7.4.3 Dynamic Performance Study in Case of Load Torque Step-Down Condition

Figures 7.47 through 7.52 (Case 3) show the performance of the adaptive MPC and CMPC driving the IPMSM with the desired speed of 143 rad/s (1365 rpm), the initial torque suddenly reduced from 6 Nm to 0.5 Nm, and a VSI voltage of 300 V.

As shown in Figure 7.47, the adaptive MPC has a settling time of about 0.365 s, less than that of CMPC, which is 0.487 s due to the accelerated speed depending on the $\omega_{F\text{-ref}}$ calculation. The torque moment suddenly decreases as shown in Figure 7.47, which indicates the importance of the TOGI estimation in the adaptive MPC for its comparatively lower speed overshoot than those of other conventional methods.

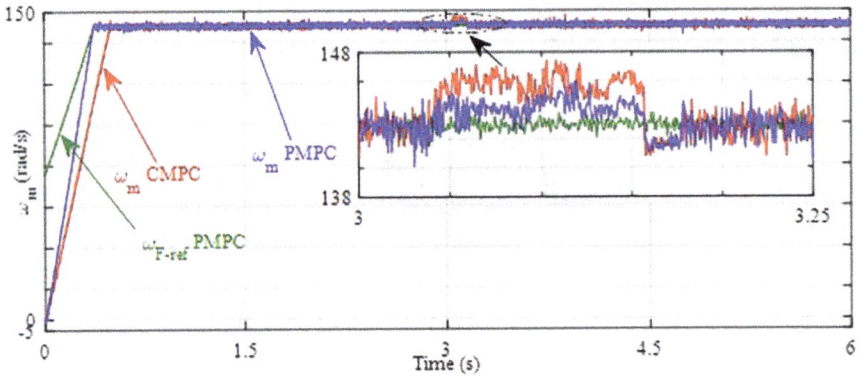

**FIGURE 7.47**    Motor velocity (Case 3).

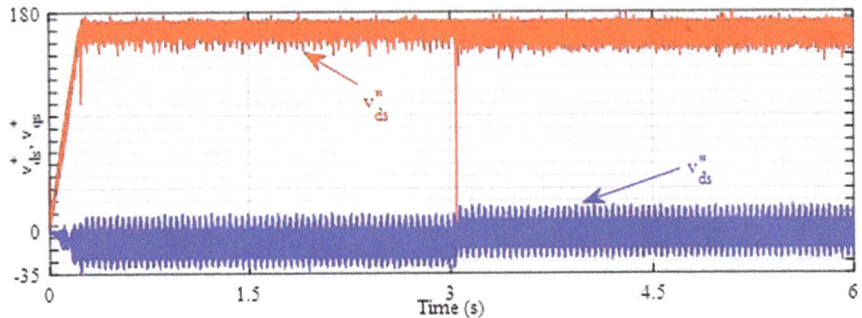

**FIGURE 7.48**    Optimized $d$-$q$ axes stator voltages of the adaptive MPC (Case 3).

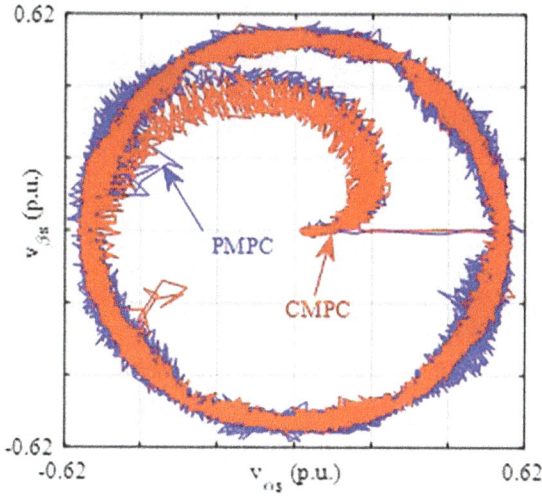

**FIGURE 7.49**   Voltage trajectory, base value 300 V (Case 3).

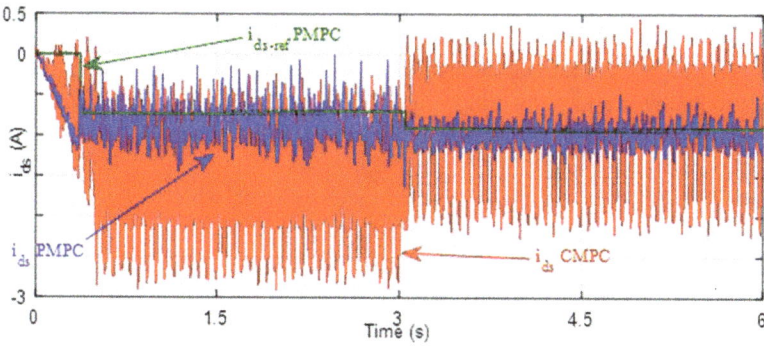

**FIGURE 7.50**   The $d$-axis currents (Case 3).

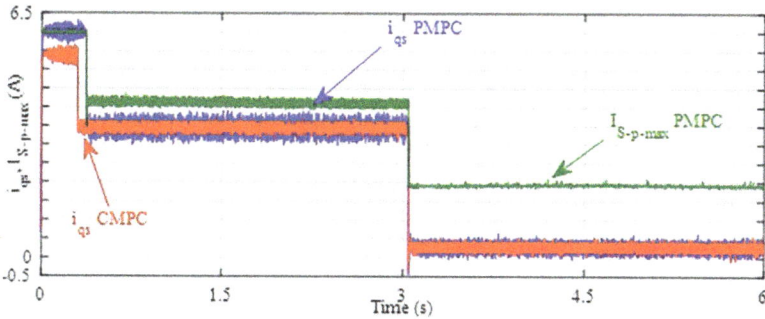

**FIGURE 7.51**   The $q$-axis currents and magnitude of stator current (Case 3).

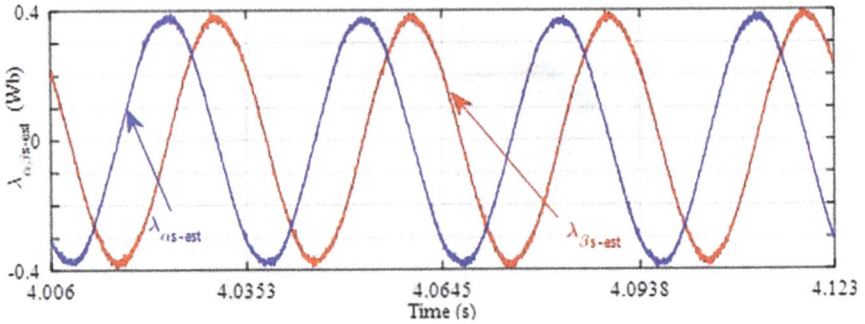

**FIGURE 7.52**   Estimation fluxes in the stationary frame based on TOGI observer (Case 3).

The reference voltages of the $d$-$q$ axes of the adaptive MPC, shown in Figure 7.48, are optimized by the new AVPSO algorithm. The range of search areas for these manipulated variables varies from -173.2 to 173.2 V. The voltage trajectory in Figure 7.49 shows the $d$-$q$ axes of the voltage vectors in the flux-weakening region placed on the maximum voltage limits that the inverter will supply to the 173.2 V.

Figures 7.50 and 7.51 show the $d$-$q$ axes current of both MPC control strategies. The $\Theta_{\text{Ids-Is}}$ of Figure 7.50 under both strategies is about 139°, and the average value of $i_{\text{ds-ref}}$ is about -0.94 A, as shown in Figure 7.50. The values of $i_{\text{ds}}$ are changed during loading disturbance under adaptive MPC and CMPC and can reach about -0.91 and -0.99 A, respectively.

Figure 7.51 displays the q-axis current measurement, which exhibits a low transient current. Therefore, the maximum static current definition is utilized for prediction purposes. In this figure, the average value of the $q$-axis current under the adaptive MPC and CMPC is changed from about 3.44 to 0.31 A. The rotor position estimated in Figure 7.54 demonstrates the TOGI flux observer power to estimate $\alpha$-$\beta$ axes fluxes and electromagnetic torque, as shown in Figures 7.52 and 7.53, respectively.

The overall efficiency is weak due to the motor running in the flux-weakening region and under unrated torque. The efficiency of both algorithms decreases as the torque changes from about 82.83% to 23.68%. The output power of each of the two MPC strategies before torque varies is 0.858 kW. After reducing the torque, the output power is 0.0714 kW. Figures 7.55 through 7.58 show the performance of the conventional FOC strategy of the Case 3 scenario. The FOC strategy has a settling time and steady-state error of about 1.1 s and 4.24%, separately, as shown in Figure 7.55.

It also causes a significant increase in torque ripple peak, which is about 30% and 250.6% before and after the load torque drop, respectively, as shown in Figure 7.56, while the torque ripple peak of the proposed MPC is less than that and reaches 5.3% and 92%, before and after the load torque drop, respectively, as shown in Figure 7.53. Figures 7.47 and 7.48 present the $d$-$q$ axes stator currents having mean values of about -3 A, 3.5 A before load drop, and 0.3 A after load drop, respectively.

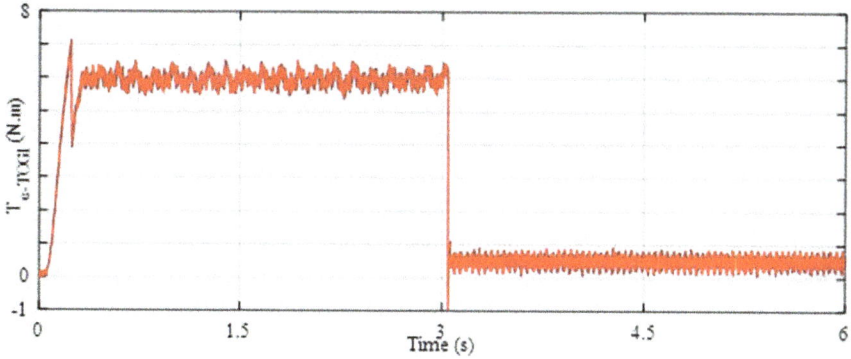

**FIGURE 7.53**  Estimated $T_{\text{e-TOGI}}$ (Case 3).

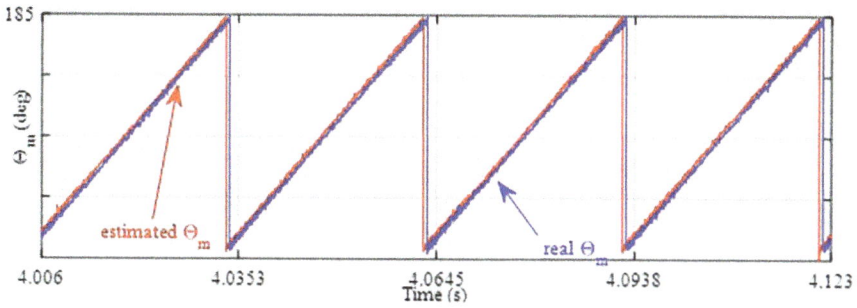

**FIGURE 7.54**  Estimated and real rotor positions (Case 3).

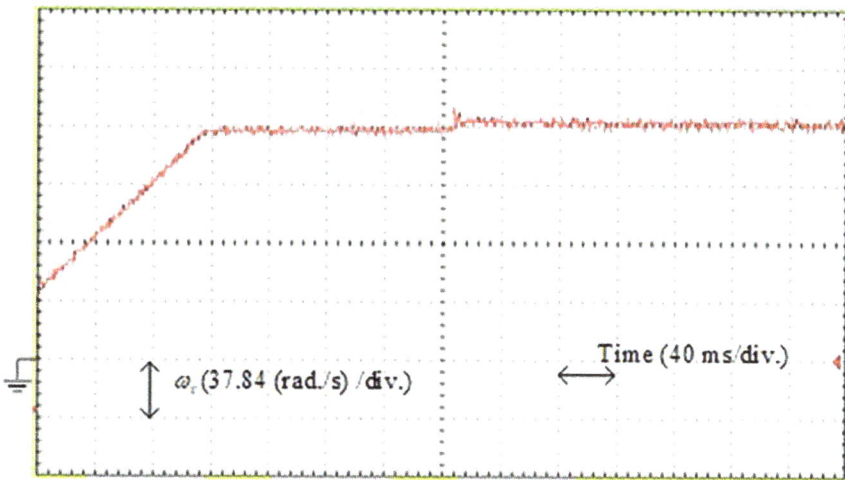

**FIGURE 7.55**  Motor speed of FOC strategy (Case 3).

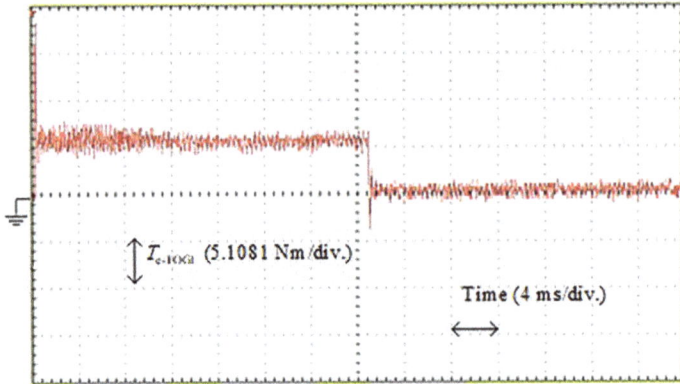

**FIGURE 7.56**   Estimated torque of FOC strategy (Case 3).

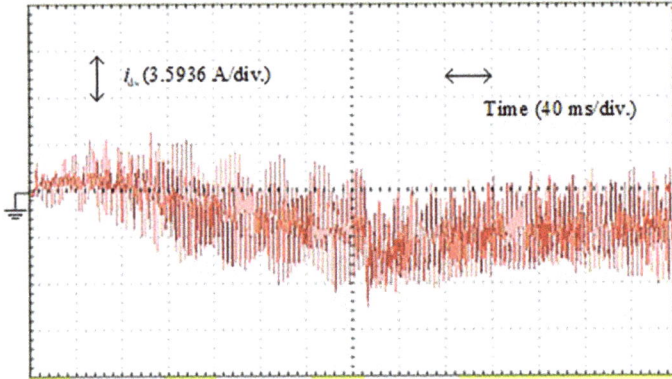

**FIGURE 7.57**   Measured $d$-axis stator current of FOC strategy (Case 3).

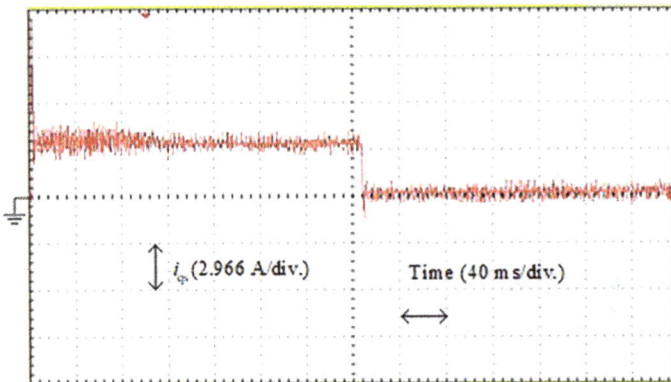

**FIGURE 7.58**   Measured $q$-axis stator current of FOC strategy (Case 3).

### 7.4.4 THE DISCRETIZATION MATRICES OF THE DLPM

Figures 7.59 to 7.76 present the instant time value of the $A$, $B$, and $E$ discretization matrices for all previous cases. As noted in this figure, their instant values differ in each case, which fully depicts the objective function on the real conditions of IPMSM. It is known that adaptive MPC is more robust in real-life changes in IPMSM conditions.

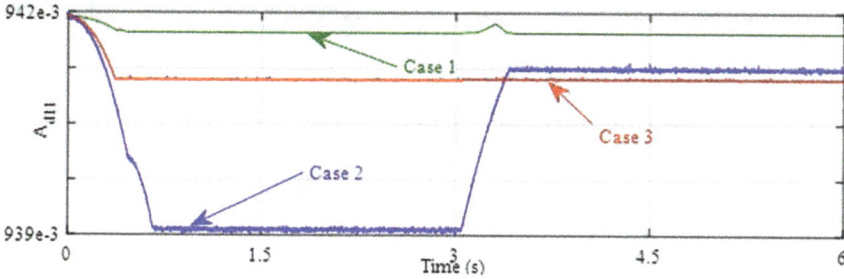

**FIGURE 7.59**   Discretization matrices of $A_{d11}$.

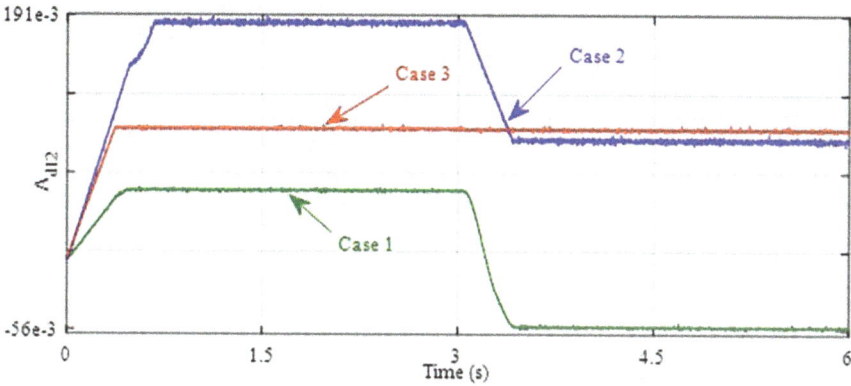

**FIGURE 7.60**   Discretization matrices of $A_{d12}$.

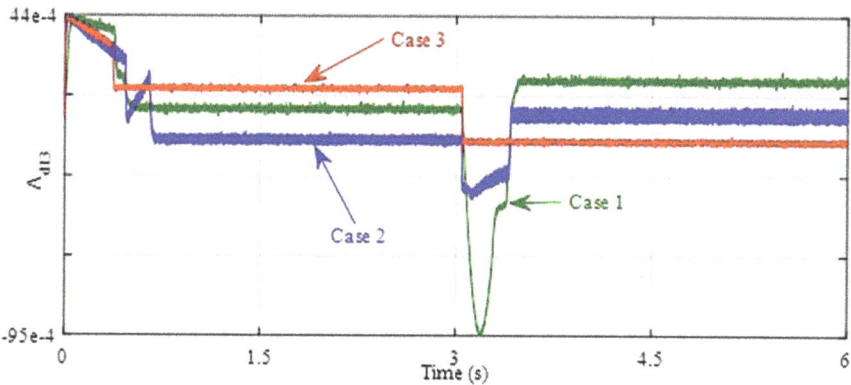

**FIGURE 7.61**   Discretization matrices of $A_{d13}$.

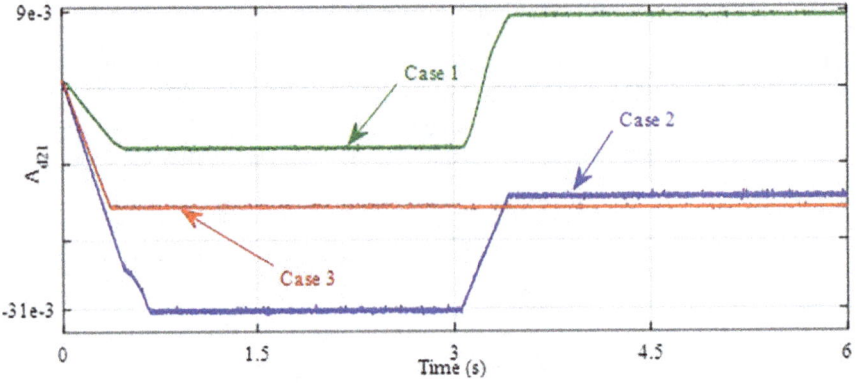

**FIGURE 7.62** Discretization matrices of $A_{d21}$.

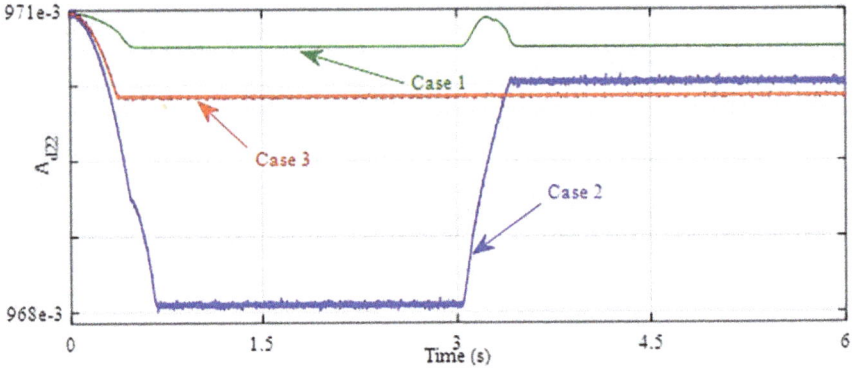

**FIGURE 7.63** Discretization matrices of $A_{d22}$.

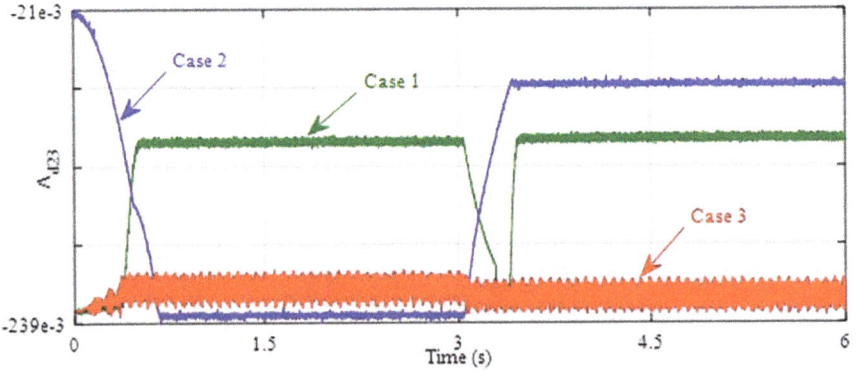

**FIGURE 7.64** Discretization matrices of $A_{d23}$.

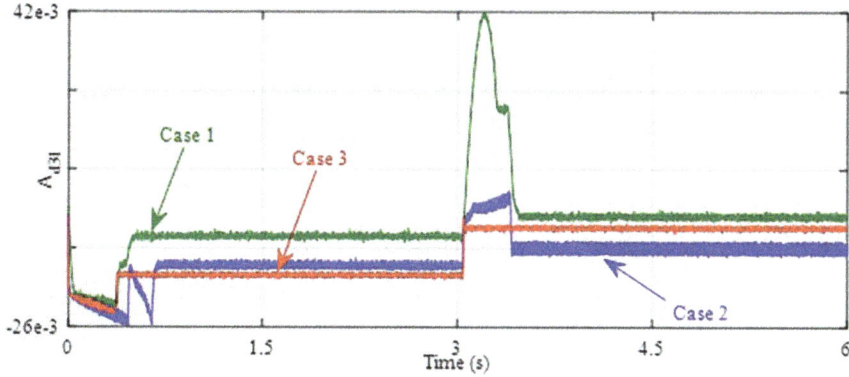

**FIGURE 7.65**   Discretization matrices of $A_{d31}$.

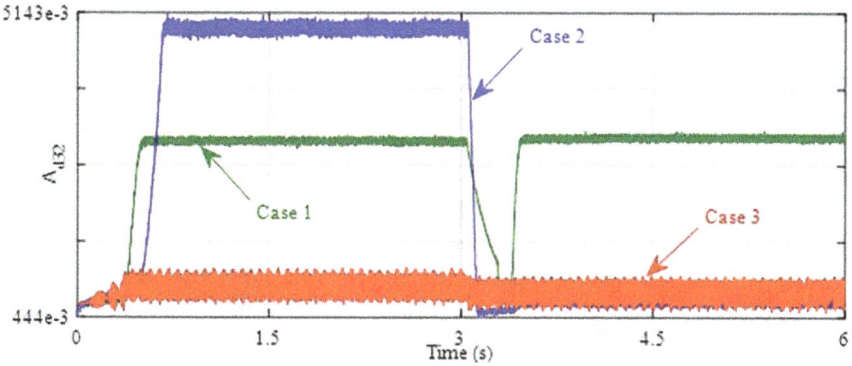

**FIGURE 7.66**   Discretization matrices of $A_{d32}$.

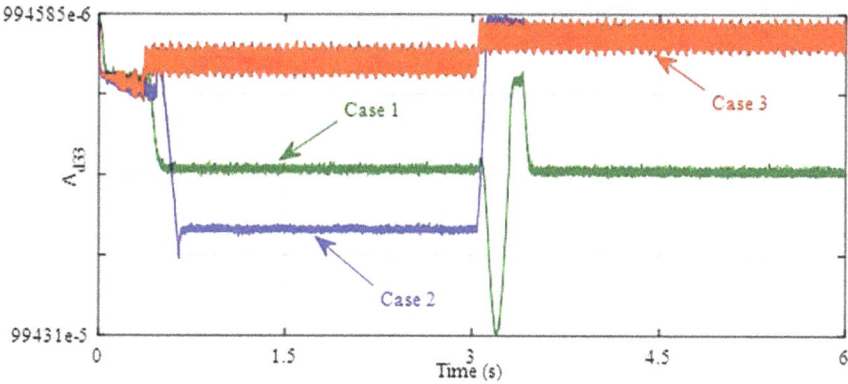

**FIGURE 7.67**   Discretization matrices of $A_{d33}$.

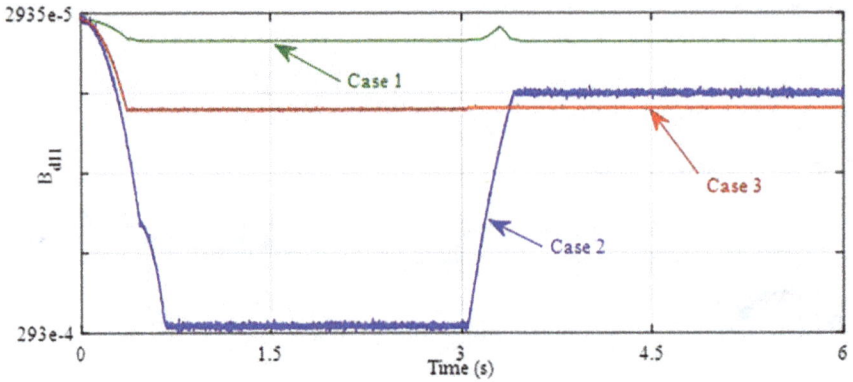

**FIGURE 7.68**    Discretization matrices of $B_{d11}$.

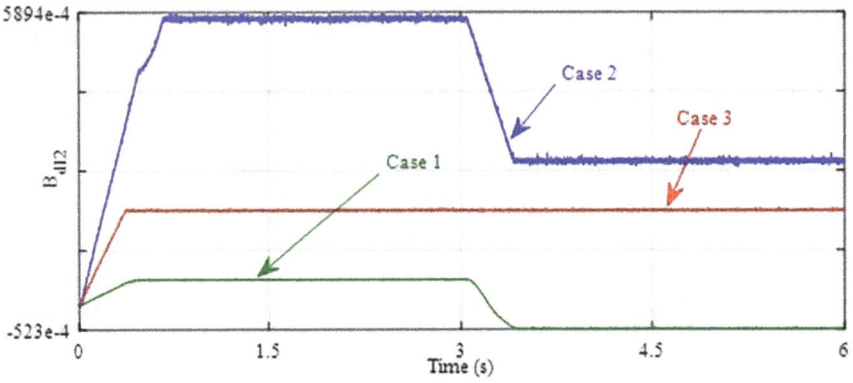

**FIGURE 7.69**    Discretization matrices of $B_{d12}$.

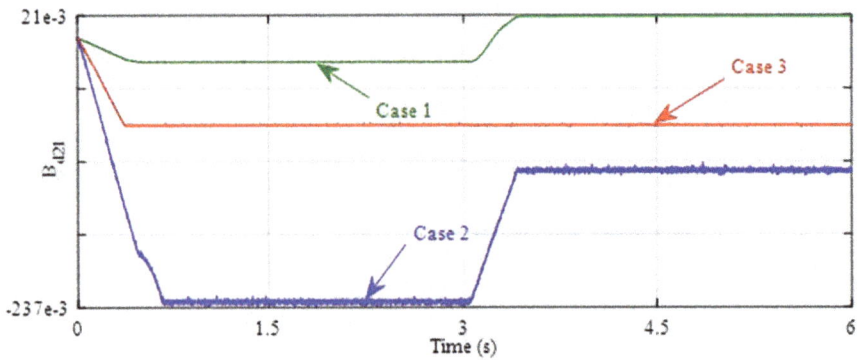

**FIGURE 7.70**    Discretization matrices of $B_{d21}$.

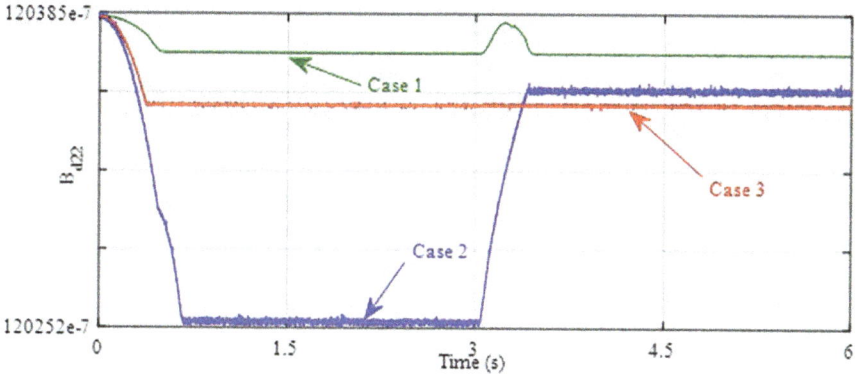

**FIGURE 7.71**   Discretization matrices of $B_{d22}$.

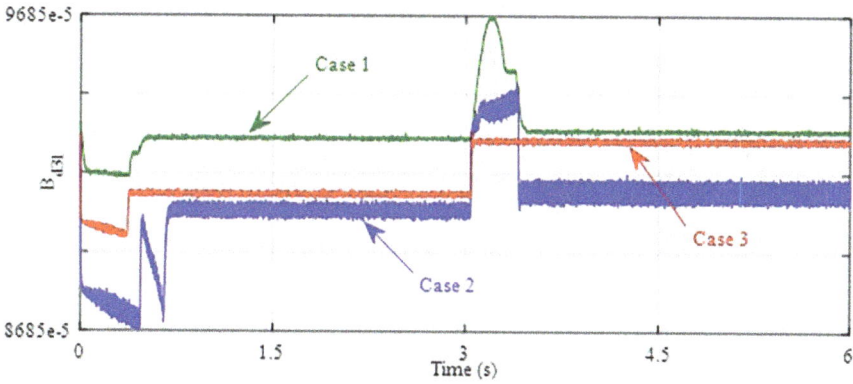

**FIGURE 7.72**   Discretization matrices of $B_{d31}$.

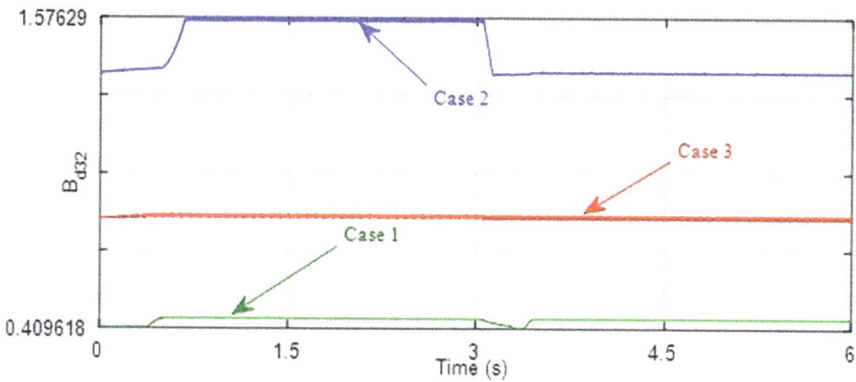

**FIGURE 7.73**   Discretization matrices of $B_{d32}$.

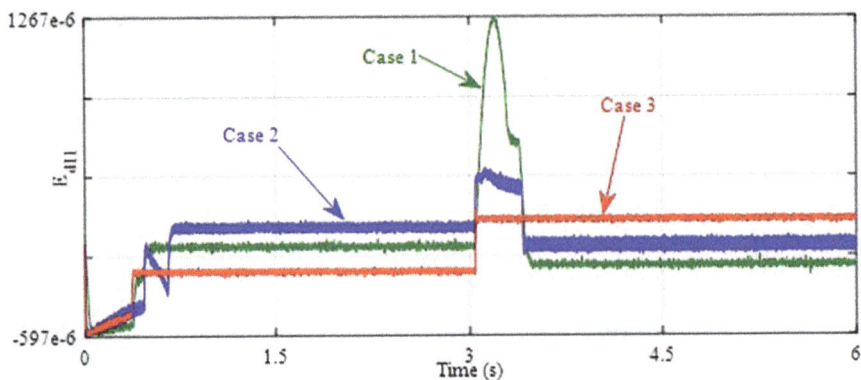

**FIGURE 7.74**  Discretization matrices of $E_{d11}$.

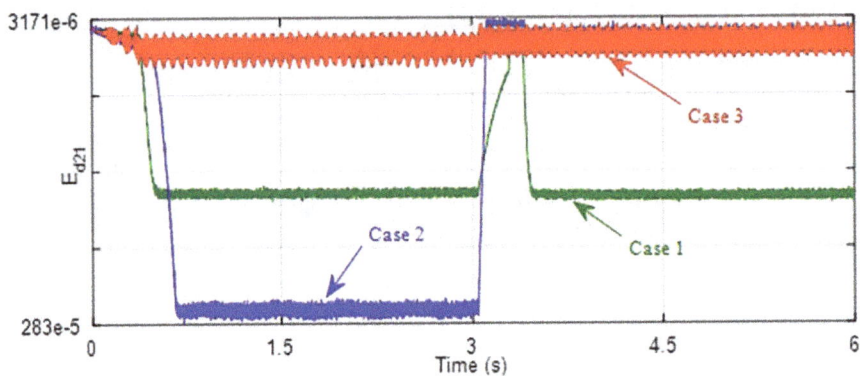

**FIGURE 7.75**  Discretization matrices of $E_{d21}$.

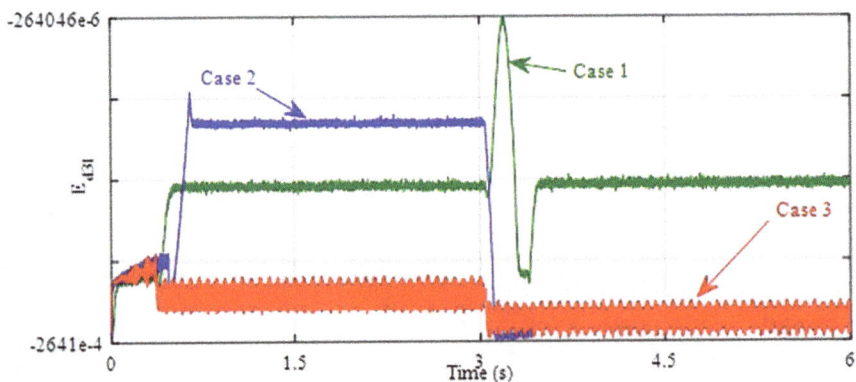

**FIGURE 7.76**  Discretization matrices of $E_{d31}$.

## 7.5 SUMMARY

In this chapter, we introduce a new adaptive DLPM to enhance the linear representation of the IPMSM and accelerate the performance of the AVPSO technique. The key points presented in this chapter are as follows:

(1) The new adaptive DLPM has three state vector variables compared to the seven variables in the CPM. The DLPM operating point equations can be updated for any DC-link voltage, load torque, and motor speed.
(2) An MPC is introduced, which overcomes the limitation of the load torque that should be previously known. It uses the DLPM with the TOGI flux observer to estimate the torque, which is unknown in an Electric Vehicle system and could result in a fatal mistake.
(3) An improved algorithm is proposed to avoid overspeed and high current by manipulating the reference speed and setting the maximum prediction stator current. This algorithm does not depend on uncertain constraints or variable parameters, and it replaces the traditional maximum approximate rate of manipulated variables.
(4) The AVPSO algorithm optimizes the compensated $d$-$q$ axes voltages as initial solutions concluded from the FOC strategy. The optimal parameters of the PI controller are set in this FOC strategy. Thus, the adaptive MPC is compared to the traditional FOC strategy, and its linearity is essential to facilitate the work of the optimizer.

TThe adaptive MPC has been validated in comprehensive experiments of DC-link voltage, rotational speed, and loading, which have demonstrated excellent dynamic performance and strong robustness. The adaptation in the adaptive MPC controller is essential to the MPC algorithm's efficiency, and failure to refresh the operating points will lead to an inaccurate and inefficient linear model.

## BIBLIOGRAPHY

[1]  G. Schoonhoven and M. N. Uddin. MTPA- and FW-based robust nonlinear speed control of IPMSM drive using Lyapunov stability criterion. IEEE Transactions on Industry Applications, 2016, 52(5):4365–4374.

[2]  T. Geyer. Model predictive control of high power converters and industrial drives, John Wiley & Sons, November 2016.

[3]  P. Chen, M. Yang, and T. Sun. editors. PSO-based on-line tuning PID controller for setpoint changes and load disturbance. 2011 IEEE Congress of Evolutionary Computation (CEC), 5–8 June 2011.

[4]  J. Pena, A. Upegui, and E. Sanchez. editors. Particle swarm optimization with discrete recombination: an online optimizer for evolvable hardware. First NASA/ESA Conference on Adaptive Hardware and Systems (AHS'06), 15–18 June 2006.

[5]  A. Song, W. N. Chen, T. Gu, H. Zhang, and J. Zhang. A Constructive particle swarm optimizer for virtual network embedding. IEEE Transactions on Network Science and Engineering, August 2020, 7(3):1406–1420.

[6]   B. Ufnalski and L. M. Grzesiak. editors. Particle swarm optimization of an online trained repetitive neurocontroller for the sine-wave inverter. IECON 2013–39th Annual Conference of the IEEE Industrial Electronics Society, 10–13 November 2013.

[7]   W. Xu, M. M. Ismail, Y. Liu, and M. R. Islam. Parameter optimization of adaptive flux-weakening strategy for permanent-magnet synchronous motor drives based on particle swarm algorithm. IEEE Transactions on Power Electronics, April 2019, 34(12):12128–12140.

[8]   G. De Souza, D. Odloak, and A. C. Zanin. Real time optimization (RTO) with model predictive control (MPC). Computers & Chemical Engineering, December 2010, 34(12):1999–2006.

[9]   K. Graichen and B. Käpernick. A real-time gradient method for nonlinear model predictive control, INTECH Open Access Publisher, February 2012.

[10]  M. Klaučo, M. Kalúz, and M. Kvasnica. Real-time implementation of an explicit MPC-based reference governor for control of a magnetic levitation system. Control Engineering Practice, March 2017, 60:99–105.

[11]  Park, Hyeongjun, Jing Sun, Steven Pekarek, Philip Stone, Daniel Opila, Richard Meyer, Ilya Kolmanovsky, and Raymond DeCarlo.

[12]  R. Van Parys, M. Verbandt, J. Swevers, and G. Pipeleers. Real-time proximal gradient method for embedded linear MPC. Mechatronics, May 2019, 59:1–9.

[13]  J. Liu, C. Gong, Z. Han, and H. Yu. IPMSM model predictive control in flux-weakening operation using an improved algorithm. IEEE Transactions on Industrial Electronics, March 2018, 65(12):9378–9387.

[14]  L. Wang. Model predictive control system design and implementation using MAT-LAB®, Springer Science & Business Media, February 2009.

# 8 Conclusions and Future Work

## 8.1 CONCLUSIONS

The scope of this book is divided into two major dimensions: improved field-oriented control strategy and improved model predictive control strategy for PMSM drives employed in light electric vehicles. The problem formulation and objectives highlighted in Chapter 1 are mainly concerned with the internal cogging vibration affecting motor performance and the necessity of operating the PMSMs in the flux-weakening region. The key contents of the book can be summarized as follows:

(1) To achieve high-performance PI regulation, the anti-windup proportional and integral (AWPI) technique has been considered to block speed overshoot and overcurrent in the flux-weakening region. The AWPI technique plays an essential role in the PI regulator of the flux-weakening control loop; it reduces the integral error value. For the $d$-axis PI controller, it increases the negative mean value of the $d$-axis reference stator current as that of the demagnetizing current. Therefore, the demagnetizing current is not weakened by velocity or load torque changes.

(2) The control loop has been considered for reference duty-cycle magnitude feedback due to its advantages of simplicity and strength against parameter variation and automatic operation in the flux-weakening region. The low-pass filter rejects the high variation of the demagnetizing current caused by the rapid change of the reference vector of the duty cycle. This filter applies a more negative average value, which greatly affects the utilization of the DC-link voltage source.

(3) The robust stability of the dynamic performance of the surface-mounted PMSM in the flux-weakening region depends on the novel method for adjusting the PI gains of PI regulators. This method calculates the error matrix in terms of the PI parameters based on the linearized control loops of the advanced drive scheme and motor mathematical model. The operating points of the linearization process are determined from the location of the flux-weakening region. Hence, an optimization algorithm has been presented to optimize the PI gain that reduces the evaluation of the error matrix. This procedure has been done offline for the implemented optimization technique based on the adaptive velocity particle swarm optimization (AVPSO) algorithm.

(4) The results presented in Chapter 3 confirm the success of the surface-mounted permanent magnet synchronous motor (PMSM) performance under the advanced control strategy compared to the conventional strategy,

DOI: 10.1201/9781003320128-8

for the same control conditions and load torque. The advanced control strategy reduces the speed overshoot at the constant-torque region by 66.7% and eliminates it completely in the flux-weakening region. Moreover, the motor speed under the advanced strategy can transition quickly and more flexibly between the constant-torque and flux-weakening regions. The advanced strategy also reduces the settling time by 50% when the load torque experiences a sudden step-down.

(5) Chapter 4 examines the method of improving the PI gains for the current regulators. The focus is on the q-axis regulator, which works in conjunction with the VSFPWM control algorithm to reduce the ripples of the stator current and torque. The advanced strategy prioritizes the minimization of the PI regulator errors through weight factors, employing an advanced objective function that includes stability constraints and terms. Furthermore, a genetic algorithm is considered to minimize the improved objective function and obtain optimal PI parameters.

The improved advanced strategy can reduce the peak of ripple values of the torque and $d$-$q$ axes stator currents, as well as the switching losses, by 80.88%, 70.37%, 79.58%, and 5%, respectively. In the case of machine speed oscillation, the advanced strategy reduces the peak of torque ripple and switching losses by 62.3% and 14.38%, respectively. Additionally, the advanced strategy has a lower copper loss due to the lower amplitude of the stator current compared to the conventional strategy.

(6) Chapter 5 presents modifications to the field-oriented control strategy discussed in Chapter 4, aiming to match the desired control performance of the salient-type PMSM drive system. The modified control strategy achieves high-performance drive during a sudden change in the reference torque and speed.

(7) Chapter 6 presents the finite control set model predictive control for a salient PMSM drive. The optimal stator voltage components are selected based on the finite state control method to cancel the inverter modulator. The cost function directly controls the motor speed and the required stator current in the flux-weakening region. Therefore, the advanced predictive control strategy achieves high drive performance during sudden changes in the reference torque and speed.

(8) The adaptive model predictive control (MPC) strategy is considered for a salient PMSM drive, incorporating features not present in the vector control strategy, such as multi-variable control and online optimization that consider all the constraints before selecting the manipulated variables. Specifically, the real-time adaptive velocity particle swarm optimization (AVPSO) algorithm is employed to select the best $d$-$q$ axes voltages. A new discrete linear plant model is employed, in which the state-space matrices are mathematically calculated. The functions of the current operating points are updated every time step for prediction. Additionally, the model is supported by a TOGI flux observer, which is used for estimating electromagnetic torque. The advanced algorithm for computing the reference speed and maximum stator current magnitude effectively reduced speed overshoot and overcurrent. The comprehensive results

confirmed the excellent dynamic performance of the interior PMSM with the advanced adaptive MPC strategy. The performance of the advanced MPC was compared to that of traditional MPC and FOC strategies under the same operating conditions and load torque. The results indicated that the advanced MPC controller was capable of achieving the fastest transient period and highest efficiency.

This book presents a variety of optimization algorithms, including PSO and GA, particularly in the context of PMSM drives. In Chapter 3, AVPSO was introduced as an offline mode optimization process that works with MATLAB simulation software. In the offline mode, the swarm size reduction, total number of iterations, and calculation time are insignificant. In Chapters 4 and 5, the GA optimization procedure was presented in the offline mode, while in Chapter 7, the AVPSO algorithm was presented as an online mode. Therefore, it was essential to reduce the number of swarm sizes and the total number of iterations to reduce the calculation time. To this end, the improved FOC strategy presented in Chapter 3 was employed to enhance the initial solutions of the real-time AVPSO algorithm for optimizing the $d$-q axes reference voltages. Consequently, the $d$-$q$ axes reference voltages calculated by the FOC strategy were considered initial solutions close to the optimal solutions.

## 8.2 COVERAGE OF EMERGING TECHNOLOGIES

This book aimed to present advanced techniques to enhance the overall performance of PMSMs. These techniques can be summarized as follows:

(1) A parameter tuning algorithm was presented for adjusting the PI parameter regulators in the vector control strategy. This algorithm improves stability performance in the flux-weakening region by reducing peak ripples.
(2) An adaptive MPC strategy was presented based on an adaptive predictive model of the PMSM plant. A new performance control algorithm was also presented to prevent speed overshoot and overcurrent, which can enhance drive performance under heavy speed and load disturbances.
(3) A novel AVPSO algorithm was discussed for the adaptive MPC strategy to optimize manipulated variables. This optimization algorithm can reduce operating time and memory size by decreasing the generation number and particle swarm size and modifying the initial population with the computed voltages of $d$-$q$ axes at each prediction time step.

## 8.3 SUGGESTIONS FOR FUTURE DEVELOPMENTS

There are still some deficiencies in the related research work presented in this book, such as testing the advanced strategies in industrial load systems such as electric vehicles. Therefore, future work is expected as follows:

1) An appropriate method should be implemented online to identify machine parameters, PI tuning methods, and the adaptive linear MPC strategy.

(2) The PI parameters of the current regulator should be adjusted online using the improved FOC strategy. This strategy can include a sliding-mode speed controller to eliminate torque and current ripple.

(3) The equation used to determine the demagnetization current in the MPC strategy presented in this book should be replaced by a novel algorithm to avoid the negative impact of changes in the machine parameters. This algorithm should be based on the maximum stator current size of the machine and inverter, the rated electromagnetic torque, and the static current reduction of the predictive $d$-$q$ axes.

# Index

For Product Safety Concerns and Information please contact our EU representative GPSR@taylorandfrancis.com
Taylor & Francis Verlag GmbH, Kaufingerstraße 24, 80331 München, Germany

www.ingramcontent.com/pod-product-compliance
Lightning Source LLC
Chambersburg PA
CBHW060348220326
41598CB00023B/2841